Pitman Research Notes in Mathematics Series

Submission of proposals for consideration
Suggestions for publication, in the form of outlines and representative samples, are invited by the Editorial Board for assessment. Intending authors should approach one of the main editors or another member of the Editorial Board, citing the relevant AMS subject classifications. Alternatively, outlines may be sent directly to the publisher's offices. Refereeing is by members of the board and other mathematical authorities in the topic concerned, throughout the world.

Preparation of accepted manuscripts
On acceptance of a proposal, the publisher will supply full instructions for the preparation of manuscripts in a form suitable for direct photo-lithographic reproduction. Specially printed grid sheets are provided and a contribution is offered by the publisher towards the cost of typing. Word processor output, subject to the publisher's approval, is also acceptable.

Illustrations should be prepared by the authors, ready for direct reproduction without further improvement. The use of hand-drawn symbols should be avoided wherever possible, in order to maintain maximum clarity of the text.

The publisher will be pleased to give any guidance necessary during the preparation of a typescript, and will be happy to answer any queries.

Important note
In order to avoid later retyping, intending authors are strongly urged not to begin final preparation of a typescript before receiving the publisher's guidelines and special paper. In this way it is hoped to preserve the uniform appearance of the series.

Longman Scientific & Technical
Longman House
Burnt Mill
Harlow, Essex, UK
(tel (0279) 26721)

Titles in this series

1 Improperly posed boundary value problems
A Carasso and A P Stone
2 Lie algebras generated by finite dimensional ideals
I N Stewart
3 Bifurcation problems in nonlinear elasticity
R W Dickey
4 Partial differential equations in the complex domain
D L Colton
5 Quasilinear hyperbolic systems and waves
A Jeffrey
6 Solution of boundary value problems by the method of integral operators
D L Colton
7 Taylor expansions and catastrophes
T Poston and I N Stewart
8 Function theoretic methods in differential equations
R P Gilbert and R J Weinacht
9 Differential topology with a view to applications
D R J Chillingworth
10 Characteristic classes of foliations
H V Pittie
11 Stochastic integration and generalized martingales
A U Kussmaul
12 Zeta-functions: An introduction to algebraic geometry
A D Thomas
13 Explicit *a priori* inequalities with applications to boundary value problems
V G Sigillito
14 Nonlinear diffusion
W E Fitzgibbon III and H F Walker
15 Unsolved problems concerning lattice points
J Hammer
16 Edge-colourings of graphs
S Fiorini and R J Wilson
17 Nonlinear analysis and mechanics: Heriot-Watt Symposium Volume I
R J Knops
18 Actions of fine abelian groups
C Kosniowski
19 Closed graph theorems and webbed spaces
M De Wilde
20 Singular perturbation techniques applied to integro-differential equations
H Grabmüller
21 Retarded functional differential equations: A global point of view
S E A Mohammed
22 Multiparameter spectral theory in Hilbert space
B D Sleeman
24 Mathematical modelling techniques
R Aris
25 Singular points of smooth mappings
C G Gibson
26 Nonlinear evolution equations solvable by the spectral transform
F Calogero
27 Nonlinear analysis and mechanics: Heriot-Watt Symposium Volume II
R J Knops
28 Constructive functional analysis
D S Bridges
29 Elongational flows: Aspects of the behaviour of model elasticoviscous fluids
C J S Petrie
30 Nonlinear analysis and mechanics: Heriot-Watt Symposium Volume III
R J Knops
31 Fractional calculus and integral transforms of generalized functions
A C McBride
32 Complex manifold techniques in theoretical physics
D E Lerner and P D Sommers
33 Hilbert's third problem: scissors congruence
C-H Sah
34 Graph theory and combinatorics
R J Wilson
35 The Tricomi equation with applications to the theory of plane transonic flow
A R Manwell
36 Abstract differential equations
S D Zaidman
37 Advances in twistor theory
L P Hughston and R S Ward
38 Operator theory and functional analysis
I Erdelyi
39 Nonlinear analysis and mechanics: Heriot-Watt Symposium Volume IV
R J Knops
40 Singular systems of differential equations
S L Campbell
41 N-dimensional crystallography
R L E Schwarzenberger
42 Nonlinear partial differential equations in physical problems
D Graffi
43 Shifts and periodicity for right invertible operators
D Przeworska-Rolewicz
44 Rings with chain conditions
A W Chatters and C R Hajarnavis
45 Moduli, deformations and classifications of compact complex manifolds
D Sundararaman
46 Nonlinear problems of analysis in geometry and mechanics
M Atteia, D Bancel and I Gumowski
47 Algorithmic methods in optimal control
W A Gruver and E Sachs
48 Abstract Cauchy problems and functional differential equations
F Kappel and W Schappacher
49 Sequence spaces
W H Ruckle
50 Recent contributions to nonlinear partial differential equations
H Berestycki and H Brezis
51 Subnormal operators
J B Conway

52 Wave propagation in viscoelastic media
F Mainardi
53 Nonlinear partial differential equations and their applications: Collège de France Seminar. Volume I
H Brezis and J L Lions
54 Geometry of Coxeter groups
H Hiller
55 Cusps of Gauss mappings
T Banchoff, T Gaffney and C McCrory
56 An approach to algebraic K-theory
A J Berrick
57 Convex analysis and optimization
J-P Aubin and R B Vintner
58 Convex analysis with applications in the differentiation of convex functions
J R Giles
59 Weak and variational methods for moving boundary problems
C M Elliott and J R Ockendon
60 Nonlinear partial differential equations and their applications: Collège de France Seminar. Volume II
H Brezis and J L Lions
61 Singular systems of differential equations II
S L Campbell
62 Rates of convergence in the central limit theorem
Peter Hall
63 Solution of differential equations by means of one-parameter groups
J M Hill
64 Hankel operators on Hilbert space
S C Power
65 Schrödinger-type operators with continuous spectra
M S P Eastham and H Kalf
66 Recent applications of generalized inverses
S L Campbell
67 Riesz and Fredholm theory in Banach algebra
B A Barnes, G J Murphy, M R F Smyth and T T West
68 Evolution equations and their applications
F Kappel and W Schappacher
69 Generalized solutions of Hamilton-Jacobi equations
P L Lions
70 Nonlinear partial differential equations and their applications: Collège de France Seminar. Volume III
H Brezis and J L Lions
71 Spectral theory and wave operators for the Schrödinger equation
A M Berthier
72 Approximation of Hilbert space operators I
D A Herrero
73 Vector valued Nevanlinna Theory
H J W Ziegler
74 Instability, nonexistence and weighted energy methods in fluid dynamics and related theories
B Straughan
75 Local bifurcation and symmetry
A Vanderbauwhede
76 Clifford analysis
F Brackx, R Delanghe and F Sommen
77 Nonlinear equivalence, reduction of PDEs to ODEs and fast convergent numerical methods
E E Rosinger
78 Free boundary problems, theory and applications. Volume I
A Fasano and M Primicerio
79 Free boundary problems, theory and applications. Volume II
A Fasano and M Primicerio
80 Symplectic geometry
A Crumeyrolle and J Grifone
81 An algorithmic analysis of a communication model with retransmission of flawed messages
D M Lucantoni
82 Geometric games and their applications
W H Ruckle
83 Additive groups of rings
S Feigelstock
84 Nonlinear partial differential equations and their applications: Collège de France Seminar. Volume IV
H Brezis and J L Lions
85 Multiplicative functionals on topological algebras
T Husain
86 Hamilton-Jacobi equations in Hilbert spaces
V Barbu and G Da Prato
87 Harmonic maps with symmetry, harmonic morphisms and deformations of metrics
P Baird
88 Similarity solutions of nonlinear partial differential equations
L Dresner
89 Contributions to nonlinear partial differential equations
C Bardos, A Damlamian, J I Díaz and J Hernández
90 Banach and Hilbert spaces of vector-valued functions
J Burbea and P Masani
91 Control and observation of neutral systems
D Salamon
92 Banach bundles, Banach modules and automorphisms of C*-algebras
M J Dupré and R M Gillette
93 Nonlinear partial differential equations and their applications: Collège de France Seminar. Volume V
H Brezis and J L Lions
94 Computer algebra in applied mathematics: an introduction to MACSYMA
R H Rand
95 Advances in nonlinear waves. Volume I
L Debnath
96 FC-groups
M J Tomkinson
97 Topics in relaxation and ellipsoidal methods
M Akgül
98 Analogue of the group algebra for topological semigroups
H Dzinotyiweyi
99 Stochastic functional differential equations
S E A Mohammed

100 Optimal control of variational inequalities
V Barbu
101 Partial differential equations and dynamical systems
W E Fitzgibbon III
102 Approximation of Hilbert space operators. Volume II
C Apostol, L A Fialkow, D A Herrero and D Voiculescu
103 Nondiscrete induction and iterative processes
V Ptak and F-A Potra
104 Analytic functions – growth aspects
O P Juneja and G P Kapoor
105 Theory of Tikhonov regularization for Fredholm equations of the first kind
C W Groetsch
106 Nonlinear partial differential equations and free boundaries. Volume I
J I Díaz
107 Tight and taut immersions of manifolds
T E Cecil and P J Ryan
108 A layering method for viscous, incompressible L_p flows occupying R^n
A Douglis and E B Fabes
109 Nonlinear partial differential equations and their applications: Collège de France Seminar. Volume VI
H Brezis and J L Lions
110 Finite generalized quadrangles
S E Payne and J A Thas
111 Advances in nonlinear waves. Volume II
L Debnath
112 Topics in several complex variables
E Ramírez de Arellano and D Sundararaman
113 Differential equations, flow invariance and applications
N H Pavel
114 Geometrical combinatorics
F C Holroyd and R J Wilson
115 Generators of strongly continuous semigroups
J A van Casteren
116 Growth of algebras and Gelfand–Kirillov dimension
G R Krause and T H Lenagan
117 Theory of bases and cones
P K Kamthan and M Gupta
118 Linear groups and permutations
A R Camina and E A Whelan
119 General Wiener–Hopf factorization methods
F-O Speck
120 Free boundary problems: applications and theory, Volume III
A Bossavit, A Damlamian and M Fremond
121 Free boundary problems: applications and theory, Volume IV
A Bossavit, A Damlamian and M Fremond
122 Nonlinear partial differential equations and their applications: Collège de France Seminar. Volume VII
H Brezis and J L Lions
123 Geometric methods in operator algebras
H Araki and E G Effros
124 Infinite dimensional analysis–stochastic processes
S Albeverio
125 Ennio de Giorgi Colloquium
P Krée
126 Almost-periodic functions in abstract spaces
S Zaidman
127 Nonlinear variational problems
A Marino, L Modica, S Spagnolo and M Degiovanni
128 Second-order systems of partial differential equations in the plane
L K Hua, W Lin and C-Q Wu
129 Asymptotics of high-order ordinary differential equations
R B Paris and A D Wood
130 Stochastic differential equations
R Wu
131 Differential geometry
L A Cordero
132 Nonlinear differential equations
J K Hale and P Martinez-Amores
133 Approximation theory and applications
S P Singh
134 Near-rings and their links with groups
J D P Meldrum
135 Estimating eigenvalues with *a posteriori/a priori* inequalities
J R Kuttler and V G Sigillito
136 Regular semigroups as extensions
F J Pastijn and M Petrich
137 Representations of rank one Lie groups
D H Collingwood
138 Fractional calculus
G F Roach and A C McBride
139 Hamilton's principle in continuum mechanics
A Bedford
140 Numerical analysis
D F Griffiths and G A Watson
141 Semigroups, theory and applications. Volume I
H Brezis, M G Crandall and F Kappel
142 Distribution theorems of L-functions
D Joyner
143 Recent developments in structured continua
D De Kee and P Kaloni
144 Functional analysis and two-point differential operators
J Locker
145 Numerical methods for partial differential equations
S I Hariharan and T H Moulden
146 Completely bounded maps and dilations
V I Paulsen
147 Harmonic analysis on the Heisenberg nilpotent Lie group
W Schempp
148 Contributions to modern calculus of variations
L Cesari
149 Nonlinear parabolic equations: qualitative properties of solutions
L Boccardo and A Tesei
150 From local times to global geometry, control and physics
K D Elworthy

151 A stochastic maximum principle for optimal control of diffusions
U G Haussmann
152 Semigroups, theory and applications. Volume II
H Brezis, M G Crandall and F Kappel
153 A general theory of integration in function spaces
P Muldowney
154 Oakland Conference on partial differential equations and applied mathematics
L R Bragg and J W Dettman
155 Contributions to nonlinear partial differential equations. Volume II
J I Díaz and P L Lions
156 Semigroups of linear operators: an introduction
A C McBride
157 Ordinary and partial differential equations
B D Sleeman and R J Jarvis
158 Hyperbolic equations
F Colombini and M K V Murthy
159 Linear topologies on a ring: an overview
J S Golan
160 Dynamical systems and bifurcation theory
M I Camacho, M J Pacifico and F Takens
161 Branched coverings and algebraic functions
M Namba
162 Perturbation bounds for matrix eigenvalues
R Bhatia
163 Defect minimization in operator equations: theory and applications
R Reemtsen
164 Multidimensional Brownian excursions and potential theory
K Burdzy
165 Viscosity solutions and optimal control
R J Elliott
166 Nonlinear partial differential equations and their applications. Collège de France Seminar. Volume VIII
H Brezis and J L Lions
167 Theory and applications of inverse problems
H Haario
168 Energy stability and convection
G P Galdi and B Straughan
169 Additive groups of rings. Volume II
S Feigelstock
170 Numerical analysis 1987
D F Griffiths and G A Watson
171 Surveys of some recent results in operator theory. Volume I
J B Conway and B B Morrel
172 Amenable Banach algebras
J-P Pier
173 Pseudo-orbits of contact forms
A Bahri
174 Poisson algebras and Poisson manifolds
K H Bhaskara and K Viswanath
175 Maximum principles and eigenvalue problems in partial differential equations
P W Schaefer
176 Mathematical analysis of nonlinear, dynamic processes
K U Grusa
177 Cordes' two-parameter spectral representation theory
D F McGhee and R H Picard
178 Equivariant K-theory for proper actions
N C Phillips
179 Elliptic operators, topology and asymptotic methods
J Roe
180 Nonlinear evolution equations
J K Engelbrecht, V E Fridman and E N Pelinovski
181 Nonlinear partial differential equations and their applications. Collège de France Seminar. Volume IX
H Brezis and J L Lions
182 Critical points at infinity in some variational problems
A Bahri
183 Recent developments in hyperbolic equations
L Cattabriga, F Colombini, M K V Murthy and S Spagnolo
184 Optimization and identification of systems governed by evolution equations on Banach space
N U Ahmed
185 Free boundary problems: theory and applications. Volume I
K H Hoffmann and J Sprekels
186 Free boundary problems: theory and applications. Volume II
K H Hoffmann and J Sprekels
187 An introduction to intersection homology theory
F Kirwan
188 Derivatives, nuclei and dimensions on the frame of torsion theories
J S Golan and H Simmons
189 Theory of reproducing kernels and its applications
S Saitoh
190 Volterra integrodifferential equations in Banach spaces and applications
G Da Prato and M Iannelli
191 Nest algebras
K R Davidson
192 Surveys of some recent results in operator theory. Volume II
J B Conway and B B Morrel
193 Nonlinear variational problems. Volume II
A Marino and M K Murthy
194 Stochastic processes with multidimensional parameter
M E Dozzi
195 Prestressed bodies
D Iesan
196 Hilbert space approach to some classical transforms
R H Picard
197 Stochastic calculus in application
J R Norris
198 Radical theory
B J Gardner
199 The C* – algebras of a class of solvable Lie groups
X Wang

200 Stochastic analysis, path integration and dynamics
D Elworthy
201 Riemannian geometry and holonomy groups
S Salamon
202 Strong asymptotics for extremal errors and polynomials associated with Erdös type weights
D S Lubinsky
203 Optimal control of diffusion processes
V S Borkar
204 Rings, modules and radicals
B J Gardner
205 Numerical studies for nonlinear Schrödinger equations
B M Herbst and J A C Weideman
206 Distributions and analytic functions
R D Carmichael and D Mitrović
207 Semicontinuity, relaxation and integral representation in the calculus of variations
G Buttazzo
208 Recent advances in nonlinear elliptic and parabolic problems
P Bénilan, M Chipot, L Evans and M Pierre
209 Model completions, ring representations and the topology of the Pierce sheaf
A Carson
210 Retarded dynamical systems
G Stepan
211 Function spaces, differential operators and nonlinear analysis
L Päivärinta
212 Analytic function theory of one complex variable
Y Komatu, K Niino and C C Yang
213 Elements of stability of visco-elastic fluids
J Dunwoody
214 Jordan decompositions of generalised vector measures
K D Schmidt
215 A mathematical analysis of bending of plates with transverse shear deformation
C Constanda
216 Ordinary and partial differential equations Vol II
B D Sleeman and R J Jarvis
217 Hilbert modules over function algebras
R G Douglas and V I Paulsen
218 Graph colourings
R Wilson and R Nelson
219 Hardy-type inequalities
A Kufner and B Opic
220 Nonlinear partial differential equations and their applications. College de France Seminar Volume X
H Brezis and J L Lions

Analytic function theory of one complex variable

Yûsaku Komatu, Kiyoshi Niino and
Chung-chun Yang (Editors)

Kanazawa Institute of Technology /Kanazawa University, Japan/
Naval Research Laboratory Washington DC

Analytic function theory of one complex variable

Longman Scientific & Technical

Copublished in the United States with
John Wiley & Sons, Inc., New York

Longman Scientific & Technical,
Longman Group UK Limited,
Longman House, Burnt Mill, Harlow
Essex CM20 2JE, England
and Associated Companies throughout the world.

Copublished in the United States with
John Wiley & Sons, Inc., 605 Third Avenue, New York, NY 10158

First published 1989

AMS Subject Classification: 30XX, 30EXX, 30FXX

ISSN 0269-3674

British Library Cataloguing in Publication Data
Yang, C. C.
Analytic function theory of one complex variable.
1. Calculus. Functions of one complex variable
I. Title II. Niino, K. III. Komatu, Y.
515.9′3

ISBN 0-582-03780-8

Library of Congress Cataloging-in-Publication Data
Analytic function theory of one complex variable/(edited by) C. C. Yang, K. Niino, and Y. Komatu.
p. cm.—(Pitman research notes in mathematics series: 212)
ISBN 0-470-21336-1
1. Functions of complex variables. 2. Analytic functions.
I. Yang, C. C. II. Niino, K. (Kiyoshi), 1941- . III. Komatu, Y. (Yûsaku), 1914–
IV. Series.

QA331.A6544 1989
515′.93–dc20 89–8073 CIP

Printed and bound in Great Britain
by Biddles Ltd, Guildford and King's Lynn

Contents

PREFACE

CONTRIBUTORS TO THIS VOLUME

ON THE MYRBERG TYPE PHENOMENON 1
Mikihiro Hayashi and Mitsuru Nakai

ON A CLASS OF RIEMANN SURFACES CHARACTERIZED BY PERIOD REPRODUCING DIFFERENTIALS ... 13
Naondo Jin and Yukio Kusunoki

WEIERSTRASS POINTS ON TRIGONAL RIEMANN SURFACES 21
Takao Kato

ON SUBORDINATION FOR BMOA 38
Shoji Kobayashi

DISTORTION PROPERTIES OF INTEGRAL OPERATORS RELATED TO FRACTIONAL CALCULUS 48
Yûsaku Komatu

A COEFFICIENT INEQUALITY FOR MEROMORPHIC UNIVALENT FUNCTIONS 77
Yoshihisa Kubota

AN INCLUSION LATTICE OF AUTOMORPHISM GROUPS OF COMPACT RIEMANN SURFACES OF GENUS 5 86
Akikazu Kuribayashi and Hideyuki Kimura

VARIATIONAL FORMULAS UNDER QUASICONFORMAL DEFORMATION 105
Fumio Maitani

MEROMORPHIC FUNCTIONS WITH A PERFECT SET AS THE SET OF SINGULARITIES 123
Kikuji Matsumoto and Toshiko Kurokawa

GAP SERIES 149
Takafumi Murai

MEROMORPHIC SOLUTIONS OF SOME POLYNOMIAL DIFFERENCE EQUATIONS 178
Yoshikuni Nakamura and Niro Yanagihara

ZEROS OF CERTAIN DIFFERENTIAL POLYNOMIALS 199
Mitsuru Ozawa

SURVEYS OF ANALYTIC MAPPINGS BETWEEN TWO RIEMANN SURFACES 226
Mitsuru Ozawa and Kiyoshi Niino

INTERPOLATION PROBLEMS OF PICK-NEVANLINNA TYPE 253
Saburou Saitoh

DISCRETENESS OF REAL TWO-GENERATOR FREE GROUPS 263
Hiroki Sato

CONFORMAL EMBEDDINGS OF AN OPEN RIEMANN SURFACE INTO CLOSED SURFACES OF THE SAME GENUS 287
Masakazu Shiba

DIRICHLET PRINCIPLE AND THE ANALYTIC PROLONGATION THEOREM ONTO ANALYTIC BOUNDARY .. 299
Keiichi Shibata

ON A DISTORTION THEOREM OF THE LOGARITHMIC CAPACITY 325
Nobuyuki Suita

PINCHING DEFORMATION OF ARBITRARY RIEMANN SURFACES AND VARIATIONAL FORMULAS FOR ABELIAN DIFFERENTIALS 330
Masahiko Taniguchi

ON ALGEBROID SOLUTIONS OF SOME ALGEBRAIC DIFFERENTIAL EQUATIONS 346
Nobushige Toda and Masakimi Kato

SOME ADVANCES IN THE THEORY OF FACTORIZATION OF ENTIRE OR MEROMORPHIC FUNCTIONS ... 356
Hironobu Urabe

POLYNOMIAL DIFFERENCE EQUATIONS WHICH HAVE MEROMORPHIC SOLUTIONS OF FINITE ORDER .. 368
Niro Yanagihara

Preface

Although the theory of analytic functions of one complex variable is a classical field, it has always been a flourishing and applicable one in mathematical research. The purpose of this volume is to offer the reader an overview of new research directions and achievements in this field contributed by Japanese mathematicians in recent years.

The contents consist of expository surveys, surveys with new results as well as original research papers on diverse topics written by respected authors. While the collection may fail to cover all the topics in this field thoroughly, it, nevertheless, will be able to provide the reader with an opportunity to keep abreast of what is going on in the current research of complex analysis in Japan.

We are grateful to the contributors for their enthusiasm and collaboration towards this project. The manuscript was skillfully typed by Karen Gallo. Her assistance is much appreciated. Finally we are grateful to the Publishers for their support during the preparation of this book.

Yusaku Komatu
Kiyoshi Niino
Chung-chun Yang

Contributors to this volume

Mikihiro Hayashi
Department of Mathematics, Hokkaido University, Sapporo 060, Japan

Naondo Jin
Department of Mathematics, Gakushuin University, Tokyo 171, Japan

Masakimi Kato
Department of Mathematics, Faculty of Liberal Arts, Shizuoka University, Ohya, Shizuoka 422, Japan

Takao Kato
Department of Mathematics, Yamaguchi University, Yoshida, Yamaguchi 753, Japan

Hideyuki Kimura
Department of Mathematics, Tokyo Institute of Technology, Oh-okayama, Meguro-ku, Tokyo 152, Japan

Shōji Kobayashi
The Technological University of Nagaoka, Kamitomioka, Nagaoka 949-54, Japan

Yûsaku Komatu
Kanazawa Institute of Technology, 4-29-40 Takadanobaba, Shinjuku-ku, Tokyo 169, Japan

Yoshihisa Kubota
Department of Mathematics, Tokyo Gakugei University, Koganei-shi, Tokyo 184, Japan

Akikazu Kuribayashi
Department of Mathematics, Chuo University, 1-13-27, Kasuga, Bunkyo-ku, Tokyo 112, Japan

Toshiko Kurokawa
Department of Mathematics, Faculty of Education, Mie University, Kamihama-cho, Tsu 514, Japan

Yukio Kusunoki
Department of Mathematics, Kyoto University, Kyoto 606, Japan

Fumio Maitani
Department of Mathematics, Kyoto Institute of Technology, Matsugazaki, Sakyo-ku, Kyoto 606, Japan

Kikuji Matsumoto
Department of Mathematics, College of General Education, Nagoya University, Chikusa-ku, Nagoya 464, Japan

Takafumi Murai
Department of Mathematics, Faculty of Science,
Nagoya University, Chikusa-ku, Nagoya 464, Japan

Mitsuru Nakai
Department of Mathematics, Nagoya Institute of Technology,
Gokiso, Showa-ku, Nagoya 466, Japan

Yoshikuni Nakamura
Department of Mathematics, College of General Education,
Chiba University, Yayoi, Chiba 260, Japan

Kiyoshi Niino
Faculty of Technology, Kanazawa University,
2-40-20, Kodatsuno, Kanazawa 920, Japan

Mitsuru Ozawa
Department of Mathematics, Science University of Tokyo,
Noda, Chiba 278, Japan

Saburou Saitoh
Department of Mathematics, Faculty of Engineering,
Gunma University, Kiryu 376, Japan

Hiroki Sato
Department of Mathematics, Faculty of Science,
Shizuoka University, Ohya, Shizuoka 422, Japan

Masakazu Shiba
Department of Mathematics, Faculty of Science,
Hiroshima University, Hiroshima 730, Japan

Keiichi Shibata
Faculty of Science, Okayama University of Science,
Ridai-cho, Okayama 700, Japan

Nobuyuki Suita
Department of Mathematics, Tokyo Institute of Technology,
Oh-okayama, Meguro-ku, Tokyo 152, Japan

Masahiko Taniguchi
Department of Mathematics, Kyoto University, Kyoto 606, Japan

Nobushige Toda
Department of Mathematics, Nagoya Institute of Technology,
Gokiso, Showa-ku, Nagoya 466, Japan

Hironobu Urabe
Department of Mathematics, Kyoto University of Education,
Fushimi-ku, Kyoto 612, Japan

Niro Yanagihara
Department of Mathematics, Faculty of Science,
Chiba University, Yayoi, Chiba 260, Japan

ON THE MYRBERG TYPE PHENOMENON

Mikihiro Hayashi and Mitsuru Nakai

We denote by $H^\infty(W)$ the set of bounded holomorphic functions on a Riemann surface W. Let $\tilde{W}$ be an unlimited possibly infinitely branched covering surface of a Riemann surface W with the projection map φ. We say that the *Myrberg type phenomenon* occurs for $\tilde{W}$ if we have

$$H^\infty(\tilde{W}) = H^\infty(W)\circ\varphi. \tag{1}$$

Denote by $\Delta(c,r)$ the open disk in the complex plane $\mathbb{C}$ with center c and radius r. Let Δ_0 be the punctured unit disk $\Delta(0,1) \setminus \{0\}$ and $\tilde{\Delta}_0$ the two sheeted unlimited covering surface of Δ_0 with branch points over 2^{-k} $(k = 1, 2, \cdots)$. We denote by φ the projection map of $\tilde{\Delta}_0$ onto Δ_0.

Choose a sequence $\{r_k\}$ of positive numbers r_k satisfying $2^{-k-1} + r_{k+1} < 2^{-k} - r_k$. This simply means that $\{\Delta_k\}$ is the sequence of mutually disjoint closed disks $\Delta_k = \overline{\Delta}\left(2^{-k}, r_k\right)$ $(k = 1, 2, \cdots)$ contained in the base surface Δ_0 of the covering surface $(\tilde{\Delta}_0, \Delta_0, \varphi)$. Consider a plane region R and its two sheeted covering surface $\tilde{R}$ given by the following:

$$R = \Delta_0 \setminus \bigcup_{k=1}^{\infty} \Delta_k, \quad \tilde{R} = \varphi^{-1}(R). \tag{2}$$

The region R is usually referred to as an L-*domain* in the sense of Zalcman. Such regions with arbitrary sequences of centers instead of $\{2^{-k}\}$, i.e. general L-domains, are also systematically studied from various view points by Behrens and many others. We stress that the covering surface $(\tilde{R}, R, \varphi)$ is *unlimited* and *smooth.*

The *purpose* of this paper is twofold. In part 1, we will give a simpler alternative proof to a result obtained in our former paper [1] that gives a condition for the occurrence of the Myrberg type phenomenon to the above covering surface ($\tilde{R}$, R, φ). In part 2, we will give a genuinely weaker condition than that formerly given in [1] to ($\tilde{R}$, R, φ) by which the Myrberg type phenomenon does not occur.

1. P. J. Myrberg's famous remark is that the Myrberg type phenomenon occurs for the covering surface ($\tilde{\Delta}_0$, Δ_0, φ). The first purpose of this paper is to give an elementary and self-contained simple alternative proof to the following result originally obtained in [1]: The Myrberg type phenomenon can occur for the covering surface ($\tilde{R}$, R, φ) in spite of the entire absence of branch points if $\{r_k\}$ decreases so rapidly that

$$\sum_{k=1}^{\infty} k/\log(1/r_k) < \infty. \tag{3}$$

This is nothing but the Wiener criterion : (3) is necessary and sufficient for the point $z = 0$ to be an irregular boundary point of the region R in the sense of the potential theory. We restate our assertion as

Theorem. *Under the condition* (3) *the Myrberg type phenomenon occurs for the unlimited smooth covering surface* ($\tilde{R}$, R, φ).

Proof. Take an arbitrary g in $H^{\infty}(\tilde{R})$ with which we associate an f in $H^{\infty}(R)$ given by the following relation:

$$f(z) = \left(g(z_+) - g(z_-)\right)^2 \tag{4}$$

for every z in R where $\varphi^{-1}(z) = \{z_+, z_-\}$. We have to show that $f \equiv 0$ on R. Contrariwise we assume that f is not identically zero on R and we will derive a contradiction after a series of long discussions given in the sequel. Without loss of generality we may assume that $|g| \leq 1/2$ so that $|f| \leq 1$. The condition (3) is equivalent to the existence of a sequence $\{n(k)\}_{k=1}^{\infty}$ of positive numbers n(k) such that

$$r_k = 2^{-n(k)k}, \quad \sum_{k=1}^{\infty} 1/n(k) < \infty. \tag{5}$$

Needless to say $n(k) \to \infty$ $(k \to \infty)$. For sufficiently large positive integers k we

consider annuli

$$\Delta_k^* = \left\{r_k < | z - 2^{-k} | < 2^{-k-2}\right\}.$$

We choose a coordinate w on $\varphi^{-1}(\Delta_k^*)$ so as to satisfy the relation $\varphi(w) = w^2 + 2^{-k}$.

Then $\varphi^{-1}(\Delta_k^*)$ is represented as

$$\varphi^{-1}(\Delta_k^*) = \left\{\sqrt{r_k} < | w | < 2^{-k/2-1}\right\}$$

and $w = \varphi^{-1}(z)$ as $w = \sqrt{z - 2^{-k}}$.

Observe that γ_k: $| w | = \left(\sqrt{r_k} 2^{-k/2-1}\right)^{1/2}$ is a concentric circle in the annulus $\varphi^{-1}(\Delta_k^*)$. We will estimate the diameter $\operatorname{diam}\left(g(\gamma_k)\right)$ of the image curve $g(\gamma_k)$ of γ_k under the mapping g. For the sake of simplicity we put $\sqrt{r_k} = a$ and $2^{-k/2-1} = b$. Then $\varphi^{-1}(\Delta_k^*)$ is $\{a < | w | < b\}$ and γ_k: $|w| = \sqrt{ab}$. Take w in γ_k and use the

Cauchy

integral formula to deduce

$$| g'(w) | = \left| \frac{1}{2\pi i}\left(\int_{|\omega|=b} \frac{g(\omega)}{(\omega-w)^2} d\omega - \int_{|\omega|=a} \frac{g(\omega)}{(\omega-w)^2} d\omega \right) \right|$$

$$\leq \frac{1}{2\pi i}\left(\int_{|\omega|=b} \frac{1/2}{(b-\sqrt{ab})^2} | d\omega | + \int_{|\omega|=a} \frac{1/2}{(\sqrt{ab}-a)^2} | d\omega| \right)$$

$$= \frac{1}{(\sqrt{b} - \sqrt{a})^2}.$$

In view of this we see that

$$\operatorname{diam}\left(g(\gamma_k)\right) \leq \frac{1}{2}\int_{\gamma_k} | g'(w) | | dw | \leq \frac{\pi\sqrt{ab}}{(\sqrt{b} - \sqrt{a})^2}$$

$$= \pi \frac{\sqrt{a/b}}{(1 - \sqrt{a/b})^2} \le \pi \frac{\left(\sqrt{r_k}2^{k/2+1}\right)^{1/2}}{\left(1 - \left(\sqrt{r_k}2^{k/2+1}\right)^{1/2}\right)^2}.$$

Therefore we have estimates

$$\operatorname{diam}\left(g(\gamma_k)\right) \le 2^{-(n(k)/8)\,k}$$

for all sufficiently large k. We set

$$\rho_k = 2^{-k} + \sqrt{r_k}2^{-k/2-1}.$$

The point ρ_k is in Δ_k^* for every sufficiently large k and $(\rho_k)_+$ and $(\rho_k)_-$ belong to γ_k where we set $\varphi^{-1}(\rho_k) = \left\{(\rho_\kappa)_+, (\rho_k)_-\right\}$. Hence we have

$$\begin{aligned} | f(\rho_\kappa) | &= | g((\rho_\kappa)_+) - g((\rho_k)_-) |^2 \\ &\le \left(\operatorname{diam}(g(\gamma_k))\right)^2 \le 2^{-(n(k)/4)\,k} \end{aligned}$$

Since $\rho_k^n > 2^{-nk}$ for every $n = 0, 1, \cdots$, by the above estimate of $| f(\rho_k) |$, we see that $| f(\rho_k) / \rho_k^n | \le 2^{-(n(k)/4-n)\,k}$ and we can conclude that

$$\lim_{m\to\infty} f(\rho_m) / \rho_m^n = 0 \qquad (n = 0, 1, \cdots). \tag{6}$$

For each $n = 0, 1, \cdots$ we define an auxiliary function G_n defined by the following equation

$$G_n(z) = \sum_{k=0}^{\infty} \frac{1}{2\pi i} \int_{\partial\Delta_k} \frac{f(\zeta)}{\zeta^n(\zeta - z)}\, d\zeta$$

for every z in $\mathbb{C}$ as far as it is meaningful, where the negative (positive, resp.) direction is given to $\partial\Delta_k$ for $k > 0$ ($k = 0$, resp.) in each integral. By (5) this is well defined for z in R. Again by (5) this is also well defined for $z = 0$:

$$G_n(0) = \sum_{k=0}^{\infty} \frac{1}{2\pi i} \int_{\partial\Delta_k} \frac{f(\zeta)}{\zeta^{n+1}}\, d\zeta.$$

We now maintain that the following limits exist:

$$\lim_{m \to \infty} G_n(\rho_m) = G_n(0) \quad (n = 0, 1, \cdots). \tag{7}$$

For the purpose we estimate the absolute value of the integrand in the integral defining $G_n(\rho_m)$ on every integrating domain $\partial\Delta_k$ separately for sufficiently large m. First for k = m

$$\max_{\zeta \epsilon \partial\Delta_k} \left| \frac{f(\zeta)}{\zeta^n(\zeta - \rho_m)} \right| \leq \frac{1}{(s^{-k} - r_k)^n \left(\sqrt{r_k} 2^{-k/2-1} - r_k\right)} \leq 2^{-k}/r_k$$

since k = m is sufficiently large. Next for sufficiently large k with k ≠ m we have

$$\max_{\zeta \epsilon \partial\Delta_k} \left| \frac{f(\zeta)}{\zeta^n(\zeta - \rho_m)} \right| \leq \frac{1}{\left(2^{-k} - r_k\right)^n \cdot \sqrt{r_k}} \leq 2^{-k}/ r_k.$$

Thus the absolute value of the integrand in the integral defining $G_n(\rho_m)$ is dominated by an integrable step function independent of m. Therefore the Lebesgue dominated convergence theorem allows us to take the limit under the integral sign to conclude the required relation (7).

Using relations (7) and (8) obtained above we derive the following representation formula of f(z) which is crucial in our proof:

$$f(z) = z^n G_n(z) \quad (n = 0, 1, \cdots). \tag{8}$$

In order to derive this it is convenient to derive the following auxiliary information simultaneously:

$$G_n(0) = 0 \quad (n = 0, 1, \cdots). \tag{9}$$

We will prove both of (8) and (9) at the same time by the mathematical induction. Using the Cauchy integral formula with the aid of (5) we at once see that (8) is true for $n = 0$. In particular we have $f(\rho_m) = G_0(\rho_m)$. On letting $m \to \infty$ in this identity and using (6) and (7) for the case $n = 0$ we conclude that $G_0(0) = 0$. Namely we have shown (8) and (9) for $n = 0$. Now we have to show that the validity of (8) and (9) for n implies that for $n + 1$ instead of n. Starting from $f(z) = z^n G_n(z)$ and $G_n(0) = 0$ we proceed as follows:

$$
\begin{aligned}
f(z) &= z^n G_n(z) = z^n\left(G_n(z) - G_n(0)\right) \\
&= z^n \sum_{k=0}^{\infty} \frac{1}{2\pi i}\left(\int_{\partial\Delta_k} f(\zeta)\left(\frac{1}{\zeta^n(\zeta - z)} - \frac{1}{\zeta^{n+1}}\right) d\zeta\right) \\
&= z^{n+1} \sum_{k=0}^{\infty} \frac{1}{2\pi i}\int_{\partial\Delta_k} \frac{f(\zeta)}{\zeta^{n+1}(\zeta - z)}\, d\zeta = z^{n+1} G_{n+1}(z),
\end{aligned}
$$

which shows that (8) for $n + 1$ holds. In particular we have $f(\rho_m) / \rho_m^n = G_{n+1}(\rho_m)$. On making $m \to \infty$ in this identity and using (6) and (7) for $n + 1$ we see that $G_{n+1}(0) = 0$ or (9) for $n + 1$ is valid. The induction is herewith complete.

We take another sequence of closed disks $\Delta_k' = \overline{\Delta}\left(2^{-k}, \sqrt{r_k}\right)$ $(k = 1, 2, \cdots)$. These disks are mutually disjoint for sufficiently large k. By changing the radii suitably, if necessary, for the first finite number of disks Δ_k' under the restriction $\Delta_k' \supset \Delta_k$, we may assume that Δ_k' are all contained in Δ_0 and mutually disjoint. Consider the region

$$
Y = \Delta_0 \setminus \bigcup_{k=1}^{\infty} \Delta_k' .
$$

Although well known, for the sake of completeness, here we prove the existence of a strictly positive harmonic function $v(z)$ on Y with boundary values zero on $\partial Y - \{0\}$ such that

$$v(z) \le \log \frac{1}{|z|} \tag{10}$$

for every z in Y. Observe that the condition (3) is equivalent to $\Sigma_{k=1}^{\infty} k/\log(1/\sqrt{r_k}) < \infty$. Therefore the point $z = 0$ is also an irregular boundary point for the smaller region Y in the sense of the Dirichlet problem. Let $v(z, \zeta)$ be the harmonic Green's function for the region Y. The Bouligand theorem says that

$$\lim_{\zeta \epsilon Y} \sup_{\zeta \to 0} \quad v(z, \zeta) > 0$$

for a and hence for every z in R. By the Harnack principle, there exists a sequence $\{\zeta_n\}$ in Y convergent to 0 such that $\{v(z, \zeta_n)\}$ converges to a strictly positive harmonic function $v(z)$ on Y uniformly on every compact subset of Y. Since each $v(z, \zeta_n)$ has vanishing boundary values on $\partial Y - \{0\}$ and each component of $\partial Y - \{0\}$ is a circle, the limit function $v(z)$ must have boundary values zero on $\partial Y - \{0\}$. Moreover observe that

$$v(z, \zeta_n) < \log \left| \frac{1-\bar{\zeta}_n z}{z - \zeta_n} \right|$$

on Y. On making $n \to \infty$, we obtain (10).

The condition (5) assures that

$$C_n = \sup_{z \epsilon Y} | G_n(z) | \le \sum_{k=0}^{\infty} \frac{1}{2\pi} \int_{\partial \Delta_k} \frac{1}{|\zeta|^n (\sqrt{r_k} - r_k)} |d\zeta| < \infty$$

for every $n = 0, 1, \cdots$. Hence by (8) we have $|f(z)| \le C_n |z|^n$ for z in $\overline{Y} - \{0\}$ or equivalently $\log(1/|f(z)|)$ dominates the function $n \log(1/|z|) - \log C_n$ on $\overline{Y} - \{0\}$ for every $n = 0, 1, \cdots$. Thus in particular by (10) we obtain the following

$$\log \frac{1}{|f(z)|} \ge n\, v(z) - \log C_n \quad (n = 0, 1, \cdots) \tag{11}$$

for every z in $Y - \{0\}$. Consider a superharmonic function

$$s(z) = s_{n,\epsilon}(z) = \log \frac{1}{|f(z)|} + \epsilon \log \frac{1}{|z|} - n\, v(z)$$

on $\overline{Y} - \{0\}$ for an arbitrarily fixed positive number ϵ. Since $|f| \le 1$ on Y and $v = 0$ on

$\partial Y - \{0\}$, we see that the boundary values of s on $\partial Y - \{0\}$ are nonnegative. By (11) and the effect of the term $\epsilon \log(1 / |z|)$, we have $\lim_{z \epsilon Y,\, z \to 0} s(z) = \infty$. Therefore we can conclude that

$$\lim_{z \epsilon Y} \inf_{z \to \zeta} s(z) \geq 0 \qquad (\zeta \in \partial Y).$$

By the minimum principle for superharmonic functions we see that $s(z) = s_{n,\epsilon}(z) \geq 0$ for every z in Y. On letting ϵ tend to zero we finally conclude that

$$\log \frac{1}{|f(z)|} \geq n\, v(z) \quad (n = 0, 1, \cdots)$$

for any z in Y, which is a desired contradiction.

The proof of the theorem is herewith complete.

2. We turn to the second purpose of this paper of giving the condition by which the Myrberg type phenomenon does not occur for the covering surface $(\tilde{R}, R, \varphi)$. Roughly speaking the Myrberg type phenomenon occurred for $(\tilde{R}, R, \varphi)$ if the sequence $\{r_k\}$ of radii decreased *enough rapidly*. We now show, again roughly speaking, that the Myrberg type phenomenon does not occur for $(\tilde{R}, R, \varphi)$ if the sequence $\{r_n\}$ of radii decreases *enough slowly*. The precise meaning of this is given in the following

Theorem. *If the sequence* $\{r_n\}$ *of radii decreases so slowly that*

$$\inf_{n \epsilon N} 2^{n(n-d)/2} \cdot r_n > 0 \tag{12}$$

for some constant $d > 1$, *then the Myrberg type phenomenon does not occur for the covering surface* $(\tilde{R}, R, \varphi)$.

In our former paper [1] we have proved the theorem under the genuinely stronger assumption that $\inf_{n \epsilon N} 2^n r_n > 0$. Suppose the theorem is valid under the assumption that $\inf_{n \epsilon N} 2^{N(n)} r_n > 0$ with $N(n) \to \infty$ $(n \to \infty)$ instead of (12). Then we can see (cf. the reduction to (15) below) that $\Sigma_{n=1}^{\infty} n/N(n) = \infty$ since otherwise the origin 0 is an irregular boundary point of R and the theorem is invalid by our first result in part 1. Therefore if $N(n) = a_0 n^{\nu} + a_1 n^{\nu - 1} + \cdots + a_{\nu}$ ($a_0 > 0$) is a polynomial of n, then $\deg N \le 2$. Thus the condition (12) in the theorem is the best possible in this very restricted sense.

Proof. It is convenient to consider auxiliary quantities τ and σ given by

$$\tau = 2^{(d+1)/2}, \quad \sigma = \max\left((\tau + 2) / (\tau - 2), 4\right). \tag{13}$$

We need to introduce one more constant

$$K = \max\left(\sup_{n \in \mathbb{N}} \left(2^{-(n-1)/2} \tau\right)^n \sigma, \sup_{n \in \mathbb{N}} \left(2^{-n(n-d)/2} / r_n\right) \right) \tag{14}$$

which is seen to be finite by virtue of (12). This in particular implies that

$$r_n \ge K^{-1} \tau^n 2^{-n(n+1)/2} \quad (n \in \mathbb{N}).$$

If we replace each r_n by a smaller radius, then $\tilde{R}$ is replaced by a larger subregion of $\tilde{\Delta}_0$ for which we only have to prove the invalidity of the Myrberg type phenomenon. Therefore we may and thus will assume that

$$r_n = K^{-1} \tau^n 2^{-n(n+1)/2} \quad (n \in \mathbb{N}). \tag{15}$$

Expressing r_n as $r_n = K^{-1}\left(2^{-n(n-1)/2} \tau\right)^n 2^{-n}$ and using (14) we see that

$$r_n \le 2^{-n} / \sigma \quad (n \in \mathbb{N}). \tag{16}$$

We consider a function p(z) given by the infinite product

$$p(z) = \prod_{n=1}^{\infty} \left(1 + q_n(z)\right)$$

formed by using rational functions

$$q_n(z) = \frac{2^{-n}}{z - 2^{-n}} \quad (n \in \mathbb{N}).$$

By the simple estimate $|q_n(z)| \le 2^{-(n-1)} / |z|$ ($|z| \ge 2^{-(n-1)}$), we see that the above infinite product converges almost uniformly on $\hat{\mathbb{C}} \setminus \{0\}$ and defines a meromorphic function p(z) on $\hat{\mathbb{C}} \setminus \{0\}$, where $\hat{\mathbb{C}}$ is the extended complex plane $\mathbb{C} \cup \{\infty\}$. Since p(z) can also be expressed as

$$p(z) = \prod_{n=1}^{\infty} \frac{z^2}{(z - 2^{-(2n - 1)})(z - 2^{-2n})},$$

we can conclude that the square root $f(z) = \sqrt{p(z)}$ defines a single valued meromorphic function on the covering surface $\tilde{\Delta}_0$ of Δ_0 but not on Δ_0. Hence we can conclude that f belongs to $H^{\infty}(\tilde{R}) \setminus H^{\infty}(R) \circ \varphi$ if we can prove that $p \in H^{\infty}(R)$.

For the proof of $p \in H^{\infty}(R)$ we first estimate quantities

$$Q_{nk} = \sup_{z \in \partial\Delta_k} | 1 + q_n(z) | \quad (n, k \in \mathbb{N}).$$

If $n < k$, then for $z \in \partial\Delta_k$ we have the following by (16):

$$| 1 + q_n(z) | = | \frac{z}{z - 2^{-n}} | \leq \frac{2^{-k} + r_k}{2^{-n} - 2^{-k} - r_k}$$

$$\leq \frac{2^{-k} + 2^{-k} / \sigma}{2^{-n} - 2^{-k} - 2^{k} / \sigma}$$

$$\leq \frac{(1 + 1/\sigma)2^{-k}}{2^{-n} - (1 + 1/\sigma)2^{-n}/2} = \frac{2(\sigma + 1)}{\sigma - 1} 2^{n - k}.$$

By (13) we see that $2(\sigma + 1) / (\sigma - 1) \leq \tau$ and we obtain the following estimate:

$$Q_{nk} \leq \tau \cdot 2^{n - k} \quad (n < k). \tag{17}$$

If $n = k$, then for $z \in \partial\Delta_k$ we proceed by using (16) as follows:

$$| 1 + q_n(z) | =| \frac{z}{z - 2^{-k}} | \leq \frac{2^{-k} + r_k}{r_k}$$

$$\leq \frac{2^{-k} + 2^{-k} / \sigma}{r_k} \leq \frac{\sigma + 1}{\sigma} \cdot \frac{2^{-k}}{r_k}.$$

By (13) we see that $(\sigma + 1) / \sigma \leq \tau$ and we have

$$Q_{kk} \leq \tau \cdot (2^{-k} / r_k). \tag{18}$$

Finally, if $k < n$, then for $z \in \partial\Delta_k$ we have again by using (16) the following:

$$| 1 + q_n(z) | \leq 1 + | q_n(z) | \leq 1 + \left| \frac{2^{-n}}{z - 2^{-n}} \right|$$

$$\leq 1 + \frac{2^{-n}}{2^{-k} - r_k - 2^{-n}}$$

$$\leq 1 + \frac{2^{-n}}{2^{-k} - 2^{-k}/\sigma - 2^{-k}/2} \leq 1 + \frac{2\sigma}{\sigma - 2} 2^{-(n - k)}.$$

Since $\sigma \geq 4$ by (13), we have $2\sigma / (\sigma - 2) \leq 4$ and therefore

$$Q_{nk} \leq 1 + 4 \cdot 2^{-(n - k)} \quad (k < n). \tag{19}$$

Fix an arbitrary $m \in \mathbb{N}$ and let $z \in \partial\Delta_k$ $(1 \leq k \leq m)$. By using (17), (18), (19) and then (15) we proceed as follows:

$$\left| \prod_{n=1}^{m} (1 + q_n(z)) \right| = \left(\prod_{n=1}^{k-1} | 1 + q_n(z) | \right) \cdot | 1 + q_k(z) | \cdot \left(\prod_{n=k+1}^{m} |1 + q_n(z) | \right)$$

$$\leq \left(\prod_{n=1}^{k-1} Q_{nk} \right) \cdot Q_{kk} \cdot \left(\prod_{n=k+1}^{m} Q_{nk} \right)$$

$$\leq \left(\prod_{n=1}^{k-1} (\tau \cdot 2^{n-k}) \right) \cdot \tau(2^{-k}/r_k) \cdot \left(\prod_{n=k+1}^{m} (1 + 4 \cdot 2^{-(n-k)}) \right)$$

$$= \left(\tau^k 2^{-k(k + 1)/2} / r_k \right) \cdot \prod_{n=1}^{m-k} (1 + 4 \cdot 2^{-n}) \leq M$$

where $M = K \cdot \prod_{n=1}^{\infty} (1 + 4 \cdot 2^{-n})$ is a finite constant.

The function $p_m(z) = \Pi_{n=1}^{m} (1 + q_n(z))$ is holomorphic on $R_m = \hat{\mathbb{C}} \setminus \cup_{k=1}^{m} \Delta_k$ and continuous on its closure $\overline{R}_m$ and, as we have seen above, $| p_m(z) | \leq M$ on its boundary ∂R_m. Therefore by the maximum principle we see that $| p_m(z) | \leq M$ on R_m and in particular on R. By making $m \to \infty$ in $| p_m(z) | \leq M$ $(z \in R)$ we conclude that $| p(z) | \leq M$ $(z \in R)$, or $p \in H^\infty(R)$, as was to be proved.

The proof of the theorem is herewith complete.

Acknowledgements. To complete the present work the first (second, resp.) named author was supported in part by Grant-in-Aid for Scientific Research No. 61540071 (61540094, resp.), Japanese Ministry of Education, Science and Culture.

References

[1] Hayashi, M., and M. Nakai, Point separation by bounded analytic functions of a covering Riemann surface, Pacific Jour. Math., 134 (1988), pp. 261 - 273.

Mikihiro Hayashi
Department of Mathematics
Hokkaido University

Mitsuru Nakai
Department of Mathematics
Nagoya Institute of Technology

ON A CLASS OF RIEMANN SURFACES CHARACTERIZED BY PERIOD REPRODUCING DIFFERENTIALS

Naondo Jin and Yukio Kusunoki

1. Introduction.

In the papers [3, 8] M. Watanabe (née Mori) introduced the following condition, which we call here (W) *condition (or property)*,

$$\Gamma_{he}(R) \cap {}^*\Gamma_{hse}(R) \subset {}^*\Gamma_{he}(R) \tag{W}$$

for Riemann surface R (cf. Section 2.1 for the notations), and obtained some equivalent conditions and interesting consequences. The space $\Gamma_{he}(R) \cap {}^*\Gamma_{hse}(R)$ is trivial by definition if and only if R belongs to class O_{KD}. Hence R in O_{KD} has property (W). While, in [4] B. Rodin showed that if R is a Riemann surface of finite positive genus with property (W), then R belongs to class O_{AD}. For surfaces of finite genus, it is known that $O_{AD} = O_{KD}$, hence in this case, $R \in O_{AD}$ if and only if R satisfies the condition (W). The Rodin's proof makes use of Torelli's theorem and a few facts in Teichmüller space theory.

In this paper we shall give a new characterization of the property (W) for general Riemann surfaces in terms of specific period reproducing differentials, and further give another proof to the Rodin's result without using big tool.

2. Preliminaries.

For the sake of convenience we recall some definitions and known facts in the theory of differentials on (open) Riemann surfaces.

2.1 Let Γ (resp. Γ_h) be the Hilbert space of square integrable real measurable (resp. harmonic) differentials on Riemann surface R, where the inner product is given by

$$(\omega_1, \omega_2) = (\omega_1, \omega_2)_R = \iint_R \omega_1 \wedge *\omega_2,$$

$*\omega_2$ being the conjugate differential of ω_2. The space Γ has the well-known orthogonal decomposition $\Gamma = \Gamma_h + \Gamma_{eo} + {}^*\Gamma_{eo}$, where Γ_{eo} is the closure in Γ of $\Gamma_{eo}^{\infty} = \{df \mid f$ is a C^{∞}-function with compact support on $R\}$ and ${}^*\Gamma_{eo} = \{*\omega \mid \omega \in \Gamma_{eo}\}$.

Let Γ_{he} (resp. Γ_{hse}) be the subspace of Γ_h whose elements ω are exact (resp. semi-exact) on R, that is, $\int_\gamma \omega = 0$ for every (resp. every dividing) 1-cycle γ on R. Given a closed subspace Γ_x of Γ_{he}, the orthogonal complement of Γ_x in Γ_h is denoted by $\Gamma_x{}^{\perp}$. For the spaces $\Gamma_{ho} = ({}^*\Gamma_{he})^{\perp}$ and $\Gamma_{hm} = ({}^*\Gamma_{hse})^{\perp}$, the following inclusion relations hold; $\Gamma_h \supset \Gamma_{hse} \supset \Gamma_{he} \supset \Gamma_{hm}$, $\Gamma_{hse} \supset \Gamma_{ho} \supset \Gamma_{hm}$. The Γ_{hm} is known as the space of harmonic measure differentials.

<u>2.2</u>. Let R be an open Riemann surface. By a neighborhood of the ideal boundary of R we mean the exterior of a compact set of R having no relatively compact complements. Let ω be a real differential defined in a neighborhood of the ideal boundary of R and Γ_x be any closed subspace of Γ_{he}. Then ω is said to have Γ_x-*behavior* if the following representation holds in some neighborhood of the ideal boundary of R:

$$\begin{cases} \omega = \omega_1 + df, \ \omega_1 \in \Gamma_x \\ *\omega = \omega_2 + dg, \ \omega_2 \in \left({}^*\Gamma_x\right)^{\perp} \end{cases} \tag{1}$$

where f and g are C^{∞}-functions on R such that df and dg belong to Γ_{eo}. We say that a function w has Γ_x-behavior if dw does. Obviously in this case the first equation in (1) may be replaced by $w = v + f$ with some harmonic function v on R such that $dv \in \Gamma_x$. The following basic property for special Γ_x will be used later.

<u>Lemma 1.</u> *Let V be a neighborhood of the ideal boundary of R such that the relative*

boundary ∂V consists of a finite number of mutually disjoint analytic Jordan curves. Let u be a harmonic function on $\overline{V} = V \cup \partial V$. If u have Γ_{he}-behavior, then

$$(du,\ dw)_V = \int_{\partial V} w * du \quad \textit{for every } w \in HD(V). \tag{2}$$

If u has Γ_{hm}-behavior, then

$$(du,\ dw)_V = \int_{\partial V} u * dw \quad \textit{for every } w \in KD(V), \tag{3}$$

where $HD(V)$ is the set of harmonic functions on $\overline{V}$ with finite Dirichlet integral over V and $KD(V) = \{u \in HD(V) \mid *du$ is semi-exact in $V\}$.

The above properties give actually a characterization of those behaviors (For the proof see Kusunoki [1, 2] and Yoshida [9]).

2.3. For a given 1-cycle c on R and a closed subspace Γ_y of Γ_h there exists uniquely the period reproducing differential $\sigma_y(c)$ in Γ_y such that

$$\int_c \omega = (\omega,\ \sigma_y(c)) \quad \text{for every } \omega \in \Gamma_y.$$

We are interested of course in the case $\sigma_y(c) \not\equiv 0$. It is known (cf. [2] [9]) that (i) $*\sigma_y(c)$ is exact on $R - c$, (ii) $*\sigma_y(c)$ has $({}^*\Gamma_y)^{\perp}$-behavior, (iii) $\int_{c'} *\sigma_y(c) = c \times c'$ for any 1-cycle c' where the intersection number $c \times c'$ is taken as $+1$ when c' crosses c from its right shore to the left with respect to the positive direction of c.

3. New characterization of property (W).

3.1. The main result of this paper is the following

Theorem. Let R be an arbitrary Riemann surface and $\sigma_{hse}(c)$ (resp. $\sigma_{ho}(c)$) be the period reproducing differential in the space Γ_{hse} (resp. Γ_{ho}) for 1-cycle c on R. Then the following properties are equivalent;

(a) *R has property (W).*

(b) $\| \sigma_{hse}(c) \| = \|\sigma_{ho}(c) \|$ *(equivalently, $\sigma_{hse}(c) = \sigma_{ho}(c)$) for every 1*

-cycle c on R.

Furthermore, if R is of finite positive genus, the next properties are also equivalent;

(c) *R belongs to class O_{AD}.*

(d) $\| \sigma_{hse}(c) \| = \| \sigma_{ho}(c) \|$ *(equivalently,* $\sigma_{hse}(c) = \sigma_{ho}(c)$*) for some non-dividing 1-cycle c on R.*

The proof will be given in the following sections after providing two lemmas.

3.2. For a fixed 1-cycle c, consider the integrals

$$s_c^0(p) = \int^p *\sigma_{ho}(c), \quad s_c^1(p) = \int^p *\sigma_{hse}(c), \ p \in R,$$

then s_c^0 and s_c^1 are single-valued harmonic functions on R − c and have Γ_{he}- and Γ_{hm}-behavior respectively. Let

$$s_c(p) = \begin{cases} \dfrac{s_c^0(p) - s_c^1(p)}{\| \sigma_{hse}(c) - \sigma_{ho}(c) \|}, & \text{if } \sigma_{hse}(c) \not\equiv \sigma_{ho}(c) \\ 0, & \text{otherwise}, \end{cases} \tag{4}$$

then from the intersection property (iii) we find that $s_c(p)$ is a single-valued harmonic function on R and $\int_\gamma *ds_c = 0$ for every dividing curve γ on R. Hence s_c belongs to the following class

$$KD_1(R) = \{ u \in KD(R) \mid \| du \| \le 1 \}.$$

Lemma 2. *For a fixed 1-cycle on c on R, the above function s_c is the unique function in $KD_1(R)$, up to an additive constant, which maximizes the functional*

$$u \to \left| \int_c *du \right|, \quad u \in KD_1(R).$$

The maximum value is equal to $\| \sigma_{hse}(c) - \sigma_{ho}(c) \|$.

Proof. It suffices to treat the case of non-dividing curve c. For any $u \in KD_1(R)$ we have $\int_c *du = \left(*du, \sigma_{hse}(c)\right)$. Since $*du$ belong to $^*\Gamma_{he}$, $\left(*du, \sigma_{ho}(c)\right) = 0$. It follows that

$$\int_c *du = \left(*du, \sigma_{hse}(c) - \sigma_{ho}(c)\right). \tag{5}$$

Thus we may assume that $\sigma_{hse}(c) \not\equiv \sigma_{ho}(c)$. By Schwarz inequality

$$\begin{aligned} \left|\int_c *du\right| &\leq \| du \| \, \| \sigma_{hse}(c) - \sigma_{ho}(c) \| \\ &\leq \| \sigma_{hse}(c) - \sigma_{ho}(c) \|. \end{aligned}$$

Since from (5) we have

$$\int_c *ds_c = \| \sigma_{hse}(c) - \sigma_{ho}(c) \| , \tag{6}$$

it holds that $\left|\int_c *du\right| \leq \int_c *ds_c$ and the equality occurs only when $u = s_c +$ const.,

q.e.d.

Since $\left(\sigma_{ho}(c), \sigma_{hse}(c) - \sigma_{ho}(c)\right) = \left(*\sigma_{ho}(c), d(s_c^0 - s_c^1)\right) = 0$, we have

$$\| \sigma_{hse}(c) - \sigma_{ho}(c) \|^2 = \| \sigma_{hse}(c) \|^2 - \|\sigma_{ho}(c) \|^2. \tag{7}$$

Corollary. *$\sigma_{hse}(c) = \sigma_{ho}(c)$, equivalently $\| \sigma_{hse}(c) \| = \| \sigma_{ho}(c) \|$ if and only if $\int_c *du = 0$ for every $u \in KD(R)$.*

Since the property (W): $\Gamma_{he} \cap {}^*\Gamma_{hse} \subset {}^*\Gamma_{he}$ for R is equivalent to that for every $u \in KD(R)$, $*du$ is exact on R. Hence we proved the first part of our theorem by Corollary.

3.3. To prove the remaining part of Theorem we assume that R is a Riemann surface of finite genus g. Then there exists a compact continuation of R, that is, a pair (R_1, f_1) of compact Riemann surface R_1 of genus g and a conformal mapping f_1 of R into R_1. We use the following

Lemma 3. *Let R be a Riemann surface of finite genus. Then the following properties are equivalent;*

1) *R belongs to class O_{AD}.*

2) *For every compact continuation (R_1, f_1) of R, $R_1 - f_1(R)$ is a set of two dimensional measure zero.*

Proof. (cf. Sario-Nakai [6], pp. 46 - 48). Here we add some comments. Take a regular subregion R_0 of R with connected (analytic) boundary ∂R_0 such that $R - \overline{R}_0$ is a planar region. Map $R - R_0$ conformally into the unit disk $U = \{ \mid z \mid \leq 1\}$ so that ∂R_0 corresponds to ∂U, and denote by E the inner boundary of the image of $R - R_0$, a realization of the ideal boundary of R. Now let (R_1, f_1) be any compact continuation of R, then U and E can be considered as the conformal images of $R_1 - f_1(R_0)$ and $R_1 - f_1(R)$ respectively. Now if R belongs to O_{AD}, then E is AD-null, hence the area $\mid E \mid$ of E is zero.

Conversely, suppose the condition 2) holds. It suffices to show that the set E above is AD-null. To see this consider an arbitrary conformal mapping F of $\Omega = \hat{\mathbb{C}} - E$ and show that $E' = \hat{\mathbb{C}} - F(\Omega)$ has vanishing area. Let Δ be a simply connected domain bounded by an analytic curve $F(\partial U)$ which contains the set E'. Let ψ be a conformal mapping of Δ onto $V = \{ \mid w \mid \leq 1\}$ and set $E'' = \psi(E')$. By conformal correspondence we may replace U by V and obtain a new compact continuation of R. Hence $\mid E'' \mid = 0$ and therefore $\mid E' \mid = 0$, q.e.d.

The next lemma is crucial in our argument.

Lemma 4. *Let R be a Riemann surface of finite genus. Suppose there exists a non-constant harmonic function u in a neighborhood of the ideal boundary of R which has both Γ_{he}- and Γ_{hm}-behavior. Then R must belong to class O_{AD}.*

Proof. Let V be a neighborhood of the ideal boundary of R such that u is harmonic on $\overline{V} = V \cup \partial V$ and the relative boundary ∂V consists of a finite number of analytic dividing curves on R. Let (R_1, f_1) be an arbitrary compact continuation of R. We may identify $f_1(R)$ with R. Then $R - V$ is a compact subset of R_1, and $V_1 = R_1 - (R - V)$ contains V, and ∂V_1 coincides with ∂V. Denote by w the solution (harmonic function) of Dirichlet problem on V_1 with boundary values $u|_{\partial V_1}$. Since the boundary function is a C^∞-function, Dirichlet integral of w over V_1 is finite. Moreover w has a harmonic extension across ∂V_1, because u is harmonic on $\overline{V}$ and $u - w$ is zero on analytic boundary. Since R and R_1 have the same genus, a dividing curve on R is also dividing on R_1, actually homologous to zero on R_1. Hence $w \in KD(V)$. As u has Γ_{hm}-behavior, by Lemma 1 we have

$$(du, dw)_V = \int_{\partial V} u *dw = \int_{\partial V_1} w *dw = \| dw \|^2_{V_1},$$

moreover, since u has Γ_{he}-behavior,

$$(du, dw)_V = \int_{\partial V} w *du = \int_{\partial V} u *du = \| du \|^2_V.$$

It follows that

$$\begin{aligned} 0 \leq \| du - dw \|^2_V &= \| du \|^2_V + \| dw \|^2_V - 2(du, dw)_V \\ &= \| dw \|^2_V - \| dw \|^2_{V_1} \\ &= - \| dw \|^2_{R_1 - R} \leq 0, \end{aligned}$$

therefore $u \equiv w$ on V and $\| dw \|_{R_1 - R} = 0$. From the hypothesis u is non-constant, so the area of $R_1 - R$ must be zero. Since (R_1, f_1) is arbitrary, we conclude R belongs to O_{AD} by Lemma 3, q.e.d.

Proof of the latter half of Theorem. (*d*) $\Rightarrow$ (*c*); If R is of finite positive genus and $\sigma_{hse}(c) = \sigma_{ho}(c)$ for a non-dividing curve c on R, then $s_c^0 = s_c^1$ is a harmonic function with both Γ_{he}-and Γ_{hm}-behavior and is non-constant by intersection property (iii), hence $R \in O_{AD}$ by Lemma 4. Now (c) $\Rightarrow$ (a) ($\Leftrightarrow$ (b)) (cf. Introduction) and (b) $\Rightarrow$ (d) is trivial, which completes the proof.

3.4. Remark. By Lemma 4 we know that if R is of finite genus and has vanishing "K-span "(cf. Sario-Oikawa [7]) at some point on R, then R belongs to O_{AD}. This is also seen from the fact that K-span is represented by Dirichlet norm of the difference of two functions, one of which has Γ_{he}-behavior and the other had Γ_{hm}-behavior. For further results of the classification by span, see also M. Sakai [5].

References

[1] Kusunoki, Y., Characterizations of canonical differentials. J. Math. Kyoto Univ. 5 (1966), pp. 197 - 207.

[2] Kusunoki, Y., Riemann surfaces and conformal mappings, (Japanese). Asakura, Tokyo, 1973.

[3] Mori, M., Contributions to the theory of differentials on open Riemann surfaces. J. Math. Kyoto Univer. 4 (1964), pp. 77 - 97.

[4] Rodin, B., On a paper of M. Watanabe. J. Math. Kyoto Univ. 6 (1967), pp. 393 - 395.

[5] Sakai, M., On the vanishing of the span of a Riemann surface. Duke Math. J. 41 (1974), pp. 497 - 510.

[6] Sario, L., and M. Nakai, Classification theory of Riemann surfaces. Springer, Berlin-Heidelberg-New York, 1970.

[7] Sario, L., and K. Oikawa, Capacity functions. Springer, Berlin-Heidelberg-New York, 1969.

[8] Watanabe, M., A remark on the Weierstrass points on open Riemann surfaces. J. Math. Kyoto Univ. 5 (1966), pp. 185 - 192.

[9] Yoshida, M., The method of orthogonal decomposition for differentials on open Riemann surfaces. J. Sci. Hiroshima Univ. Sci. A - I 32 (1968), pp. 181 - 210.

Department of Mathematics, Faculty of Science
Kyoto University

WEIERSTRASS POINTS ON TRIGONAL RIEMANN SURFACES

Takao Kato

1. Introduction

A compact Riemann surface which admits a two-sheeted covering of $\mathbb{P}^1$ is said to be hyperelliptic. The Weierstrass gap sequences at the ramification points of the two-sheeted coverings of the sphere are well known [11]. In various points of view, there are many investigation concerning with the hyperelliptic loci in the moduli space (for eg. [2, 23, 19, 1, 17, 4]).

In the present paper, we shall survey some results on the Weierstrass gap sequences at the ramification points of trigonal Riemann surfaces and on subspaces of trigonal Riemann surfaces in the moduli space ([14, 7, 8, 16, 9, 15]). Except for Theorem 4.4, all the results of this paper are not new.

Let S be a trigonal Riemann surface of genus $g \geq 5$ and let $x\colon S \to \mathbb{P}^1$ be a trigonal covering. Let $D = (x)_\infty$ be the polar divisor of a meromorphic function x on S. For a positive integer k, let $\ell(kD)$ be the affine dimension of the space of meromorphic functions on S whose divisors are multiples of kD. Since $\ell(D) = 2$ and $\ell(kD) = 3k + 1 - g$ for $k > (2g - 2)/3$, there is an integer n such that $\ell(nD) = n + 1$ and $\ell((n + 1) D) \geq n + 3$. Obviously, n is uniquely determined by S. In this case we say that S is of the n-th kind [8]. The number $n - 1$ is also said to be the Maroni invariant [20] which is named after Maroni's study [18]. By the Riemann-Roch theorem it is easy to see that $(g - 1)/3 \leq n \leq g/2$ [7, Lemma 2.4, Corollary 2.11].

Assume S is of the n-th kind. We say that a point P on S is a total (resp. an ordinary) ramification point if the ramification index of x at P is equal to three (resp. two). If P is a total ramification point, then the gap sequence at P is either

$$(1, 2, 4, 5, ..., 3n - 2, 3n - 1, 3n + 1, 3n + 4, ..., 3(g - n - 1) + 1),$$

or

$$(1, 2, 4, 5, ..., 3n - 2, 3n - 1, 3n + 2, 3n + 5, ..., 3(g - n - 1) + 2).$$

We say that P is of type I in the former case and of type II in the latter case (cf. [14, 7]). If P is an ordinary ramification point, then the gap sequence at P is either

$$(1, 2, 3, ..., 2n - 1, 2n, 2n + 1, 2n + 3, ..., 2g - 2n -1),$$

or

$$(1, 2, 3, ..., 2n - 1, 2n, 2n + 2, 2n + 4, ..., 2g - 2n).$$

We say that P is of type I in the former case and of type II in the latter case (cf. [8, 16]).

If $n = (g - 1)/3$, then there is neither a total nor an ordinary ramification point of type II (cf. [16, 9]) and if $n = g/2$, then types I and II are equal.

In section 2, we give a defining equation of a trigonal Riemann surface and show that the kind and the types of the ramification points are determined by the given equation.

In section 3, we discuss the existence and the nonexistence of trigonal Riemann surfaces of given genus, kind and the numbers of various types of ramification points.

Let $M_{g,3,n}$ be the set of trigonal Riemann surfaces of genus, g, and of the n-th kind.

For $n \neq g/2$, let $M_{g,3,n}(\rho_1, \rho_2)$ be the subset of $M_{g,3,n}$ consisting of Riemann surfaces which have ρ_1 total ramification points of type I and ρ_2 total ramification points of type II. As remarked above, since there is no difference between types I and II in case $n = g/2$, we mean $M_{g,3,g/2}(\rho_1, \rho_1)$ to be the subset of $M_{g,3,g/2}$ consisting of Riemann surfaces which have p_1 total ramification points.

For $n \neq g/2$, Let $M_{g,3,n}(\rho_1, \rho_2, \rho_3, \rho_4)$ be the subset of $M_{g,3,n}(\rho_1, \rho_2)$ consisting of Riemann surfaces which have p_3 ordinary ramification points of type I and ρ_4 ordinary ramification points of type II.

In section 4, we compute the dimension of $M_{g,3,n}(\rho_1, \rho_2)$'s and $M_{g,3,n}(\rho_1, \rho_2, \rho_3, \rho_4)$'s in the moduli space.

In section 5, we consider the incidence relations of these spaces and again we discuss the existence theorems.

2. Defining Equation for Trigonal Surface

Assume S is a trigonal Riemann surface of genus g and of the n-th kind. Let x: S $\to \mathbb{P}^1$ be a trigonal covering and let $D = (x)_\infty$. Since $\ell((n + 1) D) \geq n + 3$, there exists a meromorphic function y on S such that $(y)_\infty = (n + 1)D$ and the full meromorphic function field on S is generated by x and y. Then we have:

Lemma 2.1 ([16, Lemma 2.1′]). *The Riemann surface S is defined by an algebraic equation*

$$y^3 + Q(x)y + R(x) = 0. \tag{2.1}$$

where $\deg Q = 2n + 2$, $\deg R = 3n + 3$ *and* $\deg(4Q^3 + 27R^2) = 6n + 6$.

A trigonal Riemann surface S is said to be cyclic if there is a conformal automorphism φ of S such that the order of φ is three and $S/\langle\varphi\rangle$ is the sphere, where $\langle\varphi\rangle$ is a cyclic group generated by φ. Then we have:

Lemma 2.2 ([16, Lemma 2.2]). *For a trigonal Riemann surface S of genus g (≥ 5) and of the n-th kind, the following three conditions are equivalent:*

i) *S is cyclic;*

ii) *All of the ramification points are total, $2g - 3n + 1$ of them are of* *type I and the other* $3n - g + 1$ *points are of type II;*

iii) *In the equation in Lemma 2.1,* $Q(x)$ *is identically zero and* $R(x)$ *has* $2g - 3n + 1$ *simple zeros and* $3n - g + 1$ *double zeros.*

Henceforth, we assume S is not cyclic i.e. Q(x) is not identically zero.

If there is a common zero α of Q(x) and R(x) such that the order of zero of Q(x) at $x = \alpha$ is greater than one and that of R(x) is greater than two, then the polar divisor of $y/(x - \alpha)$ is nD. Hence, $\ell(nD) \geq n + 2$. A contradiction. Thus there is no such a common zero of Q(x) and R(x). Then, Q(x) and R(x) satisfy the following equations by suitable polynomials Γ_i (i = 1, 2, 3) and Π_j(j = 1, 2, 3, 4) in x:

$$Q(x) = \Gamma_1(x)\, \Pi_1(x)\Pi_2(x)^2\Pi_4(x), \tag{2.2}$$

$$R(x) = \Gamma_2(x)\Pi_1(x)\Pi_2(x)^2\Pi_4(x)^2 \tag{2.3}$$

and

$$\Pi_3(x)\Gamma_3(x)^2 = 4\Gamma_1(x)^3\Pi_1(x)\Pi_2(x)^2 + 27\Gamma_2(x)^2\Pi_4(x). \tag{2.4}$$

Here, $\Gamma_1\Pi_1\Pi_2$ and $\Gamma_2\Pi_4$ have no common zero,

$$\Pi_j(x) = \prod_{i=1}^{\rho_j}(x - a_{i,j}), \quad (j = 1, ..., 4) \tag{2.5}$$

for nonnegative integer, ρ_i (j = 1, ..., 4) and mutually distinct complex numbers $a_{i,j}$ (i = 1, ..., ρ_j, j = 1, ... 4),

$$\deg \Gamma_1 = 2n + 2 - \rho_1 - 2\rho_2 - \rho_4, \tag{2.6}$$

$$\deg \Gamma_2 = 3n + 3 - \rho_1 - 2\rho_2 - 2\rho_4, \tag{2.7}$$

$$2 \deg \Gamma_3 = 6n + 6 - 2\rho_1 - 4\rho_2 - \rho_3 - 3\rho_4 \tag{2.8}$$

and

$$2\rho_1 + 2\rho_2 + \rho_3 + \rho_4 - 4 = 2g. \tag{2.9}$$

In [16, Theorem 4.3], we proved that a_{1j}'s correspond to total ramification points of type

I, $a_{2,j}$'s correspond to total ramification points of type II. $a_{3,j}$'s correspond to ordinary ramification points of type I and $a_{4,j}$'s correspond to ordinary ramification points of type II. This result is contained in Theorem 2.5 below.

Conversely, we first give a defining equation of a trigonal surface S by (2.1), where Q(x) and R(x) satisfy (2.2) – (2.8). According to the Cardano formula for the cubic equations, we have an explicit solution of (2.1), i.e. y is written explicitly in terms of x. Considering the Puiseux series at $x = a_{i,j}$, we have:

Lemma 2.3 ([15, Theorem A]). *Assume S is a trigonal Riemann surface defined by the equation* (2.1), *where* $Q(x)$ *and* $R(x)$ *satisfy* (2.2) – (2.8). *Then,*

i) *There is a total ramification point over* $x = a_{i,j}$, $i = 1, \ldots, p_j$ *if and only* $j = 1$ *or* 2,

ii) *There is an ordinary ramification point over* $x = a_{i,j}$, $i = 1, \ldots, p_j$ *if and only if* $j = 3$ *or* 4,

iii) *There is no other ramification point.*

Next, we are concerned with holomorphic differentials on S. Using a similar argument as in [6, p. 273] and observing the orders of zeros of y, $3y^2 + Q(x)$ and dx, we have:

Lemma 2.4 ([15, Lemma 1]). *Assume S is a trigonal Riemann surface defined by* (2.1), *where* $Q(x)$ *and* $R(x)$ *in* (2.1) *satisfy* (2.2) – (2.9). *Then, every holomorphic differential on S is given by*

$$\Omega(D, E) = \frac{D(x)y + E(x)\Pi_2(x)\Pi_4(x)}{3y^2 + \Gamma_1(x)\Pi_1(x)\Pi_2(x)^2\Pi_4(x)}dx, \tag{2.10}$$

where $D(x)$ *and* $E(x)$ *are suitable polynomials in x so that* $\deg D \leq n - 1$ *and deg* $E \leq 2n - \rho_2 - \rho_4$ *and every zero of* Γ_3 *is also a zero of* $D(x)y + E(x)\Pi_2(x)\Pi_4(x)$ *with the same multiplicity for a suitable branch of* y.

For an arbitrary polynomial D(x) of degree less than or equal to $n - 1$, there is an E(x) of degree less than or equal to $2n - \rho_2 - \rho_4$ so that (2.10) is holomorphic. By (2.8) and (2.9), we have $2n - \rho_2 - \rho_4 - \deg\Gamma_3 = g - n - 1 > 0$. Hence, any $g - n - 1$ coefficients of E(x) can be chosen arbitrarily. Considering this fact, and again observing the orders of zeros of y, $3y^2 + Q(x)$ and dx, precisely, we have:

Theorem 2.5 ([16, Theorem 4.3], [15, Theorem 1]). *Assume* S *is a trigonal Riemann surface defined by the equation* (2.1), *where* $Q(x)$ *and* $R(x)$ *satisfy* (2.2) – (2.8). *Then,*

i) The genus g *of* S *is given by*

$$\frac{2\rho_1 + 2\rho_2 + \rho_3 + \rho_4 - 4}{2}$$

ii) If $g \geq 5$ *and* $2n \leq g \leq 3n + 1$, *then* S *is of the* n*-th kind and*

1. *There is a total ramification point of type I over* $x = a_{i,1}$, $i = 1, \ldots, \rho_1$,
2. *There is a total ramification point of type II over* $x = a_{i,2}$, $i = 1, \ldots, \rho_2$,
3. *There is an ordinary ramification point of type I over* $x = a_{i,3}$, $i = 1, \ldots, \rho_3$,
4. *There is an ordinary ramification point of type II over* $x = a_{i,4}$, $i = 1, \ldots, \rho_4$,
5. *There is no other ramification point.*

iii) If $g \geq 5$ *and* $(3n - 3\rho - 1)/2 \leq g \leq 2n - 2\rho$, *where* $\rho = \deg \Gamma_3$, *then* S *is of the* $(g - n + \rho)$*-th kind and*

1. *There is a total ramification point of type I over* $x = a_{i,2}$,

$i=1,\ldots,\rho_2,$

2. *There is a total ramification point of type II over* $x = a_{i,1},$

$i=1,\ldots,\rho_1,$

3. *There is an ordinary ramification point of type I over* $x = a_{i,3},$

$i=1,\ldots,\rho_3,$ *and* $x = a_{i,4},$ $i = 1, \ldots, \rho_4,$

4. *There is no other ramification point.*

In this case there is no ordinary ramification point of type II.

In connection with Lemma 2.4, we have the following lemma due to Maroni. For the notion and the notation of rational normal scrolls, the reader should refer to [13].

Lemma 2.6 ([18], [20, §1], [9, §1]). *For a trigonal Riemann surface S of genus g and of the n-th kind, the canonical embedding of S into* $\mathbb{P}^{g-1}$ *lies on a rational normal scroll* $S_{n-1,g-n-1}$ *such that the ruling of* $S_{n-1,g-n-1}$ *cuts out the linear series* g_3^1 *on S.*

Coppens' study is mainly based on the fact mentioned in the above lemma and the intersection-dimension theorem, while the author and Horiuchi's study depends upon Lemma 2.1. It seems to be convenient to employ Lemma 2.6 when we study our problems from the algebraic view points.

3. Existence theorem

Observing the equations (2.6), (2.9) and ii) of Theorem 2.5, we have:

Theorem 3.1. *If S is a non-cyclic trigonal Riemann surface of genus g and of the n-th kind having* ρ_1 *total ramification points of type I,* ρ_2 *total ramification points of type II,* ρ_3 *ordinary ramification points of type I and* ρ_4 *ordinary ramification points of type II, then the following inequalities hold:*

$$\rho_1 + 2\rho_2 + \rho_4 \leq 2n + 2, \tag{3.1}$$

$$\rho_1 + 2\rho_2 + 2\rho_4 \leq 3n + 3, \tag{3.2}$$

$$\rho_2 + \rho_4 \leq 3n + 1 - g. \tag{3.3}$$

From this theorem we have easily the following corollaries.

Corollary 3.2 ([8, Theorem 4.1]). *Let S be a trigonal Riemann surface of genus g and of the n-th kind having* ρ_1 *total ramification points of type I and* ρ_2 *total ramification points of type II. If* $\rho_1 + 2\rho_2 \geq 2n + 3$, *then S is cyclic and* $\rho_1 = 2g + 1 - 3n$ *and*

$\rho_2 = 3n - g + 1$.

Corollary 3.3 ([16, Theorem 6.1, Corollary 6.2]). *If S is non-cyclic, then* $2\rho_2 + \rho_4 \leq \rho_1$. *Hence, there exists an ordinary ramification point of type I on every non-cyclic trigonal Riemann surface.*

As to the existence theorem, appealing to the intersection-dimension theorem, Coppens proved the following:

Theorem 3.4 ([7, Theorem 5.1]). *Let* (g, n, ρ_1, ρ_2) *be given four-tuple satisfying that* $g \geq 5$, $(g - 1)/3 \leq n \leq g/2$, $\rho_1 + 2\rho_2 \leq 2n + 2$ *and* $\rho_2 \leq 3n - g + 1$. *If* $n \neq g/2$, *then there exists a trigonal Riemann surface of genus g and of the n-th kind having* ρ_1 *total ramification points of type I and* ρ_2 *total ramification points of type II. If* $n = g/2$, *then there exists a trigonal Riemann surface of genus g and of the n-th kind having* $\rho_1 + \rho_2$ *total ramification points.*

For a general $M_{g,3,n}(\rho_1, \rho_2, \rho_3, \rho_4)$, we do not know whether it is empty or not. However, in case $\rho_2 + \rho_4 = 3n + 1 - g$, we can construct an example of a Riemann surface as follows:

Let S be defined by

$$y^3 + \Gamma_1(x)\Pi_1(x)\Pi_2(x)^2\Pi_4(x)y + \Gamma_2(x)\Pi_1(x)\Pi_2(x)^2\Pi_4(x)^2 = 0,$$

where $\Pi_1(x) = x^{p_1} + 2$, $\Pi_2(x) = x^{p_2} + 3$, $\Pi_4(x) = x^{p_4} + 5$ $\Gamma_1(x) =$

$\alpha(x - 1)^{2n+2-\rho_1-2\rho_2-\rho_4}$ and $\Gamma_2(x) = (x + 1)^{3n+3-\rho_1-2\rho_2-2\rho_4}$. For a suitable choice of α, S is in $M_{g,3,n}(\rho_1, \rho_2, \rho_3, \rho_4)$.

Hence, we have:

Theorem 3.5 ([16, Theorem 5.1]). *For every six-tuple* $(g, n, \rho_1, \rho_2, \rho_3, \rho_4)$ *satisfying that* $g \geq 5$, $(g - 1)/3 \leq n < g/2$, (3.1), (3.2), (2.9) *and that* $\rho_2 + \rho_4 = 3n + 1 - g$, *there exists a trigonal Riemann surfaces of genus g and of the n-th kind which has* p_1 *total ramification points of type I,* ρ_2 *total ramification points of type II,* ρ_3 *ordinary ramification points of type I and* ρ_4 *ordinary ramification points of type II.*

Combining the above example with iii) of Theorem 2.5, we have:

Corollary 3.6. *For* $g \geq 5$, $(g- 1)/3 \leq n < g/2$, *if* $\rho_1+\rho_2 \leq n + 1$, $\rho_2 \leq 3n-g + 1$ *and* $2g + 4 = 2\rho_1+ 2\rho_2 + \rho_3$, *then there exists a trigonal Riemann surface of genus g and of the n-th kind which has* ρ_1 *total ramification points of type I,* ρ_2 *total ramification points of type II and* ρ_3 *ordinary ramification points of type I.*

4. Dimension of $M_{g,3,n}$'s

In [9] Coppens considered $M_{g,3,n}$'s as subsets of the coarse moduli space M_g. On the other hand, in [15] the author considered them in the Teichmüller space T_g. Since there exists a natural map of T_g to M_g which is proper analytic and locally finite to one [21], the difference between T_g and M_g has no influence on the dimension and the incidence relation of $M_{g,3,n}$'s e. t. c.

Let

$$M'_{g,3,n}(\rho_1, \rho_2) = \bigcup_{t_1+t_2\geq\rho_1+\rho_2,\ t_2\geq p_2} M_{g,3,n}(t_1, t_2)$$

By Corollary 3.2, if $\rho_1 + 2\rho_2 \geq 2n + 3$, then $\rho_1 = 2g + 1 - 3n$, $\rho_2 = 3n - g + 1$ and $\rho_3 = \rho_4 = 0$. In this case, from [14] it follows that $M_{g,3,n}(2g + 1 - 3n, 3n - g + 1)$ is irreducible of dimension $g - 1$.

In the sequel we may assume that $\rho_1 + 2\rho_2 \leq 2n + 2$ and $p_2 \leq 3n - g + 1$ for $n \neq g/2$ and that $\rho_1 < 2n + 2$ for $n = g/2$. In this case, by Theorem 3.4, $M_{g,3,n}(\rho_1, \rho_2)$ is not empty.

In this connection, Coppens proved:

Theorem 4.1 ([9, Theorems 2.4 and 2.9]). *If* $(g - 1)/3 \leq n < g/2$, *then*

$$dim(M'_{g,3,n}(\rho_1, \rho_2)) = g + 2n + 2 - \rho_1 - 2\rho_2$$

and there exists an irreducible component of $M_{g,3,n}(\rho_1, \rho_2)$ *of dimension* $g + 2n + 2 - \rho_1 - 2\rho_2$.

If $n = g/2$, *then*

$$dim(M'_{g,3,n}(\rho_1, \rho_2)) = dim(M_{g,3,n}(\rho_1, \rho_1)) = 2g + 1 - \rho_1.$$

Moreover, Coppens studied on the structure of $M'_{g,3,n}(\rho_1, \rho_2)$ and he proved

Theorem 4.1 ([5, §2], [9, Theorem 3.3]).

i) *For* $(g - 1)/3 \leq n \leq g/2$, $M'_{g,3,n}(1, 0)$ *is irreducible and unirational.*

ii) *For* $(g - 1)/3 < n < g/2$, $M'_{g,3,n}(0, 1)$ *is irreducible and unirational.*

Next, we consider the dimension of $M_{g,3,n}(\rho_1, \rho_2, \rho_3, \rho_4)$. In general, it is not known whether $M_{g,3,n}(\rho_1, \rho_2, \rho_3, \rho_4)$ is empty or not for given six-tuple (g, n, ρ_1, ρ_2,

$\rho_3, \rho_4)$ (cf. §5). So we assume $M_{g,3,n}(\rho_1, \rho_2, \rho_3, \rho_4)$ is non-empty.

At first, we prepare a lemma concerning with the resultant of two polynomials.

Lemma 4.3. *Let* $A(x) = a_m x^m + \cdots + a_1 x + a_0$ *and* $B(x) = b_n x^n + \cdots + b_1 x + b_0$ *be two polynomials with* $a_m b_n \neq 0$ *and let*

$$R(A,B) = \begin{vmatrix} a_m & a_{m-1} & \cdots & \cdots & a_0 & 0 & \cdots & 0 \\ 0 & a_m & a_{m-1} & \cdots & \cdots & a_0 & 0 & \cdots \\ & & \cdots & & \cdots & & & \\ 0 & \cdots & 0 & a_m & a_{m-1} & \cdots & \cdots & a_0 \\ b_n & b_{n-1} & \cdots & \cdots & b_0 & 0 & \cdots & 0 \\ & b_n & b_{n-1} & \cdots & \cdots & b_0 & 0 & \cdots \\ & & \cdots & & \cdots & & & \\ 0 & \cdots & 0 & b_n & b_{n-1} & \cdots & \cdots & b_0 \end{vmatrix}$$

If $A(x)$ *and* $B(x)$ *have* k *distinct common zeros, then*

$$\mathrm{rank}(R(A, B)) \leq m + n - k.$$

Consider the ordered $m + n + 2$*-tuples* $(A,B) = (a_0, a_1, \ldots, a_m, b_0, b_1, \ldots, b_n)$ *as points of* $\mathbb{P}^{n+k+1}$. *If* $k \leq \min(m, n)$ *then the set*

$$V = \{(A,B) \in \mathbb{P}^{m+n+1} \mid \mathrm{rank}(R(A,B)) \leq m + n - k\}$$

is an algebraic set. Moreover, for each pair of $A(x)$ *and* $B(x)$ *which have* k *common zeros, there is a component* W *of* V *such that* $(A,B) \in W$ *and* $\dim W \geq m + n + 1 - k$ *and the set*

$$\{(A,B) \in W \mid A(x) \text{ and } B(x) \text{ have exactly } k \text{ common zeros}\}$$

is Zarisky open in W.

Then we have:

Theorem 4.4 *If* $M_{g,3,n}(\rho_1, \rho_2, \rho_3, \rho_4)$ *is non-empty, it is of dimension at least* $g + 2n + 2 - \rho_1 - 2\rho_2 - \rho_4$.

Proof. From (2.9), (3.1) and Corollary 3.3, it is easy to see that $\rho_1 \geq 3$ or $\rho_3 \geq 3$.

We first assume that $\rho_3 \geq 3$. Let S be an arbitrary element of $M_{g,3,n}(\rho_1, \rho_2, \rho_3, \rho_4)$. By Theorem 2.5 S is defined by the equation (2.1) which satisfies (2.2) – (2.8). We may assume that $a_{3,1} = 0$, $a_{3,2} = 1$ and $a_{3,3} = -1$. Let

$$Q(x) = b_{2n+2}x^{2n+2} + b_{2n+1}x^{2n+1} + \cdots + b_1 x + b_0,$$

and

$$R(x) = c_{3n+3}x^{3n+3} + c_{3n+2}x^{3n+2} + \cdots + c_1x + c_0.$$

Replacing y by $y/c_{3n+3}^{1/3}$, we may also assume that $c_{3n+3} = 1$. Let $\Gamma(x) = 4Q(x)^3 + 27R(x)^2$. By (2.2) – (2.4), $\Gamma(a_{3,j}) = 0$, $(j = 1, 2, 3)$. Then, $W_0 = V(f_1, f_2, f_3)$ is a $5n + 3$-dimensional variety in $\mathbb{C}^{5n+6}$ with the coordinate $(b_0, \ldots, b_{2n+2}, c_0, \ldots, c_{3n+2})$. Here, f_1, f_2, f_3 are defined by

$$f_1 = \Gamma(0) = 4b_0^3 + 27c_0^2,$$

$$f_2 = \Gamma(1) = 4(b_0 + b_1 + \cdots + b_{2n+2})^3 + 27(c_0 + c_1 + \cdots c_{3n+2} + 1)^2$$

and

$$f_3 = \Gamma(1) = 4(b_0 - b_1 + \cdots + b_{2n+2})^3$$
$$+ 27(c_0 - c_1 + \cdots (-1)^{3n+2}c_{3n+2} + (-1)^{3n+3})^2$$

Since $g \geq 5$, there exists a unique linear series g_3^1 on S. Hence, for generic S, there are exactly $\rho_3!/(\rho_3 - 3)!$ possible choices of Q(x) and R(x). Choose a pair of Q and R and let **a** be a point in $\mathbb{C}^{5k+6}$ corresponding to Q, R. Then, there is a sufficiently small neighborhood U of S in $M_{g,3,n}(\rho_1, \rho_2, \rho_3, \rho_4)$ so that there is an injective holomorphic map of U into the intersection V of a neighborhood of **a** and W_0 (refer to [12, §1], [3, §1] and the first paragraph in the proof of Theorem 3 in [15]).

By (2.2) – (2.9), generic elements of the image of U in V is characterized by

i) Γ and Γ' have exactly $\rho_1 + \rho_2 + \rho_4 + \rho$ common zeros, where $\rho = 3n + 1 - g - \rho_2 - \rho_4$,

ii) Γ and Γ'' have exactly $\rho_2 + \rho_4$ common zeros,

iii) Γ and Γ''' have exactly ρ_2 common zeros.

By Lemma 4.3, the dimensions of varieties $W_1 = V(\Gamma,\Gamma')$, $W_2 = V(\Gamma,\Gamma'')$ and $W_3 = V(\Gamma,\Gamma')$ in $\mathbb{C}^{5n+6}$ are at least $5n + 6 - (\rho_1 + \rho_2 + \rho_4 + \rho)$, $5n + 6 -(\rho_2 + \rho_4)$ and $5n + 6 - \rho_2$, respectively. Hence, by the intersection-dimension theorem,

$$\dim\left(W_0 \cap W_1 \cap W_2 \cap W_3\right) \geq 2n + 2 - \rho_1 - 2\rho_2 - \rho_4.$$

This is the desired inequality.

In the case that $\rho_3 \leq 2$ and $\rho_1 \geq 3$, except that it needs a little change of the defining equations of W_0, we are able to do the same discussion as above.

This completes the proof.

Moreover, the author guesses the following fact which agrees with Theorem 4.1 when $\rho_4 = 0$.

Guess 4.5. *For every six-tuple* $\left(g, n, \rho_1, \rho_2, \rho_3, \rho_4\right)$ *satisfying that* $g \geq 5$, $(g - 1)/3 \leq n < g/2$, (3.1) - (3.3) *and* (2.9), $M_{g,3,n}\left(\rho_1, \rho_2, \rho_3, \rho_2\right)$ *is non-empty constructible set of dimension* $g + 2n + 2 - \rho_1 - 2\rho_2 - \rho_4$.

For the definition of the constructible set, see [22].

5. Incidence relations

From Theorem 4.1, we have dim $M_{g,3,n} = g + 2n + 2$, if $(g - 1)/3 \leq n \leq (g - 2)/2$. Using the fact that n is lower semi-continuous in the space of trigonal Riemann surfaces $\cup_{(g-1)/3 \leq n \leq g/2} M_{g,3,n}$ [20, Lemma 2], we have immediately:

Theorem 5.1 ([9, Theorem 4.1]). *$M_{g,3,n}$ is included in the boundary of $M_{g,3,n+1}$, if $(g - 1)/3 \leq n \leq (g - 2)/2$.*

For the incidence relations on $M_{g,3,n}(\rho_1, \rho_2)$'s, Coppens proved:

Theorem 5.2 ([9, Theorem 4.4]). *Let $(g - 1)/3 \leq n \leq (g - 2)/2$.*

i) *$M_{g,3,n}(1, 0)$ is included in the boundary of $M_{g,3,n+1}(0, 1)$.*

ii) *If moreover $n \neq (g - 2)/2$, then $M_{g,3,n}(1, 0)$ is included in the boundary of $M_{g,3,n+1}(1, 0)$.*

As to the sets $M_{g,3,n}(\rho_1, \rho_2, \rho_3, \rho_4)$, deforming the coefficient polynomials Q, R of the defining equation (2.1) (cf. Lemmas 3-6 in [15]), we have:

Theorem 5.3 9[15, Theorem 2]). *Let $(g, n, \rho_1, \rho_2, \rho_3, \rho_4)$ be a six-tuple satisfying that $g \geq 5$, $(g - 1)/3 \leq n < g/2$, (3.1), (3.2), (2.9) and that $\rho_2 + \rho_4 = 3n + 1 - g$. Then $M_{g,3,n}(\rho_1, \rho_2, \rho_3, \rho_4)$ is included in the boundary of any one of*

i) $M_{g,3,n}(\rho_1 - 1, \rho_2, \rho_3 + 2, \rho_4)$, *if* $\rho_1 > 0$,

ii) $M_{g,3,n}(\rho_1, \rho_2 - 1, \rho_3 + 1, \rho_4 + 1)$, *if* $\rho_2 > 0$,

iii) $M_{g,3,n}(\rho_1 + 1, \rho_2 - 1, \rho_3, \rho_4)$, *if* $\rho_2 > 0$,

iv) $M_{g,3,n}(\rho_1, \rho_2 - 1, \rho_3 + 2, \rho_4)$, *if* $\rho_2 > 0$,

v) $M_{g,3,n}(\rho_1, \rho_2, \rho_3 + 1, \rho_4 - 1)$, *if* $\rho_4 > 0$,

vi) $M_{g,3,n}(p_1 - 1, p_2, p_3, p_4 + 2)$, *if* $p_1 > 0$,

vii) $M_{g,3,n}(\rho_1 - 1, \rho_2 + 1, \rho_3, \rho_4)$, *if* $\rho_1 > 0$.

This theorem induces an existence theorem which is not included in Theorem 3.5.

Using iii) - vii) of Theorem 5.3, we have:

Corollary 5.4. *Let* $(g, n, \rho_1, \rho_2, \rho_3, \rho_4)$ *be a six-tuple satisfying that* $g \geq 5$ *and* (2.9). *If one of the following holds:*

i) $g/3 \leq n < g/2$, $\rho_1 + 2\rho_2 + \rho_4 \leq 2n + 1$, $\rho_1 + 2\rho_2 + 2\rho_4 \leq 3n + 2$ *and* $\rho_2 + \rho_4 = 3n - g$,

ii) $(g + 2)/3 \leq n \leq g/2$, $\rho_1 + 2\rho_2 + \rho_4 \leq 2n + 1$, $\rho_1 + 2\rho_2 + 2\rho_4 \leq 3n + 1$ *and* $\rho_2 + \rho_4 = 3n - g - 1$,

then there exists a trigonal Riemann surface of genus g and of the n-th kind which has ρ_1 *total ramification points of type I,* ρ_2 *total ramification points of type II,* ρ_3 *ordinary ramification points of type I and* ρ_4 *ordinary ramification points of type II.*

Combining Theorem 3.4 and Theorem 4.4, we have:

Theorem 5.5. *Let* $\left(g, n, \rho_1, \rho_2\right)$ *be given four-tuple satisfying that* $g \geq 5$, $(g - 1)/3 \leq n < g/2$, $p_1 + 2\rho_2 \leq 2n + 2$ *and* $\rho_2 \leq 3n - g + 1$. *Let* $\rho_3 = 2g + 4 - 2\rho_1 - 2\rho_2$. *Then, there exists a trigonal Riemann surface of genus g and of the n-th kind having* ρ_1 *total ramification points of type I,* ρ_2 *total ramification points of type II,* ρ_3 *ordinary ramification points of type I and no ordinary ramification points of type II.*

References

[1] Accola, R. D. M., Riemann Surfaces, Theta Functions, and Abelian Automorphism Groups, Lecture Notes in Math., Springer-Verlag, Berlin, 1975

[2] Ahlfors, L. V., The complex analytic structure of the space of closed Riemann surfaces, Analytic Functions, Princeton Univ. Press, Princeton, 1960.

[3] Arbarello, E., Weierstrass points and moduli of curves, Compositio Math., 29 (1974), pp. 325 - 342.

[4] Bogomolov, F. A., Rationality of moduli of hyperelliptic curves of arbitrary genus, Canadian Math. Soc. Conf. Proc., 6 (1986), pp. 17 - 37.

[5] Canuto, C., Weierstrass points on trigonal curves of genus 5, Bull. de la Soc. Math. de France, 113 (1985), pp. 123 - 142.

[6] Coolidge, J. L., A Treatise on Algebraic Plane Curve, Dover, New York, 1959.

[7] Coppens, M., The Weierstrass gap sequences of the total ramification points of trigonal covering of $\mathbb{P}^1$. Indag. Math., 47 (1985), pp. 245 - 270.

[8] Coppens, M., The Weierstrass gap sequences of the ordinary ramification points of trigonal covering of $\mathbb{P}^1$; Existence of a kind of Weierstrass gap sequence, J. pure appl. Algebra, 43 (1986), pp. 11 - 25.

[9] Coppens, M., Weierstrass points on trigonal curves I: The ramification points, Preprint R. U. Utrecht, 430 (1986).

[10] Coppens, M., The number of Weierstrass points on some special curve I, Arch. Math., 46 (1986), pp. 453 - 465.

[11] Farkas, H. M., and I. Kra, Riemann Surfaces, Graduate texts in Math., 71, Springer-Verlag, 1980.

[12] Fulton, W., Hurwitz schemes and irreducibility of moduli of algebraic curves, Ann. Math., 90 (1969), pp. 542 - 575.

[13] Griffiths, P., and J. Harris, Principle of Algebraic Geometry, John Wiley & Sons, New York, 1978.

[14] Kato, T., On Weierstrass points whose first non-gaps are three, J. reine angew. Math., 316 (1980), pp. 99 - 109.

[15] Kato, T., On Subspaces of trigonal Riemann surfaces, to appear in Kodai Math. J.

[16] Kato, T., and R. Horiuchi, Weierstrass gap sequences at the ramification points of trigonal Riemann surfaces, J. pure appl. Algebra, 50 (1988), pp. 271 - 285.

[17] Katsylo, P. I., The rationality of moduli spaces of hyperelliptic curves, Math. USSR Izvestiya 25 (1985) pp. 45 - 50.

[18] Maroni, A., Le serie lineari speciali sulle curve trigonali, Annali di Matematica pura ed appl., 25 (1946), pp. 341 - 354.

[19] Martens, H. H., Varieties of special divisors on a curve II, J. reine angew. Math., 233 (1968) pp. 89 - 100.

[20] Martens, G., and F.-O. Schreyer, Line bundles and syzygies of trigonal curves,

Abh. Math. Sem. Univ. Hamburg, 56 (1986), pp. 169 - 189.

[21] Mumford, D., Abelian quotients of the Teichmüller modular group, J. analyse Math. 18 (1967) pp. 227 - 244.

[22] Mumford, D., Algebraic Geometry I, Complex Projective Varieties, Springer-Verlag, Berlin, 1976.

[23] Rauch, H. E., A moduli in conformal mapping, Comm. Pure Appl. Math., 41 (1955), pp. 236 - 238.

Department of Mathematics
Yamaguchi University
Yoshida, Yamaguchi 753

ON SUBORDINATION FOR BMOA

Shōji Kobayashi

1. Introduction

In this paper it is shown that the BMO norm of any function in the ball algebra A(U) is preserved under the composition by a non-inner function.

Let U denote the unit disk in the complex plane $\mathbb{C}$ and T be its boundary, the unit circle. The space of all functions f analytic in U and continuous on the closure $\overline{U} = U \cup T$ of U is called the ball algebra and denoted by A(U). For $0 < p < +\infty$, the Hardy class $H_p(U)$ is the space of all functions f analytic in U whose H_p norms $\| f \|_p$, defined by

$$\| f \|_p = \lim_{r \to 1} \left\{ \frac{1}{2\pi} \int_T | f(re^{i\theta}) |^p \, d\theta \right\}^{1/p},$$

are finite. Here, we identify T with the interval $[0, 2\pi)$, by $\xi = e^{i\theta}$ for $\xi \in T$ with $\theta \in [0, 2\pi)$. For $a \in U$, $\psi_a(z) = (z + a)/(1 + \bar{a}z)$ be the conformal map of U onto itself with $\psi_a(0) = a$ and $\psi_a{}'(0) > 0$. For any function f analytic in U, let

$$f_a(z) = f(\psi_a(z)) - f(a), \tag{1.1}$$

and the BMO norm $B_p(f)$ of f is defined by

$$B_p(f) = \sup_{a \in U} \{ \| f_a \|_p \}^p \tag{1.2}$$

It is a consequence of the John-Nirenberg theorem [4] that $B_p(f)$ is simultaneously finite for any p with $0 < p < +\infty$ (see, for example, [1, pp. 15-16]). The space of all functions f analytic in U for which $B_p(f)$ are finite is denoted by BMOA(U). Here, BMOA stands for "Analytic functions of Bounded Mean Oscillation". From now on and throughout the present paper we fix arbitrarily p with $0 < p < +\infty$, and we abbreviate the index p, that is, we write B(f) instead of $B_p(f)$.

For a map ψ of U into itself and a function f defined in U, we denote as usual by

$f \circ \psi$ the composite function of f and ψ, that is, $(f \circ \psi)(z) = f(\psi(z))$ for $z \in U$. A bounded analytic function ψ in U is called an inner function when it has boundary values of modulus 1 almost everywhere on T. It is well-known that any inner function is factorized into a Blaschke product and a singular inner function. By the subordination principle (see, for example, [3, p. 10]), it is seen that if ψ is an analytic map of U into itself with $\psi(0) = 0$, then the inequality

$$\| f \circ \psi \|_p \leq \| f \|_p$$

holds for any $f \in H_p(U)$. More generally it also follows from the principle that the inequality

$$B(f \circ \psi) \leq B(f) \tag{1.3}$$

holds for any $f \in BMOA(U)$ (§2, Proposition 1). Ryff [7] showed that if ψ is an inner function with $\psi(0) = 0$, the equality

$$\| f \circ \psi \|_p = \| f \|_p \tag{1.4}$$

holds for any $f \in H_p(U)$ and conversely that if (1.4) holds for a nonconstant $f \in H_p(U)$, then the analytic map ψ of U into itself with $\psi(0) = 0$ is an inner function. The author and Suita [6] showed more generally that the composition by any inner function preserves the least harmonic majorant of any positive subharmonic function. By using their result it is easily seen that if ψ is an inner function, then the equality

$$B(f \circ \psi) = B(f) \tag{1.5}$$

holds for any $f \in BMOA(U)$ (§2, Theorem 1). In the present paper it is shown that the converse is not valid, that is, we construct a non-inner analytic map φ of U into itself for which the equality (1.5) with $\psi = \varphi$ holds for any $f \in A(U)$ (§3, Theorem 2). Although it is plausible that the equality (1.5) with some non-inner function ψ holds for any

$f \in BMOA(U)$, we do not know as yet whether there exists such a non-inner function or not.

2. Subordination by inner functions.

In this section we show that the BMO norm $B(f)$ of any $f \in BMOA(U)$ is invariant under the composition by any inner function. First of all we show the inequality (1.3), which is a restatement of a result in the author's paper [5, Lemma 4.1, p. 165].

For a subharmonic function s in U, we denote by $\hat{s}$ the least harmonic majorant of s in U. In the case where s admits no harmonic majorants we set $\hat{s}(z) \equiv +\infty$. The author and Suita [6] proved that if ψ is an inner function, then

$$(s \circ \psi)\hat{\ }(z) \equiv \hat{s}(\psi(z)) \tag{2.1}$$

holds for any positive subharmonic function s in U.

Let $f \in BMOA(U)$. For $a \in U$, we consider the subharmonic function $s_a(z) = | f(z) - f(a) |^p$ in U. First note that the limit of the integrals

$$\lim_{r \to 1} \frac{1}{2\pi} \int_T | f_a\left(re^{i\theta}\right) |^p \, d\theta$$

is the value at a of the least harmonic majorant of s_a, in other words

$$\hat{s}_a(a) = \{ \| f_a \|_p \}^p. \tag{2.2}$$

Proposition 1. If ψ is an analytic map of U into itself, then the inequality $B(f \circ \psi) \leq B(f)$ holds for any $f \in BMOA(U)$.

Proof. Let $f \in BMOA(U)$ and set $b = \psi(a)$ for $a \in U$. Using (2.2), we see

$$\{ \| f \circ \varphi)_a \|_p \}^p = (s_b \circ \psi)\hat{\ }(a),$$

since $| (f \circ \psi)_a(z) |^p \equiv (s_b \circ \psi \circ \psi_a)(z)$. Noting that $\hat{s}_b(\psi(z))$ is a harmonic function majorating $(s_b \circ \psi)(z)$, we see

$$\left(s_b \circ \psi\right)\hat{\ }(z) \leq \hat{s}_b(\psi(z)),$$

in which by putting z = a, we obtain

$$(s_b \circ \psi)^\wedge(a) \le \hat{s}_b(\psi(a)) = \hat{s}_b(b). \tag{2.4}$$

Hence, we obtain by combining (2.2), (2.3) and (2.4)

$$\begin{aligned} B(f\circ\psi) &= \sup_{a\in U} \{ \|(f\circ\psi)_a\|_p\}^p \\ &\le \sup_{b\in\psi(U)} \hat{s}_b(b) \\ &\le \sup_{b\in U} \hat{s}_b(b) \\ &= B(f), \end{aligned}$$

since obviously $\psi(U) \subset U$. This completes the proof.

Theorem 1. If ψ is an inner function, then the equality

$$B(f\circ\psi) = B(f)$$

holds for any $f \in BMOA(U)$.

Proof. Putting $s = s_b$ with $b = \psi(a)$ for $a \in U$ in (2.1), we see

$$(s_b \circ \psi)^\wedge(z) = \hat{s}_b(\psi(z)),$$

and hence we see that equality holds in (2.4). By the same reasoning used in the proof of Proposition 1, we see

$$\begin{aligned} B(f\circ\psi) &= \sup_{a\in U} \{ \| (f\circ\psi)_a\|_p\}^p \\ &= \sup_{a\in U} (s_b\circ\psi)^\wedge(a) \\ &= \sup_{b\in\psi(U)} \hat{s}_b(b) \end{aligned}$$

$$= \sup_{b \in U} \hat{s}_b(b)$$

$$= B(f),$$

since the range $\psi(U)$ of U by ψ covers U except possibly for a set of capacity zero by Frostman's theorem [2] on the ranges of inner functions, in particular $\psi(U)$ is dense in U, and $\hat{s}_b(b)$ is obviously continuous as a function of b in U.

3. Subordination by a non-inner function.

In this section we construct a non-inner function φ such that the equality (1.5) with $\psi = \varphi$ holds for any $f \in A(U)$.

Let $\lambda(z)$ be any inner function with a singularity at a point ξ on T. It is known that any inner function with a factor of a Blaschke product whose zeros have a subsequence converging to ξ or a singular inner function whose determining singular measure has a mass at ξ is such an inner function (see, for example, [3, pp. 23 - 25]). Let ω be any non-inner function in A(U) with $|\omega(z)| < 1$ for $z \in U$ and $|\omega(\xi)| = 1$. We set $\varphi(z) = \lambda(z)\omega(z)$, which is obviously a non-inner function.

Theorem 2. If φ is the non-inner function constructed as above, then the equality

$$B(f \circ \varphi) = B(f) \tag{3.1}$$

holds for any $f \in A(U)$.

Proof. Let $f \in A(U)$ and ϵ be an arbitrarily small positive number. From the definition (1.2) of the BMO norm, we can take a point $b \in U$ for which the inequality

$$B(f) < \{\{ \| f_b \|_p\}^p + \epsilon \tag{3.2}$$

holds.

Without loss of generality, we may assume $\xi = 1$ and $\omega(\xi) = 1$. By the Frostman theorem cited in the previous section, any inner function takes in any neighborhood of a singular point on T every value in U possibly except for a set of capacity zero. Therefore, applying this theorem on λ, we can take a sequence $\{a_n\}$ of points in U such that

$$\lim_{n\to\infty} a_n = 1 \tag{3.3}$$

and

$$\lim_{n\to\infty} \lambda(a_n) = b. \tag{3.4}$$

By the continuity of ω, we see

$$\lim_{z\to 1} \omega(z) = 1 \tag{3.5}$$

Noting (3.4) and the continuity of $\hat{s}_b$,

$$| \hat{s}_b(\lambda(a_n)) - \hat{s}_b(b) | < \epsilon$$

holds for any integer $n > N_1$.

By (2.1) with $s = s_b$ and $\psi = \lambda$, we obtain

$$(s_b \circ \lambda)\,\hat{}\,(z) \equiv \hat{s}_b(\lambda(z)). \tag{3.7}$$

Let $M = \sup\{\, | f(z) | : z \in U\}$. Since the function $k(t) = t^p$ is uniformly continuous on the bounded interval $[0, 2M]$, we can take an $\epsilon_1 > 0$ such that

$$| t_1{}^p - t_2{}^p | < \epsilon \ \text{ if } | t_1 - t_2 | < \epsilon_1. \tag{3.8}$$

Noting the continuity of f, we can take a $\delta > 0$ such that for $\xi_1, \xi_2 \in \overline{U}$

$$| f(\xi_1) - f(\xi_2) | < \epsilon_1 \ \text{ if } | \xi_1 - \xi_2 | < \delta. \tag{3.9}$$

By (3.5), we can take an $\eta > 0$ such that $| \lambda(e^{i\theta}) - \varphi(e^{i\theta}) | < \delta$ if $| \theta | \leq \eta$, and hence

$$| f(\lambda(e^{i\theta})) - f(\varphi(e^{i\theta})) | < \epsilon_1 \tag{3.10}$$

for $| \theta | \leq \eta$.

By the Poisson integral formula, we see

$$(s_b \circ \lambda)\,\hat{}\,(a_n) = \frac{1}{2\pi} \int_T | f(\lambda(e^{i\theta})) - f(b) |^p \, P_n(e^{i\theta})\, d\theta, \tag{3.11}$$

where $P_n(e^{i\theta}) = (1 - |a_n|^2)/|e^{i\theta} - a_n|^2$ is the Poisson kernel for the point $a_n \in U$. Since $P_n(e^{i\theta})$ uniformly converges to 0 for $|\theta| > \eta$ as $n \to \infty$, we can take an N_2 such that for any integer $n > N_2$

$$\frac{1}{2\pi} \int_{|\theta| > \eta} P_n(e^{i\theta})\, d\theta < \frac{\epsilon}{(2M)^p},$$

and hence

$$\frac{1}{2\pi} \int_{|\theta| > \eta} |f(\lambda(e^{i\theta})) - f(b)|^p P_n(e^{i\theta})\, d\theta < \epsilon. \tag{3.12}$$

Combining (3.11) and (3.12), we see

$$\left| (S_b \circ \lambda)^\wedge (a_n) - \frac{1}{2\pi} \int_{-\eta}^{\eta} |f(\lambda(e^{i\theta})) - f(b)|^p P_n(e^{i\theta})\, d\theta \right| < \epsilon \tag{3.13}$$

for $n > N_2$.

Noting (3.8) and (3.10), we see

$$\left| \frac{1}{2\pi} \int_{-\eta}^{\eta} |f(\lambda(e^{i\theta})) - f(b)|^p P_n(e^{i\theta})\, d\theta \right.$$

$$\left. - \frac{1}{2\pi} \int_{-\eta}^{\eta} |f(\lambda(e^{i\theta})) - f(b)|^p P_n(e^{i\theta})\, d\theta \right| < \epsilon. \tag{3.14}$$

for every n.

Combining (3.3), (3.4) and (3.5), we see

$$\lim_{n \to \infty} \varphi(a_n) = b,$$

and hence we can take an N_3 such that $|\varphi(a_n) - b| < \delta$ for any integer $n > N_3$. Therefore, noting (3.9), we see $|f(\varphi(e^{i\theta})) - f(b)| \le \epsilon_1$, and hence

$$\left| \frac{1}{2\pi} \int_T |f(\varphi(e^{i\theta})) - f(b)|^p P_n(e^{i\theta})\, d\theta \right.$$

$$- \frac{1}{2\pi}\int_T |f(\varphi(e^{i\theta})) - f(\varphi(a_n))|^{\,p}\, P_n(e^{i\theta})\, d\theta \;<\; \Big| \;\; \epsilon, \tag{3.15}$$

for $n > N_3$. Since the integral of the second term of (3.15) is the value at a_n of the least harmonic majorant of $|f(\varphi(z)) - f(\varphi(a_n))|^{\,p}$, we see

$$\left\{\| (f\circ\varphi)_{a_n} \|_p\right\}^{p} = \frac{1}{2\pi}\int_T |f(\varphi(e^{i\theta}))- f(\varphi(a_n))|^{p} P_n(e^{i\theta})\, d\theta, \tag{3.16}$$

by noting (2.2).

By successively using (3.2), (2.2), (3.6), (3.7), (3.13), (3.14), (3.15), (3.16) and the definition (1.2) of B(f), we see that for every $n > N = \max\left(N_1, N_2, N_3\right)$

$$B(f) < \left\{\| f_b \|_p\right\}^{p} + \epsilon.$$

$$= \hat{s}_b(b) + \epsilon$$

$$< \hat{s}_b(\lambda(a_n)) + 2\epsilon$$

$$= (s_b \circ \lambda)^{\wedge}(a_n) + 2\epsilon$$

$$< \frac{1}{2\pi}\int_{-\eta}^{\eta} |f(\lambda(e^{i\theta})) - f(b)\,|^{\,p}\, P_n(e^{i\theta})\, d\theta + 3\,\epsilon$$

$$< \frac{1}{2\pi}\int_{-\eta}^{\eta} |f(\varphi(e^{i\theta})) - f(b)\,|^{\,p}\, P_n(e^{i\theta})\, d\theta + 4\,\epsilon$$

$$\leq \frac{1}{2\pi}\int_T |f(\varphi(e^{i\theta})) - f(b)|^{\,p}\, P_n(e^{i\theta})\, d\theta + 4\epsilon$$

$$\leq \frac{1}{2\pi}\int_T |f(\varphi(e^{i\theta})) - f(\varphi(a_n))|^{\,p}\, P_n(e^{i\theta})\, d\theta + 5\epsilon$$

$$= \left\{ \| (f \circ \varphi)_{a_n} \|_p \right\}^p + 5\epsilon$$

$$\leq B(f \circ \varphi) + 5\epsilon,$$

from which we obtain

$$B(f) \leq B(f \circ \varphi),$$

as asserted, since ϵ can be taken arbitrarily small.

4. Concluding remarks

For $0 < r < 1$ and $f \in BMOA(U)$, let f_r be, as usual, the function in A(U) defined by $f_r(z) = f(rz)$. We easily see

$$\lim_{r \to 1} B(f_r) = B(f),$$

since f_r converges to f in the H_p norm as $r \to 1$. Therefore if we were able to prove

$$\lim_{r \to 1} B(f_r \circ \varphi) = B(f \circ \varphi), \tag{4.1}$$

we would obtain (2.1) for $f \in BMOA(U)$.

We conclude the present paper by offering open problems on related topics.

Problem 1. Characterize the analytic maps ψ of U into itself for which (4.1) with $\varphi = \psi$ holds for any $f \in BMOA(U)$.

Problem 2. Characterize the analytic maps ψ of U into itself for which $B\left(f_r \circ \psi\right)$ is a nondecreasing function of r on (0, 1) for any $f \in BMOA(U)$.

It is easily seen that if a φ satisfies the conditions of Problem 2 and Theorem 2 then it also satisfies that of Problem 1. Note that inner functions satisfy both of the conditions in Problems 1 and 2. It is plausible and desired that there exists a non-inner analytic map ψ of U into itself which satisfies these conditions.

References

[1] Bearnstein, A., Analytic functions of bounded mean oscillation, Aspects of contemporary complex analysis, Academic Press, New York, 1980, pp. 3 - 36.

[2] Frostman, O., Potential d' équilibre et capacité des ensembles avec quelques application à la théorie des fonctions, Medd. Lunds Math. Sem. 3 (1935), pp. 1 - 118.

[3] Duren, P., Theory of H^p spaces, Academic Press, New York, 1970.

[4] John, F., and L. Nirenber, On functions of bounded mean oscillation, Comm. Pure Appl, Math. 14 (1961), pp. 415 - 426.

[5] Kobayashi, S., Image areas and BMO norms of analytic functions, Kodai Math. J. 8 (1985), pp. 163 - 170.

[6] Kobayashi, S., andN. Suita, On subordination of subharmonic functions, Kodai Math. J. 3 (1980), pp. 315 - 320.

[7] Ryff, J. V., Subordinate H^p functions, Duke Math. J. 33 (1966), pp. 347 - 354.

Department of Science and Mathematics
Nagaoka University of Technology

DISTORTION PROPERTIES OF INTEGRAL OPERATOR RELATED TO FRACTIONAL CALCULUS

Yûsaku Komatu

0. Introduction.

Let $\mathcal{F}$ denote the class of analytic functions f which are holomorphic in the unit disk $E = \{|z| < 1\}$ and normalized by $f(0) = f'(0) - 1 = 0$. Let σ be a probability measure supported by the unit interval $I = [0, 1]$. Then a linear integral operator $\mathcal{L}$ is defined by

$$\mathcal{L}f(z) = \int_I \frac{f(zt)}{t} \, d\sigma(t). \tag{0.1}$$

The purpose of the present note is to make a survey of operators of the form (0.1) through their distortion properties on several functionals defined on the class $\mathcal{F}$ and its subclasses. Most of the results have been published in [6] ~ [13].

As will be shown, any operator $\mathcal{L}$ commutes with the operator d/d log z. In particular, in the simplest case where σ is defined by $\sigma(t) = t$, the corresponding operator $\mathcal{L}$ becomes the integration with respect to log z. On the other hand, any operator $\mathcal{L}$ gives rise a family of operators $\{\mathcal{L}^\lambda\}_{\lambda \geq 0}$ depending on the continuous parameter λ such that the additivity relation

$$\mathcal{L}^\lambda \mathcal{L}^\mu = \mathcal{L}^{\lambda+\mu} \tag{0.2}$$

is satisfied. In particular, if the family is generated by $\sigma(t) = t$, then the member $\mathcal{L}^\lambda$ coincides with the fractional integration of order λ with respect to log z.

We observe especially, as a concrete example, the additive family generated by the measure $\sigma(t; a) = t^a$ with a parameter $a > 0$ in detail. Its member admits an integral representation of similar form as for $\mathcal{L}$ in terms of elementary functions. The relation to the ordinary integration operator is explained and a possibility of generalizing the basic measure is also suggested. Several distortion properties in this case are derived in explicit forms.

1. Distortion properties on the modulus and the real part.

We observe some functionals defined on $\mathfrak{F}$ and deal with their distortion properties concerning the effect of $\mathfrak{L}$. In the following lines the extreme measure concentrated at the single point 1 plays occasionally an exceptional role. We denote it by σ^*. Every $f \in \mathfrak{F}$ is transformed by σ^* into itself, while the particular function z is transformed by any $\mathfrak{L}$ always into itself.

First, let the functional M be defined on $\mathfrak{F}$ by

$$M[F] = \max_{|z|=r} |F(z)| \qquad (F \in \mathfrak{F}),$$

where $r \in [0, 1)$ is a fixed parameter.

Theorem 1. 1. For any $f \in \mathfrak{F}$ the monotonicity

$$M[\mathfrak{L}f] \le M[f]$$

holds where the equality sign appears for an $r \in (0, 1)$ if and only if $f(z) \equiv z$, unless $\sigma = \sigma^*$.

Proof. The maximum principle yields

$$\left|\frac{f(zt)}{zt}\right| \le \max_{|z|=r}\left|\frac{f(z)}{z}\right| = \frac{M[f]}{r} \qquad (0 \le t \le 1)$$

for any z with $|z| = r < 1$. Hence we get

$$\left|\frac{\mathfrak{L}f(z)}{z}\right| \le \left|\int_I \left|\frac{f(zt)}{zt}\right| d\sigma(t)\right| \le \frac{M[f]}{r}$$

and consequently $M[\mathfrak{L}f] \le M[f]$. The equality holds if and only if $|f(zt)/zt| = M[f]/r$ at every t with $d\sigma(t) > 0$ and hence $f(z) \equiv z$, unless $\sigma = \sigma^*$.

Theorem 1. 2. If $f, g \in \mathfrak{F}$ possess the Taylor coefficients in common up to the power k, then we have

$$M[\mathfrak{L}f - \mathfrak{L}g] \le M[f - g] \int_I t^k d\sigma(t). \tag{1.1}$$

Proof. For any fixed z with $|z| = r < 1$, the quantity $(f(zt) - g(zt))/t^{k+1}$ may be regarded as function of a complex variable t holomorphic on $\{|t| \leq 1\}$. The maximum principle applied to this function yields

$$\max_{|t|\leq 1} \left| \frac{f(zt) - g(zt)}{t^{k+1}} \right| = \max_{|z|=r} |f(z) - g(z)|$$

whence follows $|f(zt) - g(zt)| \leq t^{k+1} M[f - g]$ for $t \in I$. Consequently, we obtain

$$|\mathcal{L}f(z) - \mathcal{L}g(z)| \leq \int_I \frac{|f(zt) - g(zt)|}{t} d\sigma(t) \leq M[f - g] \int_I t^k d\sigma(t).$$

Next, let the functionals h and H be defined on $\mathcal{F}$ by

$$h[F] = \min_{|z|=r} \operatorname{Re} \frac{F(z)}{z} \quad \text{and} \quad H[F] = \max_{|z|=r} \operatorname{Re} \frac{F(z)}{z} \quad (F \in \mathcal{F}),$$

where $r \in [0, 1)$ is a fixed parameter.

For any $F \in \mathcal{F}$ we have always $-\infty < h[F] \leq 1 \leq H[F] < \infty$ $(0 \leq r < 1)$; here the equality signs appear only when $F(z) \equiv z$. It is readily shown that the monotonicity

$$h[f] \leq h[\mathcal{L}f] \leq H[\mathcal{L}f] \leq H[f] \tag{1.2}$$

holds for any $f \in \mathcal{F}$. However, this property may be stated in more precise form.

Theorem 1.3. For any $f \in \mathcal{F}$ we have

$$h[\mathcal{L}f] \geq h[f] + \Phi(1 - h[f]), \tag{1.3}$$

$$H[\mathcal{L}f] \leq H[f] - \Phi(H[f] - 1)$$

where Φ is given by

$$\Phi = \int_I \frac{1 - t}{1 + t} d\sigma(t). \tag{1.4}$$

The equality sign in either estimation appears for an $r \in (0, 1)$ if and only if $f(z) \equiv z$, unless $\sigma = \sigma^*$.

Proof. For any fixed z with $|z| = r < 1$, both $(f(zt)/zt - h[f])/(1 - h[f])$ and $(H[f] - f(zt)/zt)/(H[f] - 1)$ may be regarded as functions of a complex variable t holomorphic on $\{|t| \leq 1\}$ provided $f(z) \not\equiv z$. Since they have nonnegative real part and attain the value 1 at the origin, we have

$$\operatorname{Re} \frac{f(zt)/zt - h[f]}{1 - h[f]} \geq \frac{1 - t}{1 + t}, \quad \operatorname{Re} \frac{H[f] - f(zt)/zt}{H[f] - 1} \geq \frac{1 - t}{1 + t}$$

valid, in particular, for $t \in I$, whence follows, respectively,

$$\operatorname{Re} \frac{f(zt)}{zt} \geq h[f] + (1 - h[f]) \frac{1 - t}{1 + t}, \quad \operatorname{Re} \frac{f(zt)}{zt} \leq H[f] - (H[f] - 1) \frac{1 - t}{1 + t}.$$

The last inequalities remain to hold also when $h[f] = 1$ or $H[f] = 1$ and hence $f(z) \equiv z$.

Consequently, in view of

$$\operatorname{Re} \frac{\mathfrak{L}f(z)}{z} = \int_I \operatorname{Re} \frac{f(zt)}{zt} \, d\sigma(t) \text{ with } \sigma(1) = 1,$$

we obtain the desired estimations. If the equality sign appears in the first estimation for an $r \in (0, 1)$, then there exists a z with $|z| = r$ such that

$$\operatorname{Re} \frac{f(zt)}{zt} = h[f] + (1 - h[f]) \frac{1 - t}{1 + t}$$

for $t \in I$. In view of the analyticity and the normalization at the origin, we see that the relation

$$\frac{f(zt)}{zt} = h[f] + (1 - h[f]) \frac{1 - t}{1 + t}$$

holds for $t \in E$. Since the left-hand member is bounded there, we have $h[f] = 1$ and hence $f(z) \equiv z$. Equality assertion for the second estimation is similarly verified. Or, it may be remarked that either result in the theorem follows from another by considering $2z - f(z)$ instead of $f(z)$ and replacing correspondingly $\sigma(t)$ by $2t - \sigma(t)$.

By the way, we remark that the inequalities of the nature similar to those in Theorem 1.3 can be derived.

<u>Theorem 1.4</u>. For any $f \in \mathcal{F}$ we have

$$h[\mathfrak{L}f] \geq H[f] - (H[f] - 1) \int_I \frac{1 + t}{1 - t} \, d\sigma(t), \tag{1.5}$$

$$H[\mathfrak{L}f] \leq h[f] + (1 - h[f]) \int_I \frac{1 + t}{1 - t} \, d\sigma(t),$$

where the equality assertion is the same as in Theorem 1.3.

Proof. The proof proceeds quite as in Theorem 1.3. We have only to make use of

$$\mathrm{Re}\,\frac{f(zt)/zt - h[f]}{1 - h[f]} \le \frac{1+t}{1-t}, \quad \mathrm{Re}\,\frac{H[f] - f(zt)/zt}{H[f] - 1} \le \frac{1+t}{1-t}.$$

Now, in order to deal with a related relation, we refer to a theorem of Koebe [4] as a lemma which may be formulated as follows.

Lemma 1.1. Let g be an analytic function holomorphic in E and normalized by $g(0) = 0$. If it satisfies $-a < \mathrm{Re}\, g(\varsigma) < b$ for $\varsigma \in E$, then the distortion inequality

$$\frac{2(b+a)}{\pi} \arctan \frac{\tau - |\varsigma|}{1 - \tau|\varsigma|} \le \mathrm{Re}\, g(\varsigma) - \frac{b-a}{2} \le \frac{2(b+a)}{\pi} \arctan \frac{\tau + |\varsigma|}{1 + \tau|\varsigma|}$$

holds for $\varsigma \in E$ where $\tau = -\tan((\pi/4)(b-a)/(b+a))$. The extremal function for $0 < |\varsigma| < 1$ is given only by $g^*(\varepsilon\varsigma)$ with $|\varepsilon| = 1$, where g^* is a function mapping E onto the parallel strip $\{-a < \mathrm{Re}\,\omega < b\}$:

$$g^*(\varsigma) = \frac{2(b+a)}{\pi} \arctan \frac{\tau + \varsigma}{1 + \tau\varsigma} + \frac{b-a}{2}.$$

By making use of Lemma 1.1, we derive the monotonicity of $\mathcal{L}$ in more precise form [11].

Theorem 1.5. For any $f \in \mathfrak{F}$ we have

$$\frac{2(H-h)}{\pi} \int_I \arctan \frac{T-t}{1-Tt}\, d\sigma(t) + \frac{H+h}{2} \le h[\mathcal{L}f]$$
$$\le H[\mathcal{L}f] \le \frac{2(H-h)}{\pi} \int_I \arctan \frac{T+t}{1+Tt}\, d\sigma(t) + \frac{H+h}{2}; \tag{1.6}$$

here h[f] and H[f] are abbreviated simply by h and H, respectively, and T is defined by

$$T = T[f] \equiv -\tan\left(\frac{\pi}{4}\, \frac{H[f] + h[f] - 2}{H[f] - h[f]}\right).$$

Proof. For any fixed z with $|z| = r < 1$, the quantity $F(zt) = f(zt)/zt - 1$ may be regarded as function of a complex variable t holomorphic on $\{|t| \le 1\}$. It satisfies $F(0) = f'(0) - 1 = 0$ and $-(1-h) \le \mathrm{Re}\, F(zt) \le H - 1$ for $|t| \le 1$ and, in particular, for $t \in I$. In view of Lemma 1.1 applied to F(zt) as function of t, we get

$$\frac{2(H-h)}{\pi} \arctan \frac{T-t}{1-Tt} \le \mathrm{Re}\, F(zt) - \frac{H+h-2}{2} \le \frac{2(H-h)}{\pi} \arctan \frac{T+t}{1+Tt}$$

for $t \in I$. Since the definition of $\mathcal{L}$ yields

$$\mathrm{Re}\,\frac{\mathcal{L}f(z)}{z} - 1 = \mathrm{Re} \int_I \left(\frac{f(zt)}{zt} - 1\right) d\sigma(t) = \int_I \mathrm{Re}\, F(zt)\, d\sigma(t),$$

we obtain the desired relation after integrating each member of the above inequality with respect to σ.

It is noted that the estimation (1.6) given here sharpens the monotonicity (1.2). In fact, since the value $(H + h - 2)/(H - h)$ always belongs to $(-1, 1)$, that of T does also so. For any fixed $T \in (-1, 1)$ the quantities $(T - t)/(1 - Tt)$ and $(T + t)/(1 + Tt)$ are decreasing and increasing, respectively, with respect to $t \in I$, and hence their values are contained in $(-1, 1)$, whence readily follows (1.2).

On the other hand, we see that (1.6) remains valid even when h and H are replaced by any lower and upper bounds of Re $(f(z)/z)$, respectively. Accordingly, if we would use h alone without referring to H, (1.6) seems to reduce to (1.5). To see this, we have only to consider the limit as $H \to \infty$. We then get in turn

$$T = -\tan\left(\frac{\pi}{4}\frac{H + h - 2}{H - h}\right) = -1 + \frac{(1 - h)}{H} + o\left(\frac{1}{H}\right),$$

$$\arctan\frac{T - t}{1 - Tt} = \frac{\pi}{4} + \frac{(1 - h)}{2H}\frac{1 - t}{1 + t} + o\left(\frac{1}{H}\right),$$

whence really follows

$$\frac{2(H - h)}{\pi}\int_I \arctan\frac{T - t}{1 - Tt}\,d\sigma(t) + \frac{H + h}{2} = h + (1 - h)\int_I \frac{1 - t}{1 + t}\,d\sigma(t) + o\,(1).$$

Similarly, if we would retain H alone without referring to h, we obtain as $h \to -\infty$ the limit relation

$$\frac{2(H - h)}{\pi}\int_I \arctan\frac{T + t}{1 + Tt}\,d\sigma(t) + \frac{H + h}{2} = H - (H - 1)\int_I \frac{1 - t}{1 + t}\,d\sigma(t) + o(1).$$

In connection with the functionals h and H, we consider the oscillation of Re $(F(z)/z)$ on $\{|z| = r\}$ with a fixed $r \in [0, 1)$ which will be denoted by

$$\Delta[F] = \operatorname*{osc}_{|z|=r}\ \mathrm{Re}\,\frac{F(z)}{z} = H[F] - h[F] \qquad (F \in \mathcal{F}).$$

The monotonicity of $\mathcal{L}$ for Δ, namely $\Delta[\mathcal{L}f] \leq \Delta[f]$, is readily seen. More precise estimation follows from Theorem 1.5 as its corollary.

Corollary. For any $f \in \mathcal{F}$ we have

$$\Delta[\mathcal{L}f] \leq \frac{4\Delta[f]}{\pi} \int_I \arctan t \, d\sigma(t). \tag{1.7}$$

The equality sign appears only for $f(z) \equiv z$, unless $\sigma = \sigma^*$.

Proof. The estimation (1.6) in Theorem 1.5 implies

$$\Delta[\mathcal{L}f] \leq \frac{2\Delta}{\pi} \int_I \left(\arctan \frac{T+t}{1+Tt} - \arctan \frac{T-t}{1-Tt}\right) d\sigma(t)$$
$$= \frac{2\Delta}{\pi} \int_I \arctan \left(\frac{1-T^2}{1+T^2} \frac{2t}{1-t^2}\right) d\sigma(t)$$

where $\Delta = \Delta[f]$. In view of $h \leq 1 \leq H$ we have $-1 \leq T \leq 1$ and hence

$$\Delta[\mathcal{L}f] \leq \frac{2\Delta}{\pi} \int_I \arctan \frac{2t}{1-t^2} \, d\sigma(t) = \frac{4\Delta}{\pi} \int_I \arctan t \, d\sigma(t).$$

The equality assertion is readily seen.

Corollary just proved can be derived in a similar manner as the proof of Theorem 1.5, by making use of a formula of Neumann [16] as a basic lemma which has been restated by Koebe [4] as follows.

Lemma 1.2. Let g be an analytic function holomorphic in E and normalized by $g(0) = 0$. If it satisfies $\operatorname{osc}_E \operatorname{Re} g \leq \Delta$, then

$$\operatorname*{osc}_{|\zeta|=r} \operatorname{Re} g(\zeta) \leq \frac{4\Delta}{\pi} \arctan r$$

holds for $r \in [0, 1)$. The extremal function for $r \in (0, 1)$ is given only by $g^*(\varepsilon\zeta)$ with $|\varepsilon| = 1$ where $g^*(\zeta) = (2\Delta/\pi) \arctan \zeta$ is a function mapping E onto the parallel strip $\{|\operatorname{Re} \omega| < \Delta/2\}$.

2. Range of values and distortions on length and area.

Let $\mathcal{P}(\alpha)$ with $\alpha < 1$ denote the Carathéodory class of order α which consists of analytic functions p holomorphic in E and satisfying $p(0) = 1$ and $\operatorname{Re} p > \alpha$ in E. It is readily seen the $p \in \mathcal{P}(\alpha)$ implies $f = zp \in \mathcal{F}$ and $\mathcal{L}f(z)/z \in \mathcal{P}(\alpha)$.

We now observe the subclass $\mathcal{F}(\alpha)$ of $\mathcal{F}$ which consists of functions f satisfying $f/z \in \mathcal{P}(\alpha)$. As shown by Strohhäcker [20], the class of convex mappings is a subclass of

$\mathcal{F}(1/2)$. In relation to this fact, we have derived in [8] some results on the range of values concerning $\mathcal{F}(1/2)$ and then generalized them to the class $\mathcal{F}(\alpha)$ in [9].

Theorem 2.1. Any function $\in \mathcal{F}(\alpha)$ satisfies

$$\left| \frac{\mathcal{L}f(z)}{z} - \frac{\mathcal{L}\varphi(r;\, \alpha)}{r} \right| \le \frac{\mathcal{L}\psi(r;\, \alpha)}{r} - 1 \tag{2.1}$$

for $|z| \le r < 1$, where φ and ψ are elementary functions in $\mathcal{F}$ defined by

$$\varphi(z;\, \alpha) = z\, \frac{1 + (1 - 2\alpha)z^2}{1 - z^2} \text{ and } \psi(z;\, \alpha) = z\, \frac{1 + 2(1 - \alpha)z - z^2}{1 - z^2}.$$

The extremal functions for the estimation are of the form $f(z) = \bar{\varepsilon}\chi(\varepsilon z;\, \alpha)$ with

$$\chi(z;\, \alpha) = z\, \frac{1 + (1 - 2\alpha)z}{1 - z} \text{ and } |\varepsilon| = 1,$$

unless $\sigma = \sigma^*$. Further, the range of values of $\mathcal{L}f/z$ for $\{|z| \le r\}$ induced from any extremal function f is just the closed circle expressed by the estimation.

Proof. Since $f \in \mathcal{F}(\alpha)$ implies $(f - \alpha z)/(1 - \alpha) \in \mathcal{F}(0)$, we get in view of Herglotz representation for $\mathcal{P}(0)$ the expression

$$\frac{f(z)}{z} = (1 - \alpha) \int_{-\pi}^{\pi} \frac{e^{i\theta} + z}{e^{i\theta} - z}\, d\tau(\theta) + \alpha = \int_{-\pi}^{\pi} \frac{\chi(e^{-i\theta}z;\, \alpha)}{e^{-i\theta}z}\, d\tau(\theta)$$

where τ is a probability measure supported on the interval $(-\pi, \pi]$. Since the range of values of $\chi(z;\, 0)/z \in \mathcal{P}(0)$ for $\{|z| \le r\}$ is contained in the closed disk with the segment $[\chi(-r;\, 0)/(-r),\ \chi(r;\, 0)/r]$ as a diameter, that of $\chi(z;\, \alpha)/z$ is contained in the closed disk with the segment $[\chi(-r;\, \alpha)/(-r),\ \chi(r;\, \alpha)/r]$ as a diameter, of which the center and the radius are given by

$$\frac{1}{2}\left(\frac{\chi(r;\, \alpha)}{r} + \frac{\chi(-r;\, \alpha)}{-r}\right) = \frac{\varphi(r;\, \alpha)}{r},\quad \frac{1}{2}\left(\frac{\chi(r;\, \alpha)}{r} - \frac{\chi(-r;\, \alpha)}{-r}\right) = \frac{\psi(r;\, \alpha)}{r} - 1,$$

respectively. Consequently, we get

$$\left| \frac{\chi(z;\, \alpha)}{z} - \frac{\varphi(r;\, \alpha)}{r} \right| \le \frac{\psi(r;\, \alpha)}{r} - 1$$

for $|z| \le r$. On the other hand, by remembering the definition of $\mathcal{L}$, we obtain

$$\frac{\mathcal{L}f(z)}{z} - \frac{\mathcal{L}\varphi(r;\, \alpha)}{r} = \int_I \left(\int_{-\pi}^{\pi} \frac{\chi(e^{-i\theta}zt;\, \alpha)}{e^{-i\theta}zt} d\tau(\theta) - \frac{\varphi(rt;\, \alpha)}{rt} \right) d\sigma(t)$$

$$= \int_I \left(\int_{-\pi}^{\pi} \left(\frac{\chi(e^{-i\theta}zt;\,\alpha)}{e^{-i\theta}\,zt} - \frac{\varphi(rt;\,\alpha)}{rt} \right) d\tau(\theta) \right) d\sigma(t).$$

Thus, by taking into account the above inequality, we have

$$\left| \frac{\mathcal{L}f(z)}{z} - \frac{\mathcal{L}\varphi(r;\,\alpha)}{r} \right| \leq \int_I \left(\int_{-\pi}^{\pi} \left(\frac{\psi(rt;\,\alpha)}{rt} - 1 \right) d\tau(\theta) \right) d\sigma(t)$$

$$= \int_I \left(\frac{\psi(r;\,\alpha)}{rt} - 1 \right) d\sigma(t) = \frac{\mathcal{L}\psi(r;\,\alpha)}{r} - 1.$$

Concerning the extremal functions it is readily seen that the equality sign at a point on $\{|z| \leq r\}$ and necessarily on $\{|z| = r\}$ appears if and only if τ is the point measure concentrated at a single point θ, and hence f reduces to $f(z) = \overline{\varepsilon}\,\chi(\varepsilon z;\,\alpha)$ with $\varepsilon = e^{-i\theta}$.

Next, we supplement a theorem on the range of values of $\mathcal{L}f$ in E.

<u>Theorem 2.2</u>. If $\mathcal{L}\varphi(r;\,\alpha)$ is unbounded for $r \to 1-0$, then $\mathcal{L}\,\psi(r;\,\alpha)$ is also bounded and both possess necessarily finite limits. The range of values of $\mathcal{L}f/z$ in the whole disk E is then contained in the circular disk

$$\left| \frac{\mathcal{L}f(z)}{z} - \mathcal{L}\varphi(1-0;\,\alpha) \right| \leq \mathcal{L}\psi(1-0;\,\alpha) - 1. \tag{2.2}$$

If $\mathcal{L}\varphi(r;\,\alpha)$ is unbounded for $r \to 1-0$, then $\mathcal{L}\psi(r;\,\alpha)$ is also unbounded and the range of values of $\mathcal{L}f/z$ in E is contained in the half-plane

$$\mathrm{Re}\,\frac{\mathcal{L}f(z)}{z} > B(\alpha) \tag{2.3}$$

where $B(\alpha)$ denotes the surely existent limit of $\mathcal{L}\varphi(r;\,\alpha) - \mathcal{L}\,\psi(r;\,\alpha) + 1$ as $r \to 1-0$.

<u>Proof</u>. The range-circle (2.1) for $\mathcal{L}f/z$ ($|z| \leq r$) given in Theorem 2.1 swells as point set together with $r \in [0,\,1)$. Since $\varphi(r;\,\alpha)/r$ and $\psi(r;\,\alpha)/r$ increase together with r, $\mathcal{L}\varphi(r;\,\alpha)/r$ and $\mathcal{L}\psi(r;\,\alpha)/r$ increase also. The former part of the theorem follows readily from Theorem 2.1, since $\chi(-r;\,\alpha)/(-r)$ and $\mathcal{L}\chi(-r;\,\alpha)/(-r)$ decrease as r increases and the left endpoint of the diameter on the real axis of the range-circle lies at

$$\frac{\mathcal{L}\varphi(r;\,\alpha)}{r} - \frac{\mathcal{L}\psi(r;\,\alpha)}{r} + 1 = \frac{\mathcal{L}\chi(-r;\,\alpha)}{-r} \in (\alpha,\,1].$$

The latter part follows also, since there exists the limit

$$B(\alpha) = \mathcal{L}\chi(-1+0;\ \alpha) = \lim_{r\to 1-0} (\mathcal{L}\varphi(r;\ \alpha) - \mathcal{L}\psi(r;\ \alpha) + 1)$$

and the range-circle (2.1) tends to the half-plane (2.3) as $r \to 1-0$.

Now, observing the length and the area related to the image mapped by $f(z)/z$, we deal with their distortions under the effect of $\mathcal{L}$. We begin with some classical lemmas.

Lemma 2.1. Let $L(r) = L(r;\ f)$ denote the length of image-curve of $\{|z| = r < 1\}$ by the mapping $w = f(z)/z$ with $f \in \mathcal{F}(\alpha)$. Then we have

$$L(r) \le (1 - \alpha)\frac{4\pi r}{1 - r^2}.$$

The extremal function for $r \in (0,\ 1)$ is of the form $\bar{\epsilon}\chi(\epsilon z)$ with $|\epsilon| = 1$ where χ is the function defined in Theorem 2.1.

Proof. Herglotz representation applied to $(f(z)/z - \alpha)/(1 - \alpha) \in \mathcal{P}(0)$ yields

$$g(z) \equiv \frac{f(z)}{z} = 2(1 - \alpha)\int_{-\pi}^{\pi} \frac{e^{i\varphi}}{e^{i\varphi} - z}\, d\tau(\varphi) - (1 - 2\alpha)$$

where τ is a probability measure on $(-\pi,\ \pi]$. Direct calculation shows

$$\begin{aligned} L(r) &= r\int_{-\pi}^{\pi} |g'(re^{i\theta})|\, d\theta \\ &\le 2(1 - \alpha)r\int_{-\pi}^{\pi} d\tau(\varphi)\int_{-\pi}^{\pi} \frac{1}{|e^{i\varphi} - re^{i\theta}|^2}\, d\theta \\ &= 2(1 - \alpha)r\int_{-\pi}^{\pi} \frac{2\pi}{1 - r^2} d\tau(\varphi) = (1 - \alpha)\frac{4\pi r}{1 - r^2}. \end{aligned}$$

The bound in the estimation for $r \in (0,\ 1)$ appears only if $e^{i\varphi}/(e^{i\varphi} - re^{i\theta})^2$ has for every θ the same argument for $\varphi \in (-\pi,\ \pi)$ with $d\tau(\varphi) > 0$. The associated measure τ becomes then the point measure concentrated at a single point φ_o, whence follows the form of extremal function $\bar{\varepsilon}\chi(\varepsilon z)$ with $\varepsilon = e^{-i\varphi_o}$.

Lemma 2.1 now mentioned has been derived by Rogosinski [17] for $\mathcal{F}(0)$; cf. also [5], [10].

Lemma 2.2. Let $A(r) = A(r;\ f)$ denote the area of image-domain of $\{|z| < r < 1\}$ by the mapping $w = f(z)/z$ with $f \in \mathcal{F}(\alpha)$, the multiplicity being taken into account. Then we have

$$A(r) \le (1 - \alpha)^2 \frac{4\pi r^2}{(1 - r^2)^2}.$$

The extremal function is the same as in Lemma 2.1.

Proof. Following the notations in the proof of Lemma 2.1, we have in turn

$$A(r) = \int_0^r \rho \, d\rho \int_{-\pi}^{\pi} |g'(\rho e^{i\theta})|^2 \, d\theta$$

$$\le 4(1 - \alpha)^2 \int_{-\pi}^{\pi} d\tau(\varphi) \int_0^r \rho \, d\rho \int_{-\pi}^{\pi} \frac{1}{|e^{i\varphi} - e^{i\theta}|^4} \, d\theta$$

$$= 4(1 - \alpha)^2 \int_{-\pi}^{\pi} d\tau(\varphi) \int_0^r 2\pi \frac{\rho(1 + \rho^2)}{(1 - \rho^2)^3} \, d\rho$$

$$= 4(1 - \alpha)^2 \int_{-\pi}^{\pi} \pi \frac{r^2}{(1 - r^2)^2} \, d\tau(\varphi) = (1 - \alpha)^2 \frac{4\pi r^2}{(1 - r^2)^2}.$$

The extremal assertion follows similarly as above.

While the proof has been based on direct calculation, it could be made by means of Lemma 2.1 together with the isoperimetric inequality which remains valid for non-univalent mapping. In fact, we have simply

$$A(r) \le \frac{1}{4\pi} L(r)^2 \le \frac{1}{4\pi} \left((1 - \alpha) \frac{4\pi r}{1 - r^2} \right)^2 = (1 - \alpha)^2 \frac{4\pi r^2}{(1 - r^2)^2}.$$

The equality sign in the first inequality appears only if $g \in \mathcal{P}(\alpha)$ is a linear function mapping $\{|z| < r\}$ onto a circular disk and moreover E onto $\{\mathrm{Re}\, w > \alpha\}$.

The length-distortion corresponding to Lemma 2.1 states as follows:

Theorem 2.3. For any $f \in \mathcal{F}(\alpha)$ we have

$$L(r; \mathcal{L}f) \le \int_I L(rt) \, d\sigma(t),$$

where $L(\rho) = L(\rho; f))$ satisfies the estimation given in Lemma 2.1.

Proof. For the sake of brevity we put $g(z) = f(z)/z$. Then we get

$$G(z) \equiv \frac{\mathcal{L}f(z)}{z} = \int_I \frac{f(zt)}{zt}\, d\sigma(t) = \int_I g(zt)\, d\sigma(t),$$

whence follows

$$\begin{aligned} L(r; \mathcal{L}f) &= r\int_{-\pi}^{\pi} |G'(re^{i\theta})|\, d\theta \\ &= r\int_{-\pi}^{\pi} d\theta\, \Big|\int_I tg'(rte^{i\theta})\, d\sigma(t)\Big| \le r\int_{-\pi}^{\pi} d\theta \int_I t|g'(rte^{i\theta})|\, d\sigma(t) \\ &= \int_I d\sigma(t)\, rt\int_{-\pi}^{\pi} |g'(rte^{i\theta})|\, d\theta = \int_I L(rt)\, d\sigma(t). \end{aligned}$$

The area-distortion corresponding to Lemma 2.2 states similarly:

Theorem 2.4. For any $f \in \mathcal{F}(\alpha)$ we have

$$A(r; \mathcal{L}f) \le \int_I A(rt)\, d\sigma(t),$$

where $A(\rho) = A(\rho; f)$ satisfies the estimation given in Lemma 2.2.

Proof. Direct calculation as in the proof of Theorem 2.3 yields

$$\begin{aligned} A(r; \mathcal{L}f) &= \int_0^r \rho\, d\rho \int_{-\pi}^{\pi} |G'(\rho e^{i\theta})|^2\, d\theta \\ &= \int_0^r \rho\, d\rho \int_{-\pi}^{\pi} d\theta\, \Big|\int_I tg'(\rho te^{i\theta})\, d\sigma(t)\Big|^2 \end{aligned}$$

and by making use of Schwarz inequality,

$$\begin{aligned} A(r; \mathcal{L}f) &\le \int_0^r \rho\, d\rho \int_{-\pi}^{\pi} d\theta \int_I d\sigma(t) \cdot \int_I t^2 |g'(\rho te^{i\theta})|^2\, d\sigma(t) \\ &= \int_I d\sigma(t) \int_0^r \rho t^2 d\rho \int_{-\pi}^{\pi} |g'(\rho te^{i\theta})|^2\, d\theta \\ &= \int_I d\sigma(t) \int_0^{rt} s\, ds \int_{-\pi}^{\pi} |g'(se^{i\theta})|^2\, d\theta = \int_I A(rt)\, d\sigma(t). \end{aligned}$$

While Theorem 2.4 yields together with Lemma 2.2 an inequality

$$A(r; \mathcal{L}f) \le (1-\alpha)^2\, 4\pi r^2 \int_I \frac{t^2}{(1-r^2t^2)^2}\, d\sigma(t),$$

the isoperimetric inequality yields together with Theorem 2.3 and Lemma 2.1 a sharper inequality

$$A(r; \mathcal{L}f) \le \frac{1}{4\pi} L(r; \mathcal{L}f)^2 \le (1-\alpha)^2\, 4\pi r^2 \Big(\int_I \frac{t}{1-r^2t^2}\, d\sigma(t)\Big)^2.$$

By the way, note that an inequality of inverse nature always holds. In fact, by making use of Schwarz inequality, we have

$$\frac{1}{2\pi}\int_0^r \frac{L(\rho;\, \mathcal{L}f)^2}{\rho}\, d\rho = \frac{1}{2\pi}\int_0^r \left(\int_{-\pi}^{\pi} |G'(\rho e^{i\theta})\mid d\theta\right)^2 d\rho$$

$$\leq \int_0^r \rho\, d\rho \int_{-\pi}^{\pi} |G'(\rho e^{i\theta})|^2 d\theta = A(r;\, \mathcal{L}f).$$

3. Additive family of operators.

Let an operator $\mathcal{L}$ be defined by (0.1). Then, since $f \in \mathcal{F}$ implies $\mathcal{L}f \in \mathcal{F}$, the iterating sequence $\{\mathcal{L}^n\}_{n=0}^{\infty}$ with $\mathcal{L}^0 = \mathrm{id}$ arises automatically. Further, it is shown that there exists an interpolating family $\{\mathcal{L}^\lambda\}_{\lambda \geq 0}$ depending on the continuous parameter λ such that the additivity (0.2) is satisfied. In fact, we have the following theorem; cf. [6].

Theorem 3.1. If $\{\alpha_\nu\}$ is a sequence satisfying $\alpha_1 = 1$ and $\limsup |\alpha_\nu|^{1/\nu} \leq 1$, then the family of operators $\{\mathcal{L}^\lambda\}$ defined by

$$\mathcal{L}^\lambda f(z) = \sum_{\nu=1}^{\infty} \alpha_\nu{}^\lambda c_\nu z^\nu, \tag{3.1}$$

where $f(z) = \sum_{\nu=1}^{\infty} c_\nu z^\nu \in \mathcal{F}$, satisfies (0.2). Further, if the sequence $\{\alpha_\nu{}^\lambda\}_\nu$ is fully monotone, then there exits a probability measure σ_λ uniquely determined under the normalization $\sigma_\lambda(t) = (\sigma_\lambda(t-0) + \alpha_\lambda(t+0))/2$ such that $\mathcal{L}^\lambda$ admits the integral representation

$$\mathcal{L}^\lambda f(z) = \int_I \frac{f(zt)}{t}\, d\sigma_\lambda(t). \tag{3.2}$$

Proof. The former part is trivial. In particular, the assumption lim sum $|\alpha_\nu|^{1/\nu} \leq 1$ together with $\alpha_1 = 1$ ensures $\mathcal{L}^\lambda f \in \mathcal{F}$. Next, the full monotonicity of $\{\alpha_\nu{}^\lambda\}$ is a (necessary and) sufficient condition in order that the moment problem of Hausdorff type

$$\int_I t^{\nu-1}\, d\sigma_\lambda(t) = \alpha_\nu^\lambda \qquad (\nu = 1, 2, \ldots) \tag{3.3}$$

has a solution; cf. [3] and also [18]. The solution is then unique and in view of $\alpha_1 = 1$ it is a probability measure. Finally, we have

$$\mathcal{L}^\lambda F(z) = \sum_{\nu=1}^{\infty} \alpha_\nu{}^\lambda c_\nu z^\nu$$

$$= \int_I \sum_{\nu=1}^{\infty} c_\nu z^\nu t^{\nu-1} d\sigma_\lambda(t) = \int_I \frac{f(zt)}{t}\, d\sigma_\lambda(t).$$

Remark. It is also known that the solution of the moment problem (3.3) can be explicitly represented in the form

$$\sigma_\lambda(t) = \sum_{j=0}^{\infty} (2j + 1)q_j \int_0^t P_j(2\tau - 1)\, d\tau$$

where P_j is the Legendre polynomial of order j and q_j denotes the value

$$q_j = \int_I P_j(2\pi - 1)\, d\sigma_\lambda(\tau)$$

in which $\int_I \tau^{\nu-1} d\sigma_\lambda(\tau)$ is to be replaced by $\alpha_\nu{}^\lambda$.

There are several sequences $\{\alpha_\nu\}$ with $\alpha_1 = 1$ which are fully monotone. Among them there are two extreme cases. One is the case where $\alpha_\nu = 0$ for $\nu > 1$ while another is the case where $\alpha_\nu = 1$ for every ν. They correspond to the operators $\mathcal{L}$ satisfying $\mathcal{L}f(z) = z$ and $\mathcal{L}f(z) = f(z)$ for every $f \in \mathcal{F}$, respectively. In terms of probability measure, they correspond to σ concentrated at $t = 0$ and $t = 1$, respectively, the latter having been denoted by σ^*. On the other hand, if $\sigma \neq \sigma^*$, then $\{\alpha_\nu{}^\lambda\}_\nu$ becomes a positive decreasing sequence and hence $\mathcal{L}^\lambda$ is regarded as the inverse of a Gel'fond-Leont'ev derivative multiplied by z; cf. [2].

Now, two extreme cases play exceptional roles also with respect to the continuity for $\lambda \to +0$ and the limit behavior as $\lambda \to \infty$.

Theorem 3.2. For ever $f \in \mathcal{F}$ the limit relations

$$\lim_{\lambda \to +0} \mathcal{L}^\lambda f(z) = f(z) \quad \text{and} \quad \lim_{\lambda \to \infty} \mathcal{L}^\lambda f(z) = z$$

hold in E uniformally in the wider sense except the extreme cases where $\mathcal{L}f(z) = z$ for the former relation and $\mathcal{L}f(z) = f(z)$ for the latter.

Proof. The assertion follows from the series expansion (3.1). In fact, except the case where $\mathcal{L}f(z) \equiv z$, that is $\alpha_\nu = 0$ for every $\nu > 1$, the sequence $\{\alpha_\nu\}$ satisfies $0 < \alpha_\nu$

$\leq \alpha_1 = 1$ for every $\nu > 1$, whence follows $\mathcal{L}^{\lambda}f(z) \to f(z)$ as $\lambda \to +0$. On the other hand, except the case where $\mathcal{L}f(z) \equiv f(z)$, that is, $\alpha_\nu = 1$ for every ν, the sequence $\{\alpha_\nu\}$ satisfies $0 \leq \alpha_\nu < \alpha_1 = 1$ for every $\nu > 1$, whence follows $\mathcal{L}^{\lambda}f(z) \to z$ as $\lambda \to \infty$. Every limit is uniform in the wider sense.

In the following lines, we observe mainly the additive family $\{\mathcal{L}(a)^{\lambda}\}_{\lambda \geq 0}$ generated by the special measure $\sigma(t; a) = t^a$ which involves a real parameter $a > 0$; cf. [13]. The particular case $a = 1$ has been dealt with in [6], [7], [8], [9], [10], [11] and [12] from several points of view, while it plays occasionally an exceptional role in certain sense.

Theorem 3.3. The additive family of operators generated by the measure $\sigma(t; a) = t^a$ with $a > 0$ is given by the family of probability measure $\{\sigma_\lambda(t; a)\}_{\lambda \geq 0}$ defined by

$$\sigma_\lambda(t; a) = \int_0^t \rho_\lambda(\tau; a)\, d\tau, \quad \rho_\lambda(t; a) = \frac{a^\lambda}{\Gamma(\lambda)}\, t^{a-1}\left(\log \tfrac{1}{t}\right)^{\lambda-1}.$$

Proof. In order to verify the additivity, we put

$$\begin{aligned}\mathcal{L}(a)^{\lambda}f(z) &= \int_I \frac{f(zt)}{t}\, d\sigma_\lambda(t; a)\\ &= \int_I \frac{f(zt)}{t}\, \rho_\lambda(t; a)dt = \frac{a^\lambda}{\Gamma(\lambda)} \int_I f(zt)t^{a-2}\left(\log \tfrac{1}{t}\right)^{\lambda-1} dt.\end{aligned}$$

Then direct calculation shows

$$\begin{aligned}\mathcal{L}(a)^{\lambda}\mathcal{L}(a)^{\mu}f(z) &= \int_I \rho_\lambda(s; a)\, \frac{ds}{s} \int_I f(zs\tau)\rho_\mu(\tau; a)\, \frac{d\tau}{\tau}\\ &= \int_I \rho_\lambda(s; a)\, \frac{ds}{s} \int_0^s f(zt)\rho_\mu\left(\tfrac{t}{s}; a\right) \frac{dt}{t}\\ &= \int_I f(zt)\, \frac{dt}{t} \int_t^1 \rho_\lambda(s; a)\, \rho_\mu\left(\tfrac{t}{s}; a\right) \frac{ds}{s}.\end{aligned}$$

The integral with respect to s contained in the last member becomes

$$\frac{a^\lambda}{\Gamma(\lambda)}\, \frac{a^\mu}{\Gamma(\mu)} \int_t^1 s^{a-1} \left(\log \tfrac{1}{s}\right)^{\lambda-1} \left(\tfrac{t}{s}\right)^{a-1} \left(\log \tfrac{s}{t}\right)^{\mu-1} \frac{ds}{s}$$

$$= \frac{a^{\lambda+\mu}}{\Gamma(\lambda)\Gamma(\mu)}\, t^{a-1} \int_t^1 \left(\log \tfrac{1}{s}\right)^{\lambda-1} \left(\log \tfrac{s}{t}\right)^{\mu-1} \frac{ds}{s}$$

$$= \frac{a^{\lambda+\mu}}{\Gamma(\lambda)\Gamma(\mu)}\, t^{a-1} \left(\log \tfrac{1}{t}\right)^{\lambda+\mu-1} \int_0^1 u^{\lambda-1} (1 - u)^{\mu-1} du \quad [\log \tfrac{1}{s} = u \log \tfrac{1}{t}]$$

$$= \frac{a^{\lambda+\mu}}{\Gamma(\lambda)\Gamma(\mu)} t^{a-1} \left(\log \frac{1}{t}\right)^{\lambda+\mu-1} \cdot \frac{\Gamma(\lambda)\Gamma(\mu)}{\Gamma(\lambda+\mu)} = \rho_{\lambda+\mu}(t; a).$$

Consequently, we obtain $\mathfrak{L}(a)^{\lambda}\mathfrak{L}(a)^{\mu} = \mathfrak{L}(a)^{\lambda+\mu}$.

The assertion may be alternatively verified by considering the moments with respect to $\sigma_{\lambda}(t; a)$. In fact, we have the relation

$$\int_I t^{\nu-1}\, d\sigma_{\lambda}(t; a) = \frac{a^{\lambda}}{\Gamma(\lambda)} \int_I t^{\nu+a-2} \left(\log \frac{1}{t}\right)^{\lambda-1} dt = \left(\frac{a}{\nu+a-1}\right)^{\lambda},$$

showing that the moment with respect to σ_{λ} (t; a) is equal to the moment with respect to $\sigma(t; a)$ to the λth power for very $\nu = 1, 2, \ldots$.

In case of $\sigma(t; a)$ the limit relations in Theorem 3.2 for $\mathfrak{L}(a)^{\lambda}$ instead of $\mathfrak{L}^{\lambda}$ hold without exception. On the other hand, the behaviors of $\{\mathfrak{L}(a)^{\lambda}\}$ as $a \to +0$ and $a \to \infty$ become as follows.

Theorem 3.4. For every $f \in \mathfrak{F}$ the limit relations

$$\lim_{a\to+0} \mathfrak{L}(a)^{\lambda} f(z) = z \quad \text{and} \quad \lim_{a\to\infty} \mathfrak{L}(a)^{\lambda} f(z) = f(z)$$

hold in E uniformly in the wider sense.

Proof. Let z be restricted on any fixed compact subset in E. Both $|f(zt)/t - z|$ and $|f(zt)/t - f(z)|$ then possess for every $t \in I$ a bound M, say. First, we have

$$\mathfrak{L}(a)^{\lambda} f(z) - z = \frac{a^{\lambda}}{\Gamma(\lambda)} \int_I \left(\frac{f(zt)}{t} - z\right) t^{a-1} \left(\log \frac{1}{t}\right)^{\lambda-1} dt.$$

For any $\varepsilon > 0$ there exists $\tau \in (0, 1)$ such that $|f(zt)/t - z| < \varepsilon/2$ as $0 \le t < \tau$, and hence for $a < 1$

$$|\mathfrak{L}(a)^{\lambda} f(z) - z| < \frac{\varepsilon}{2} \frac{a^{\lambda}}{\Gamma(\lambda)} \int_0^{\tau} t^{a-1}\left(\log \frac{1}{t}\right)^{\lambda-1} dt + M\tau^{a-1} \frac{a^{\lambda}}{\Gamma(\lambda)} \int_{\tau}^1 \left(\log \frac{1}{t}\right)^{\lambda-1} dt.$$

The first summand of this estimate is always less then $\varepsilon/2$, while the second summand becomes less than $\varepsilon/2$ provided a is sufficiently near to zero. This leads us to the first relation of the theorem. Next, we have

$$\mathcal{L}(a)^{\lambda}f(z) - f(z) = \frac{a^{\lambda}}{\Gamma(\lambda)} \int_I \left(\frac{f(zt)}{t} - f(z)\right) t^{a-1} \left(\log \frac{1}{t}\right)^{\lambda-1} dt.$$

For any $\varepsilon > 0$ there exits $\tau \epsilon$ (0, 1) such that $|f(zt)/t - f(z)| < \varepsilon/2$ as $1 - \tau < t \leq 1$, and hence for $a > 1$

$$|\mathcal{L}(a)^{\lambda}f(z) - f(z)| < M(1 - \tau)^{a-1} \frac{a^{\lambda}}{\Gamma(\lambda)} \int_0^{1-\tau} \left(\log \frac{1}{t}\right)^{\lambda-1} dt + \frac{\varepsilon}{2} \frac{a^{\lambda}}{\Gamma(\lambda)} \int_{1-\tau}^{1} t^{a-1} \left(\log \frac{1}{t}\right)^{\lambda-1} dt.$$

Since the second summand of this estimate is always less than $\epsilon/2$ and the first summand becomes less than $\epsilon/2$ for a large enough, the second relation follows.

Though the proof given here has been based on the integral representation of $\mathcal{L}(a)^{\lambda}f$, a rather brief proof may be given by referring to its series expansion.

4. Relation to integration operator.

We begin with a property of $\mathcal{L}$ with general σ.

Theorem 4.1. Any operator $\mathcal{L}$ commutes with the operator $d/d \log z$, that is,

$$\frac{d}{d \log z} \mathcal{L} = \mathcal{L} \frac{d}{d \log z}.$$

Proof. By differentiating the defining expression of $\mathcal{L}f$, we obtain

$$\frac{d}{d \log z} \mathcal{L}f(z) = \int_I \frac{ztf'(zt)}{t} d\sigma(t) = \mathcal{L}(zf'(z)) = \mathcal{L} \frac{d}{d \log z} f(z).$$

Now, in particular case generated by $\sigma(t) = \sigma(t; 1)$, the operator $\mathcal{L}(1)^{\lambda}$ can be represented in the form

$$\mathcal{L}(1)^{\lambda} f(z) = \frac{1}{\Gamma(\lambda)} \int_{\infty}^{\log z} f(\zeta) (\log z - \log \zeta)^{\lambda-1} d \log \zeta,$$

the integration being taken along the half straight line in the left half-plane $\{\mathrm{Re} \log \zeta < 0\}$ on the $\log \zeta$-plane which is parallel to the real axis. Thus, we see that it coincides with the fractional integration of order λ with respect to log z. In particular, $\mathcal{L}(1)$ is just the inverse operator of $d/d \log z$.

The last-mentioned fact is peculiar to the case $a = 1$. The corresponding property of $\mathcal{L}(a)$ with $a \neq 1$ is stated as in the following theorem.

Theorem 4.2. The operator $\mathcal{L}(a)$ with $a \neq 1$ coincides with the integration with respect to $w = a(a-1)^{-1}z^{a-1}$ followed by multiplication by $z^{-(a-1)}$, the integration path on the log ω-plane being the half straight line $\{\arg \omega = \pi - (1-a)\arg z,\ \infty > |\omega| > |w|\}$ for $0 < a < 1$ and the segment $\{\arg\ \omega = (a-1)\arg z,\ \ 0 < |\omega| < |w|\}$ for $a > 1$, respectively.

Proof. The operator $\mathcal{L}(a)$ is, by definition, given by

$$\mathcal{L}(a)f(z) = a \int_I \frac{f(zt)}{t}\, t^{a-1}dt = \frac{a}{z^{a-1}} \int_0^z f(\zeta)\zeta^{a-2}d\zeta, \tag{4.1}$$

the last integration being taken along the segment from 0 to z. We have only to change the integration variable by $d\omega = a\zeta^{a-2}d\zeta$, or more concretely, by $\omega = a(a-1)^{-1}\zeta^{a-1}$.

Theorem 4.3 For any $a > 0$ and $\lambda \geq 1$, we have

$$\frac{d}{d \log z}\, \mathcal{L}(a)^{\lambda} = a\mathcal{L}(a)^{\lambda-1} - (a-1)\mathcal{L}(a)^{\lambda},\quad \mathcal{L}(a)^{o} = \mathrm{id}.$$

Proof. By differentiating the defining equation

$$\mathcal{L}(a)^{\lambda}f(z) = \frac{a^{\lambda}}{\Gamma(\lambda)} \int_I f(zt)t^{a-2}\left(\log \frac{1}{t}\right)^{\lambda-1} dt$$

$$= \frac{a^{\lambda}}{\Gamma(\lambda)}\, \frac{1}{z^{a-1}} \int_0^z f(\zeta)\zeta^{a-2}\left(\log \frac{z}{\zeta}\right)^{\lambda-1}d\zeta$$

with respect to log z, we have

$$\frac{d}{d \log z}\, \mathcal{L}(a)^{\lambda}f(z) = z\, \frac{a^{\lambda}}{\Gamma(\lambda)} \left(\frac{1}{z^{a-1}} \int_0^z f(\zeta)\zeta^{a-2}(\lambda-1)\left(\log \frac{z}{\zeta}\right)^{\lambda-2} \frac{1}{z}\, d\zeta\right.$$

$$\left. + \frac{-(a-1)}{z^{a}} \int_0^z f(\zeta)\zeta^{a-2}\left(\log \frac{z}{\zeta}\right)^{\lambda-1}d\zeta\right)$$

$$= (a\mathcal{L}(a)^{\lambda-1} - (a-1)\mathcal{L}(a)^{\lambda})f(z).$$

Next, we attempt to derive expression for $\mathcal{L}(a)$ in terms of the ordinary integration $\mathcal{J}$

defined by

$$\mathcal{J}f(z) = \int_0^z f(\zeta)d\zeta$$

and its iterations. For the sake of brevity we make use of Pochhammer's symbol $(x)_n = \Gamma(x + n)/\Gamma(x)$ ($n = 0, 1, \ldots$), $(x)_o$ being understood to be equal to 1 even for $x = 0$.

Theorem 4.4. For any $a > 0$ we have

$$\mathcal{L}(a) = a \sum_{\kappa=1}^{\infty} \frac{(2 - a)_{\kappa-1}}{z^{\kappa}} \mathcal{J}^{\kappa}. \tag{4.2}$$

In particular, when $a = k > 1$ is an integer, the right hand expression reduces to finite sum consisting of the beginning $k - 1$ terms.

Proof. Since $|z - \zeta| < |z|$ holds on the integration path in (4.1) except at $\zeta = 0$, we have

$$\zeta^{a-2} = z^{a-2}\left(1 - \frac{z - \zeta}{z}\right)^{a-2} = z^{a-2} \sum_{\kappa=1}^{\infty} \frac{(2 - a)_{\kappa-1}}{(\kappa-1)!} \frac{1}{z^{\kappa-1}} (z - \zeta)^{\kappa-1}.$$

Substitution followed by termwise integration yields

$$\begin{aligned}\mathcal{L}(a)f(z) &= a \sum_{\kappa=1}^{\infty} \frac{(2 - a)_{\kappa-1}}{z^{\kappa}} \frac{1}{(\kappa-1)!} \int_0^z f(\zeta)(z - \zeta)^{\kappa-1}d\zeta \\ &= a \sum_{\kappa=1}^{\infty} \frac{(2 - a)_{\kappa-1}}{z^{\kappa}} \mathcal{J}^{\kappa} f(z).\end{aligned}$$

When $a=k > 1$ is an integer, then $(2 - k)_{\kappa - 1}$ vanishes for every $\kappa \geq k$. The case $a = 1$ is exceptional in the sense that every term in the summand for $\mathcal{L}(1)$ does not vanish.

It may be noted that, for integral value of a, the relation in Theorem 4.4 can also be inductively verified, by making use of integration by parts. On the other hand, it is remarked that the operator $\mathcal{L}(k)$ with an integer k has been dealt with by several authors, namely, $\mathcal{L}(1)$ by Srivastava and Owa [19], $\mathcal{L}(2)$ by Libera [14] and Livingston [15], $\mathcal{L}(k)$ with $k > 1$ by Bernardi [1], each in connection with some classes of univalent functions.

By relaxing the restriction that the referring probability measure σ is a monomial, the result just obtained can be generalized as follows.

Theorem 4.5. Let a probability measure σ be given by

$$\sigma(t) = \int_0^\infty t^a d\tau(a)$$

where τ is a probability measure on $(0, \infty)$. Then the corresponding operator $\mathfrak{L}$ is represented in the form

$$\mathfrak{L} = \sum_{\kappa=1}^{\infty} (-1)^{\kappa-1} \frac{\varphi^{(\kappa-1)}(1)}{z^\kappa} \mathfrak{g}^\kappa$$

where φ is defined by

$$\varphi(t) = \frac{\sigma'(t)}{t} = \int_0^\infty a t^{a-2} d\tau(a).$$

Proof. We have by definition

$$\mathfrak{L}f(z) = \int_I \frac{f(zt)}{t} d\sigma(t) = \int_0^\infty \mathfrak{L}(a)f(z)d\tau(a)$$

and hence, by substituting (4.2),

$$\mathfrak{L} = \int_0^\infty a \sum_{\kappa=1}^{\infty} \frac{(2-a)_{\kappa-1}}{z^\kappa} \mathfrak{g}^\kappa \, d\tau(a) = \sum_{\kappa=1}^{\infty} \frac{\Phi_\kappa}{z^\kappa} \mathfrak{g}^\kappa, \text{ say.}$$

The coefficients Φ_κ are given by

$$\Phi_\kappa = \int_0^\infty a(2-a)_{\kappa-1} d\tau(a)$$

$$= (-1)^{\kappa-1} \left[\frac{d^{\kappa-1}}{dt^{\kappa-1}} \int_0^\infty a t^{a-2} d\tau(a)\right]^{t=1} = (-1)^{\kappa-1} \varphi^{(\kappa-1)}(1),$$

whence follows the desired result.

Now, the Hadamard product $*$ of two power series $\varphi(z) = \sum_{\nu=1}^{\infty} a_\nu z^\nu$ and $\psi(z) = \sum_{\nu=1}^{\infty} b_\nu z^\nu$ is defined by

$$\varphi * \psi(z) = \sum_{\nu=1}^{\infty} a_\nu b_\nu z^\nu.$$

It is readily seen that $\varphi, \psi \in \mathfrak{F}$ implies $\varphi * \psi \in \mathfrak{F}$ and the particular function

$$\chi(z) = \frac{z}{1-z} = \sum_{\nu=1}^{\infty} z^\nu$$

plays the role of unit function with respect to $*$ in the class $\mathfrak{F}$.

On the other hand, any operator $\mathfrak{L}$ under consideration satisfies $\mathfrak{L}(\varphi * \psi) = \varphi * \mathfrak{L}\psi$, whence follows, in particular,

$$\mathcal{L}f = \mathcal{L}(f*\chi) = f*\mathcal{L}\chi.$$

Thus, the action of $\mathcal{L}$ on any $f \in \mathcal{F}$ is reduced to the Hadamard product of f with $\mathcal{L}\chi$.

If we consider, for instance, the operator $\mathcal{L}(a)$, we have the power series expansion of $\mathcal{L}(a)\chi$ in the form

$$\mathcal{L}(a)\chi(z) = a \sum_{\nu=1}^{\infty} \frac{z^\nu}{\Gamma(\nu + a)}$$

and hence, by making use of Theorem 4.4, the equality

$$\sum_{\nu=1}^{\infty} \frac{z^\nu}{\Gamma(\nu + a)} = \sum_{\kappa=1}^{\infty} \frac{(2 - a)_{\kappa-1}}{z^\kappa} \mathcal{J}^\kappa \chi(z)$$

valid for $a > 0$. Now, as readily seen directly, $\mathcal{J}^\kappa\chi$ is expressed by the expression

$$\mathcal{J}^\kappa\chi(z) = z^\kappa \sum_{\nu=1}^{\infty} \frac{z^\nu}{(\nu + 1)_\kappa}.$$

By substituting this into the above relation and comparing the coefficients of z^ν, we obtain an identity on gamma function

$$\frac{1}{\Gamma(\nu + a)} = \sum_{\kappa=1}^{\infty} \frac{(2 - a)_{\kappa-1}}{(\nu + 1)_\kappa}.$$

Finally, it is noted that $\mathcal{J}^\kappa\chi$ is for any integer $\kappa \geq 0$ an elementary function. We have, for instance,

$$\mathcal{J}\chi(z) = \log \frac{1}{1 - z} - z, \quad \mathcal{J}^2\chi(z) = -(1 - z) \log \frac{1}{1 - z} + z - \frac{z^2}{2}.$$

For any integer $\kappa \geq 1$ we can derive similar explicit expression in the form

$$\mathcal{J}^\kappa\chi(z) = \frac{(-1)^{\kappa-1}}{(\kappa - 1)!} (1 - z)^{\kappa-1} \log \frac{1}{1 - z} + \frac{(-1)^{\kappa-1}}{\kappa!} (1 - z)^\kappa$$

$$+ \frac{(-1)^{\kappa-1}}{(\kappa - 1)!} \sum_{j=2}^{\kappa-1} \frac{1}{j} \cdot (1 - z)^{\kappa-1} + \sum_{j=0}^{\kappa-2} \frac{(-1)^j}{j!} \frac{1}{(\kappa - j)!(\kappa - j - 1)} (1 - z)^j,$$

the empty sum being zero.

5. Specialization of referring measure.

In previous papers [6] and [13] we discussed some properties on the family $\{\mathcal{L}^\lambda\}$ generated by general measure σ and specialized them in the case of $\{\mathcal{L}(a)^\lambda\}$ with $a = 1$

and subsequently with any $a > 0$.

In the following lines we exclusively consider the family $\{\mathcal{L}(a)^{\lambda}\}_{\lambda \geq o}$ with a fixed a and accordingly make use of the abbreviation

$$f_{\lambda} = \mathcal{L}(a)^{\lambda} f,$$

unless any confusion will arise.

First, we state a theorem similar to Theorem 1.2.

Theorem 5.1. For any f, g ϵ $\mathcal{F}$ and any λ, $\mu \geq 0$ the quantity $M[f_{\lambda+\delta} - g_{\mu+\delta}]$ decreases with respect to $\delta \geq 0$. More precisely, for $\delta' > \delta \geq 0$ we have

$$\left(\frac{a+1}{a}\right)^{\delta'} M[f_{\lambda+\delta'} - g_{\mu+\delta'}] \leq \left(\frac{a+1}{a}\right)^{\delta} M[f_{\lambda+\delta} - g_{\mu+\delta}].$$

Proof. The distortion inequality (1.2) in Theorem 1.2 becomes, after substituting $\mathcal{L}(a)^{\delta'-\delta}$, $f_{\lambda+\delta}$ and $g_{\mu+\delta}$ instead of $\mathcal{L}$, f and g, respectively, and accordingly $\sigma_{\delta'-\delta}(t;\, a)$ instead of $\sigma(t)$,

$$M[f_{\lambda+\delta'} - g_{\mu+\delta'}] \leq M[f_{\lambda+\delta} - g_{\mu+\delta}] \int_I t d\sigma_{\delta'-\delta}(t;\, a).$$

Here the last factor in the right-hand member is equal to

$$\int_I t\, d\sigma_{\delta'-\delta}(t;\, a) = \frac{a^{\delta'-\delta}}{\Gamma(\delta'-\delta)} \int_I t^a \left(\log \frac{1}{t}\right)^{\delta'-\delta-1} dt = \left(\frac{a}{a+1}\right)^{\delta'-\delta},$$

yielding the desired result.

Next, Theorem 1.3 takes for $\mathcal{L}(a)$ the following form:

Theorem 5.2. For any f ϵ $\mathcal{F}$, $\lambda \geq 0$ and $\delta > 0$ we have

$$h[f_{\lambda+\delta}] \geq h[f_{\lambda}] + \Phi(\delta,\, a)(1 - h[f_{\lambda}]),$$

$$H[f_{\lambda+\delta}] \leq H[f_{\lambda}] - \Phi(\delta,\, a)(H[f_{\lambda}] - 1) \tag{5.1}$$

where Φ is given by

$$\Phi(\delta,\, a) = 1 - 2a^{\delta} \sum_{\nu=2}^{\infty} \frac{(-1)^{\nu}}{(\nu + a - 1)^{\delta}}. \tag{5.2}$$

The equality sign in either estimation does not appear for any r ϵ (0, 1) unless $f(z) \equiv z$.

If, in particular, a = k is a positive integer, then

$$\Phi(\delta, k) = 1 + 2(-1)^{k-1}k^{\delta}\Big((1 - 2^{1-\delta})\zeta(\delta) + \sum_{\kappa=1}^{k} \frac{(-1)^{\kappa-1}}{\kappa^{\delta}}\Big), \tag{5.3}$$

ζ denoting Riemann zeta function.

Proof. The inequalities having been substantially shown in Theorem 1.3, it suffices to verify the expression for Φ. We first have

$$\Phi(\delta, a) = \int_I \frac{1-t}{1+t}\, d\sigma_{\delta}(t; a) = \frac{a^{\delta}}{\Gamma(\delta)} \int_I \frac{1-t}{1+t}\, t^{a-1} \Big(\log \frac{1}{t}\Big)^{\delta-1} dt$$

$$= \frac{a^{\delta}}{\Gamma(\delta)} \int_I \Big(1 - 2 \sum_{\nu=2}^{\infty} (-1)^{\nu}\, t^{\nu-1}\Big)\, t^{a-1} \Big(\log \frac{1}{t}\Big)^{\delta-1} dt$$

$$= 1 - 2a^{\delta} \sum_{\nu=2}^{\infty} \frac{(-1)^{\nu}}{(\nu + a - 1)^{\delta}}.$$

Next, in view of a formula on zeta function, we get for an integer $k > 0$

$$\sum_{\nu=2}^{\infty} \frac{(-1)^{\nu}}{(\nu + k - 1)^{\delta}} = (-1)^{k} \Big(\sum_{\kappa=1}^{\infty} - \sum_{\kappa=1}^{k}\Big) \frac{(-1)^{\kappa-1}}{\kappa^{\delta}}$$

$$= (-1)^{k}\, (1 - 2^{1-\delta})\zeta(\delta) + (-1)^{k} \sum_{\kappa=1}^{k} \frac{(-1)^{\kappa}}{\kappa^{\delta}}.$$

By substituting this, we obtain the desired expression (5.3).

We can derive another estimation of similar nature.

Theorem 5.3. The factor $\Phi(\delta, a)$ in the estimations in Theorem 5.2 may be replaced by $1 - e^{-\Phi'(0, a)}$, where Φ' is given by

$$\Phi'(0, a) = \Big[\frac{\delta}{\partial\delta}\Phi(\delta, a)\Big]^{\delta=+0} = 2 \lim_{\delta\to+0} \sum_{\nu=2}^{\infty} (-1)^{\nu} \frac{\log(\nu + a - 1) - \log a}{(\nu + a - 1)^{\delta}}. \tag{5.4}$$

If, in particular, a = k is a positive integer, then

$$\Phi'(0, k) = 2(-1)^{k-1} \log \frac{k!\, \sqrt{\pi/2}}{(2^{[k/2]} \cdot [k/2]!)^{2}} - \log k. \tag{5.5}$$

Proof. We first note that $\Phi(+0, a) = 0$. In fact, by means of integration by parts, we get

$$\Phi(\delta, a) = \frac{a^\delta}{\Gamma(\delta)} \int_I \frac{1-t}{1+t} t^{a-1} \left(\log \frac{1}{t}\right)^{\delta-1} dt$$

$$= \frac{a^\delta}{\Gamma(\delta+1)} \int_I \frac{d}{dt}\left(\frac{1-t}{1+t} t^a\right) \cdot \left(\log \frac{1}{t}\right)^\delta dt \to 0 \text{ as } \delta \to +0.$$

Thus, the first inequality in (5.1) yields

$$\frac{h[f_{\lambda+\delta}] - h[f_\lambda]}{\delta} \geq \frac{\Phi(\delta, a)}{\delta} (1 - h[f_\lambda]),$$

whence follows, as δ tends to 0^+, the inequality

$$\frac{\partial}{\partial \lambda} h[f_\lambda] \geq \Phi'(0, a) (1 - h[f_\lambda]). \tag{5.6}$$

This linear differential inequality can be readily brought into finite form. In fact, by rewriting it in the form

$$\frac{\partial}{\partial \lambda} (e^{\Phi'(0,a)\lambda} h[f_\lambda]) \geq \Phi'(0, a) e^{\Phi'(0,a)\lambda}$$

and then integrating both sides with respect to λ over the interval $(\lambda, \lambda + \delta)$, we obtain the desired estimation for h. Similar argument applies also for H. Next, (5.2) yields (5.4) readily. Finally, we have

$$\frac{\partial}{\partial \delta}\Phi(\delta, k) = 2(-1)^{k-1} k^\delta \log k \left((1 - 2^{1-\delta})\zeta(\delta) - \sum_{\kappa=1}^{k} \frac{(-1)^{\kappa-1}}{\kappa^\delta}\right)$$

$$+ (-1)^{k-1} k^\delta \left((1 - 2^{1-\delta})\zeta'(\delta) + 2^{1-\delta} \log 2 \cdot \zeta(\delta) + \sum_{\kappa=2}^{k} \frac{(-1)^{\kappa-1}}{\kappa^\delta} \log \kappa\right).$$

In view of $\zeta(0) = -1/2$ and $\zeta'(0) = -(1/2) \log 2\pi$, we get

$$\Phi'(0, k) = 2(-1)^{k-1} \log k \left(\frac{1}{2} - \sum_{\kappa=1}^{k} (-1)^{\kappa-1}\right)$$
$$+ 2(-1)^{k-1} \left(\frac{1}{2} \log 2\pi - \log 2 + \sum_{\kappa=2}^{k} (-1)^{\kappa-1} \log \kappa\right),$$

which is readily brought into the desired form (5.5).

We supplement here the monotonicity behaviors of $\Phi(\delta, a)$ and $\Phi'(0, a)$ with respect to the parameters δ and a.

Theorem 5.4. For any fixed a > 0, $\Phi(\delta, a)$ increases strictly from 0 to 1 as δ increases from 0 to ∞, while, for any fixed $\delta > 0$, $\Phi(\delta, a)$ decreases strictly from 1 to 0 as a increases from 0 to ∞. On the other hand, $\Phi'(0, a)$ decreases strictly from ∞ to 0 as a increases from 0 to ∞.

Proof. By making use of the expressions

$$\Phi(\delta, a) = \frac{a^\delta}{\Gamma(\delta)} \int_I \frac{1-t}{1+t} t^{a-1} \left(\log \tfrac{1}{t}\right)^{\delta-1} dt,$$

$$\Phi'(0, a) = \int_I \frac{d}{dt}\left(\frac{1-t}{1+t} t^a\right) \cdot \log\log \tfrac{1}{t}\, dt$$

$$= \int_I \frac{1-t}{1+t} t^{a-1}\left(\log \tfrac{1}{t}\right)^{-1} dt,$$

the assertions are verified by direct estimation; cf. [13].

Theorem 1.5 is specialized by substituting $\mathcal{L}^\lambda$, $h[f_\lambda]$, $H[f_\lambda]$, $T[f_\lambda]$, $f_{\lambda+\delta}$ and $\sigma_\delta(t; a)$ instead of $\mathcal{L}$, $h[f]$, $H[f]$, $T[f]$, $\mathcal{L}f$ and $\sigma(t)$, respectively. Corresponding to (5.6), we further obtain the following result. cf. [11].

Theorem 5.5. For any $f \in \mathcal{F}$, the system of differential inequalities

$$\frac{\partial}{\partial\lambda} h[f_\lambda] \geq \psi_\lambda(a) \frac{H[f_\lambda] - h[f_\lambda]}{2}, \quad \frac{\partial}{\partial\lambda} H[f_\lambda] \leq \Psi_\lambda(a) \frac{H[f_\lambda] - h[f_\lambda]}{2}$$

concerning the right-side derivatives is satisfied, where ψ_λ and Ψ_λ are given by

$$\left.\begin{matrix}\psi_\lambda(a)\\ \Psi_\lambda(a)\end{matrix}\right\} = \frac{4}{\pi}\int_I \log\log\frac{1}{t}\,\frac{1}{t}\frac{\partial}{\partial t}\left(t^a \arctan\frac{T[f_\lambda] \mp t}{1 \mp T[f_\lambda]t}\right) dt \mp (C + \log a),$$

$C = -\Gamma'(1)$ denoting the Euler constant.

Proof. The inequality corresponding to the first part of (1.6) is brought into

$$h[f_{\lambda+\delta}] - h[f_\lambda] \geq (\varphi_\lambda(\delta, a) + 1)\frac{H[f_\lambda] - h[f_\lambda]}{2},$$

where φ_λ is given by

$$\varphi_\lambda(\delta, a) = \frac{4a^\delta}{\pi\Gamma(\delta)}\int_I \arctan\frac{T[f_\lambda] - t}{1 - T[f_\lambda]t} \cdot t^{a-1}\left(\log\tfrac{1}{t}\right)^{\delta-1} dt.$$

Hence the first desired inequality is obtained by putting

$$\psi_\lambda(a) = \lim_{\delta\to+0}\frac{\varphi_\lambda(\delta, a) + 1}{\delta}$$

It remains thus to derive its explicit expression. Integration by parts yields

$$\varphi_\lambda(\delta, a) = \frac{4a^\delta}{\pi\Gamma(\delta+1)} \int_I \left(\log \frac{1}{t}\right)^\delta \frac{\partial}{\partial t}\left(t^a \arctan \frac{T[f_\lambda] - t}{1 - T[f_\lambda]t}\right) dt \to -1 \text{ as } \delta \to +0$$

and hence

$$\psi_\lambda(a) = [\frac{\partial}{\partial\delta}\varphi_\lambda(\delta, a)]^{\delta=+0}.$$

Again by means of integration by parts, we have

$$\varphi_\lambda(\delta, a) = \frac{4a^\delta}{\pi\Gamma(\delta+1)} \int_I \left(\log \frac{1}{t}\right)^\delta \frac{\partial}{\partial t}\left(t^a \arctan \frac{T[f_\lambda] - t}{1 - T[f_\lambda]t}\right) dt.$$

Therefore, by differentiating with respect to δ followed by putting $\delta = +0$, we obtain

$$\psi_\lambda(a) = \frac{4}{\pi} \int_I \left(\log\log \frac{1}{t} - \Gamma'(1)\right) \frac{\partial}{\partial t}\left(t^a \arctan \frac{T[f_\lambda] - t}{1 - T[f_\lambda]t}\right) dt + \varphi_\lambda(+0, a) \log a,$$

whence readily follows the desired expression for ψ_λ. The second part of the theorem is similarly verified.

Concerning Corollary 1.5, the circumstance is quite similar. The inequality (1.7) holds for $\mathcal{L}^\lambda$, $f_{\lambda+\delta}$ and $\sigma_\delta(t; a)$ instead of $\mathcal{L}$, $\mathcal{L}f$ and $\sigma(t)$, respectively, and the following theorem is derived; cf. [12] for the case $a = 1$.

Theorem 5.6. For any $f \in \mathcal{F}$ and $\delta > 0$, the inequalities

$$\frac{\partial}{\partial\lambda}\Delta[f_\lambda] \le R(a)\Delta[f_\lambda], \quad \Delta[f_\delta] \le \Delta[f] \exp(R(a)\delta)$$

are satisfied, where R is given by

$$R(a) = \frac{4}{\pi} \int_I \log\log \frac{1}{t} \frac{\partial}{\partial t}(t^a \arctan t)\, dt + C + \log a,$$

C being the Euler constant.

Proof. We start with the inequality $\Delta[f_{\lambda+\delta}] \le Q(\delta, a)\, \Delta[f_\lambda]$ where Q is given by

$$Q(\delta, a) = \frac{4a^\delta}{\pi\Gamma(\delta)} \int_I \arctan t \cdot t^{a-1}\left(\log \frac{1}{t}\right)^{\delta-1} dt.$$

Integration by parts yields

$$Q(\delta, a) = \frac{4a^\delta}{\pi\Gamma(\delta+1)} \int_I \frac{\partial}{\partial t}(t^a \arctan t) \cdot \left(\log \frac{1}{t}\right)^\delta dt,$$

whence readily follows $Q(+0, a) = 1$. Therefore, we get

$$\frac{\Delta[f_{\lambda+\delta}] - \Delta[f_\lambda]}{\delta} \le \frac{Q(\delta, a) - 1}{\delta} \Delta[f_\lambda],$$

whence follows the desired inequality with $R(a) = [\partial Q(\delta, a)/\partial\delta]^{\delta=+0}$. Now, by differentiating the last expression for Q with respect to δ followed by putting $\delta = +0$, we have the desired expression for R.

Theorems 2.1 and 2.2 concerning the range of values can also be specialized, which are stated as follows.

Theorem 5.7. Any $f \in \mathcal{F}(\alpha)$ satisfies

$$\left| \frac{f_\lambda(z)}{z} - \frac{\varphi_\lambda(f; \alpha)}{r} \right| \leq \frac{\Psi_\lambda(f; \alpha)}{r} - 1$$

for $|z| \leq r < 1$ and the extremal functions are of the same form as in Theorem 2.1. The center and the radius of the range-disk are given, respectively, by

$$\frac{\varphi_\lambda(r; \alpha)}{r} = 1 + 2(1 - \alpha)\, a^\lambda \sum_{n=1}^{\infty} \frac{r^{2n}}{(2n + a)^\lambda},$$

$$\frac{\psi_\lambda(r; \alpha)}{r} - 1 = 2(1 - \alpha)\, a^\lambda \sum_{n=1}^{\infty} \frac{r^{2n-1}}{(2n + a - 1)^\lambda}.$$

Theorem 5.8. For any $f \in \mathcal{F}(\alpha)$, if $\lambda > 1$, the range of values of $w = f_\lambda(z)/z$ in the whole disk E is contained in the circular disk

$$\left| w - \left(1 + 2(1 - \alpha)\left(\frac{a}{2}\right)^\lambda \zeta\left(\lambda; \frac{a + 2}{2}\right)\right) \right| < 2(1 - \alpha)\left(\frac{a}{2}\right)^\lambda \zeta\left(\lambda; \frac{a + 1}{2}\right),$$

where ζ denotes the generalized zeta function $\zeta(\lambda; c) = \sum_{\nu=0}^{\infty} (\nu + c)^{-\lambda}$, while, if $\lambda \leq 1$, this circular disk expands to the half-plane

$$\operatorname{Re} w > 1 - 2(1 - \alpha)\, a^\lambda \lim_{r \to 1-0} \sum_{n=1}^{\infty} \left(\frac{r^{2n-1}}{(2n + a - 1)^\lambda} - \frac{r^{2n-1}}{(2n + a)^\lambda} \right)$$

$$= 1 - 2(1 - \alpha)\, a^\lambda \sum_{\nu=1}^{\infty} \frac{(-1)^{\nu-1}}{(\nu + a)^\lambda}.$$

Here we only state the theorems. Their proofs proceed similar as those for the case $a = 1$ previously given in [9], and they will be fully published elsewhere.

References

[1] Bernardi, S. D., Convex and starlike univalent functions, Trans. Amer. Math. Soc. 136 (1969), pp.429 - 446.

[2] Gel'fond, A. O., and A. F. Leont'ev, On a generalization of Fourier series. Math. Sbornik N. S. 29 (71) (1951), pp. 477 - 500.

[3] Hausdorff, F., Summationsmethoden und Momentfolgen. Math. Z. 9 (1921), pp. 74 - 109; pp. 280 - 299.

[4] Koebe, P., Über das Schwarzsche Lemma und einige damit zusammenhängende Ungleichheitsbeziehungen der Potentialtheorie und Funktionentheorie. Math. Z. 6 (1920), pp. 52 - 84.

[5] Komatu, Y., On mean distortion for analytic functions with positive real part in a circle. Nagoya Math. J. 29 (1967), pp. 221 - 228.

[6] Komatu, Y., On a one-parameter additive family of operators defined on analytic functions regular in the unit disk. Bull. Fac. Sci.-Eng., Chuo Univ. 22 (1977), pp. 1 - 22.

[7] Komatu, Y., A one-parameter family of operators defined on analytic functions in a circle, Proc. Anal. Func./Kozubnik 1979, Lecture Notes in Math. No. 768, ed. by J. Lawrynowicz, Springer, New York (1980), pp. 292- 300.

[8] Komatu, Y., Über die Verzerrung bei konvexer Abbildung des Einheitskreises. Bull. Fac. Sci.- Eng., Chuo Univ. 24 (1981), pp. 7 - 12.

[9] Komatu, Y., On the range of analytic functions related to Carathéodory class. Ann. Polon. Math. 46 (1985), pp. 141 - 145.

[10] Komatu, Y., Über Langen-und Flächenverzerrungen für die Carathéodorysche Klasse. J. Math. Kyoto Univ. 25 (1985), pp. 627 - 633.

[11] Komatu, Y., On distortion of the real part in a class of analytic functions related to fractional integration. Complex Variables 7 (1986), pp. 97 - 106.

[12] Komatu, Y., On oscillation in a class of analytic functions related to fractional calculus. Compl. Anal. and Appl. '85, Sofia (1986), pp. 327 - 334.

[13] Komatu, Y., On a family of integral operators related to fractional calculus. Kodai Math. J. 10 (1987), pp. 20 - 38.

[14] Libera, R. J., Some classes of regular univalent functions. Proc. Amer. Math. Soc. 16 (1965), pp. 755 - 758.

[15] Livingston, A. E., On the radius of univalence of certain analytic functions. Proc. Amer. Math. Soc. 17 (1966), pp. 352 - 357.

[16] Neumann, C., Vorlesungen über Riemanns Theorie der Abelschen Integrale. 2. Aufl. Leipzig (1884).

[17] Rogosinski, W., Über positive harmonische Entwicklungen und typisch reelle Potenzreihen. Math. Z. 35 (1932), pp. 93 - 121.

[18] Shohat, J. A., and J. D. Tamarkin, The problem of moments. Math. Survey I, AMS, New York (1943), pp. 9 and 11.

[19] Srivastava, H. M., and S. A. Owa, A certain one-parameter additive family of operators defined on analytic functions. J. Math. Anal. Appl. 118 (1986), pp. 80 - 87.

[20] Strohhäcker, E., Beiträge zur Theorie der schlichten Funktionen. Math. Z. 37 (1933), pp. 356 - 380.

Kanazawa Institute of Technology

A COEFFICIENT INEQUALITY FOR MEROMORPHIC UNIVALENT FUNCTIONS

Yoshihisa Kubota

1. Let Σ denote the family of univalent functions

$$g(z) = z + \sum_{n=0}^{\infty} b_n z^{-n}$$

in $\Delta = \{z: |z| > 1\}$.

It is well known that

$$\max_{\Sigma} |b_1| = 1 \qquad ([2]),$$

$$\max_{\Sigma} |b_2| = \tfrac{2}{3} \qquad ([8]),$$

$$\max_{\Sigma} |b_3| = \tfrac{1}{2} + e^{-6} \qquad ([1]).$$

Solutions to the coefficient problems $\max_{\Sigma} |b_n|$ are not known for $n \geq 4$. There is not even an explicit general conjecture for these problems. The author [5] proved that if b_1 is real,

$$Re\, b_4 \leq \tfrac{2}{5} + \frac{3^6}{5 \cdot 2^{15}} \wedge^2,$$

where $\wedge = 1.1938...$ is the solution of a complicated equation. It seems reasonable for various reasons that

$$\max_{\Sigma} |b_4| = \tfrac{2}{5} + \frac{3^6}{5 \cdot 2^{15}} \wedge^2$$

(see [9], [10]).

Many extremal problems involving b_1, b_2 and b_3 are solved by applying Löwner's method and Jenkins' general coefficient theorem (for example, see [3], [6], [7]). However, it is difficult to give complete solutions for extremal problems involving b_n $(n \geq 4)$ by applying those methods.

In this paper we shall give an estimate for $b_4 + \alpha b_1 b_2$ by using the method in [5]:

Theorem. Let $0 \leq \alpha \leq 1$ and let

$$g(z) = z + \sum_{n=0}^{\infty} b_n z^{-n}$$

be a function in Σ. If b_1 is real, then

$$Re\{b_4 + \alpha b_1 b_2\} \le \tfrac{2}{5} + \frac{3^6}{5\cdot 2^5}\left(\frac{\alpha-1}{\alpha-2}\right)^{10} \Lambda^2, \quad \Lambda = \frac{(3-4\cos^2\psi)^3}{\cos^2\psi(9-8\cos^2\psi)^2},$$

where ψ *is the real number satisfying*

$$Im\left\{e^{i2\psi}\int_0^1 (e^{i\psi}t^3 - 2\cos\psi\cdot t^2 + e^{-i\psi}t)^{1/2}dt\right\} = 0,$$

$$0 < \psi < \pi, \ -0.44 < \cos\psi < -0.4.$$

For $0 \le \alpha < 1$, the equality holds only for the function $\tilde{g}$ which satisfies the differential equation

$$z^2\left(\frac{dw}{dz}\right)^2 \left\{w^3 - 2\left(\frac{\alpha-2}{\alpha-1}\right)\mu w + 2\left(\frac{\alpha-1}{\alpha-2}\right)\mu\right\}$$

$$= z^5 - 2\mu z^3 + 2\mu z^2 + \mu^2 z - 2(\mu^2+1) + \mu^2 z^{-1} + 2\mu z^{-2} - 2\mu z^{-3} + z^{-5},$$

$$\mu = -\frac{27}{8}\left(\frac{\alpha-1}{\alpha-2}\right)^5 \Lambda.$$

The expansion of $\tilde{g}$ at the point at infinity begins

$$z + \frac{27}{4}\frac{(\alpha-1)^4}{(\alpha-2)^5}\Lambda z^{-1} - \frac{27}{4}\frac{(\alpha-1)^5}{(\alpha-2)^6}\Lambda z^{-2} - \frac{243}{64}\frac{(\alpha-1)^8(\alpha+1)^2}{(\alpha-2)^{10}}\Lambda^2 z^{-3}$$

$$+ \left\{\frac{2}{5} + \frac{3^6}{5\cdot 2^5}\frac{(\alpha-1)^9(\alpha^2+7\alpha+2)}{(\alpha-2)^{11}}\Lambda^2\right\} z^{-5} + \cdots .$$

2. We begin by listing several lemmas which will be used later on.

Lemma 1. Let $Q(w)dw^2 = \alpha(w^3 + \beta_1 w^2 + \beta_2 w + \beta_3)\, dw^2$ be a quadratic differential on the w-sphere and let

$$g^*(z) = z + \sum_{n=1}^{\infty} b_n^* z^{-n}$$

be a function in Σ which maps Δ onto a domain admissible with respect to $Q(w)dw^2$.

If

$$g(z) = z + \sum_{n=1}^{\infty} b_n z^{-n}$$

is a function in Σ with $b_1 = b_1^*$, then

$$Re\, \alpha\{b_4 - b_4^* + \beta_1 (b_3 - b_3^*) + (\beta_2 + 2b_1^*)(b_2 - b_2^*)\} \leq 0.$$

Equality occurs only for $g(z) = g^*(z)$.

This lemma is obtained by the general coefficient theorem in its extended form [4].

Lemma 2. If

$$g(z) = z + \sum_{n=1}^{\infty} b_n z^{-n}$$

is a function in Σ, then

$$|b_1|^2 + 2|b_2|^2 + 3|b_3|^2 + 4|b_4|^2 \leq 1.$$

This lemma is a simple consequence of the area theorem. The next five lemmas were given in [5].

Lemma 3. Let

$$g(z) = z + \sum_{n=1}^{\infty} b_n z^{-n}$$

be a function in Σ. If b_1 is real, then

$$Re\,\{b_4 + b_1 b_2\} \leq \tfrac{2}{5}.$$

Lemma 4. Let Y be a non-negative real number and let

$$Q^*(w:Y)dw^2 = (w + 2Y)(w-Y)^2 dw^2.$$

If X is a real number satisfying the condition

$$80X^4 - 60X^2 + 4 \leq 3\sqrt{3}Y^{5/2} \leq 64X^5 - 40X^3, \quad \sqrt{\tfrac{5}{8}} \leq X \leq \tfrac{\sqrt{3}}{2}, \qquad (1)$$

then there is a function $g^*(z: X, Y)$ in Σ which maps Δ onto a domain admissible with respect to $Q^*(w:Y)dw^2$ and which satisfies the differential equation

$$z^2\left(\frac{dw}{dz}\right)^2\left(w^3 - 3Y^2 w + 2Y^3\right)$$
$$= z^5 - 2\mu z^3 + 2\mu z^2 + \mu^2 z - 2\left(\mu^2 + 1\right) + \mu^2 z^{-1} + 2\mu z^{-2} - 2\mu z^{-3} + z^{-5},$$

$$\mu = 16X^4 - 12X^2 + 1.$$

The expansion of $g^*(z:X, Y)$ at the point at infinity begins

$$z - (2\mu - 3Y^2)\, z^{-1} - 2(\mu - Y^3)\, z^{-2} - 3(\mu - Y^2)\,{}^2 z^{-3} + \left\{ \tfrac{2}{5} + \Psi^*(\mu, Y)\right\} z^{-4} + \cdots,$$

where

$$\Psi^*(\mu, Y) = -6\mu^2 + (8Y^3 + 6Y^2)\,\mu - \tfrac{42}{5}Y^5.$$

Lemma 5. Let

$$R_1^* = \left\{(b_1, b_2) : 0 \le b_1 \le 0.43,\ -0.49 \le b_2 \le 0\right\},$$

$$R_2^* = \left\{(b_1, b_2) : 0.43 \le b_1 \le 0.6,\ -0.38 \le b_2 \le 0\right\}.$$

If $\left(b_1, b_2\right)$ is a point in $R_1^* \cup R_2^*$, then there are real numbers X and Y, $Y \ge 0$, satisfying condition (1) such that

$$b_1 = -2\mu + 3Y^2 = -32X^4 + 24X^2 - 2 + 3Y^2,$$

$$b_2 = -2\mu + 2Y^3 = -32X^4 + 24X^2 - 2 + 2Y^3.$$

Next we consider quadratic differentials of the form

$$Q(w : r, \theta)\, dw^2 = w(w - re^{i\theta})\,(w - re^{-i\theta})\, dw^2,\ r \ge 0,\ 0 < \theta < \pi.$$

We denote by Φ the union of all trajectories of $Q(w : r, \theta)dw^2$ which have a limiting end point at a zero of $Q(w : r, \theta)$. We put

$$R(\theta) = -\frac{1}{\sqrt{2}}\int_0^1 t^{1/2}\,(1-t)^{1/2}\,[\{(1-2t\cos 2\theta + t^2)^{1/2} + (1-t)\cos\theta\}^{1/2}\cos 2\theta$$
$$+ \{(1-2t\cos 2\theta + t^2)^{1/2} - (1-t)\cos\theta\}^{1/2}\sin 2\theta]dt,$$

$$I(\theta) = -\frac{1}{\sqrt{2}}\int_0^1 t^{1/2}(1-t)^{1/2}\,[\{(1-2t\cos 2\theta + t^2)^{1/2} + (1-t)\cos\theta\}^{1/2}\sin 2\theta$$
$$- \{(1-2t\cos 2\theta + t^2)^{1/2} - (1-t)\cos\theta\}^{1/2}\cos 2\theta]dt.$$

Then it follows that $S_W - \overline{\Phi}$ consists of five end domains if and only if $I(\theta) = 0$, where S_W is the w-sphere. Moreover, it follows that there is a real number ψ such that

$$I(\psi) = 0,\quad -0.44 < \cos\psi < -0.4,\quad 0 < \psi < \pi.$$

For such a ψ we have

$$0.434 < R(\psi) < 0.487.$$

Lemma 6. Let ψ be a real number such that $I(\psi) = 0$, $-0.44 < \cos\psi < -0.4$, $0 < \psi < \pi$

and let Y be a non-negative real number. Let

$$\tilde{Q}(w:Y)dw^2 = Q(w + \tfrac{2}{3}Y \cos\psi: Y, \psi)dw^2$$
$$= \{w^3 + \tfrac{1}{3}Y^2 (3-4\cos^2\psi)\, w + \tfrac{2}{27}Y^3 \cos\psi\, (9-8\cos^2\psi)\}dw^2.$$

If X is a real number satisfying the condition

$$-256X^5 + 160X^3 \le 5R(\psi)\, Y^{5/2} \le -160X^4 + 120X^2 -8, \qquad (2)$$

then there is a function $\tilde{g}(z:X, Y)$ in Σ which maps Δ onto a domain admissible with respect to $\tilde{Q}(w:Y)\, dw^2$ and which satisfies the differential equation

$$z^2\left(\frac{dw}{dz}\right)^2\left\{w^3 + \tfrac{1}{3}Y^2(3-4\cos^2\psi)w + \tfrac{2}{27}Y^3\cos\psi(9-8\cos^2\psi)\right\}$$
$$= z^5-2\mu z^3 + 2\mu z^2 + \mu^2 z - 2(\mu^2 + 1) + \mu^2 z^{-1} + 2\mu z^{-2}-2\mu z^{-3} + z^{-5},$$

$$\mu = 16X^4 - 12X^2 + 1.$$

The expansion of $\tilde{g}(z:X, Y)$ at the point at infinity begins

$$z - \left\{2\mu + \tfrac{1}{3}Y^2(3-4\cos^2\psi)\right\} z^{-1} - 2\left\{\mu - \tfrac{1}{27}Y^3\cos\psi\, (9-8\cos^2\psi)\right\} z^{-2}$$
$$- 3\left\{\mu + \tfrac{1}{9}Y^2(3-4\cos^2\psi)\right\}^2 z^{-3} + \left\{\tfrac{2}{5} + \tilde{\Psi}\, (\mu, Y)\right\}z^{-4} + \cdots,$$

where

$$\tilde{\Psi}(\mu, Y) = -6\mu^2 - \left\{ \tfrac{2}{3}Y^2(3-4\cos^2\psi) - \tfrac{8}{27}Y^3\cos\psi(9-8\cos^2\psi)\right\} \mu$$
$$+ \tfrac{14}{405}Y^5\cos\psi\, (9-8\cos^2\psi)\, (3-4\cos^2\psi).$$

Lemma 7. Let

$$\tilde{R}_1 = \{(b_1, b_2) : -0.46 \le b_1 \le 0,\ 0 \le b_2 \le 0.425\},$$

$$\tilde{R}_2 = \{(b_1, b_2) : -0.6 \le b_1 \le -0.46,\ 0 \le b_2 \le 0.33\}.$$

If (b_1, b_2) is a point in $\tilde{R}_1 \cup \tilde{R}_2$, then there are real numbers X and Y, $0 \le Y < 2$, satisfying condition (2) such that

$$b_1 = -2\mu - \tfrac{1}{3}Y^2(3-4\cos^2\psi) = -32X^4 + 24X^2 - 2-\tfrac{1}{3}Y^2(3-4\cos^2\psi),$$

$$b_2 = -2\mu + \tfrac{2}{27}Y^3\cos\psi\, (9-8\cos^2\psi) = -32X^4 + 24^2 -2 + \tfrac{2}{27}Y^3\cos\psi\, (9-8\cos^2\psi).$$

3. We now prove the theorem. Let

$$g(z) = z + \sum_{n=0}^{\infty} b_n z^{-n}$$

be a function in Σ with real b_1. We may suppose that $b_0 = 0$. Firstly we consider the case $b_1 \leq 0$. We divide this case into several subcases.

Case (1) $-0.46 \leq b_1 \leq 0$, $0 \leq Re b_2 \leq 0.425$. In this case by Lemma 7 there are real numbers X and Y, $0 \leq Y < 2$, satisfying condition (2) such that

$$b_1 = -2\mu - \tfrac{1}{3}Y^2(3-4\cos^2\psi),\quad Re b_2 = -2\mu + \tfrac{2}{27}Y^3\cos\psi\,(9-8\cos^2\psi),$$

$$\mu = 16X^4 - 12X^2 + 1.$$

Using Lemma 1 and Lemma 6 we obtain

$$Re\, b_4 \leq Re\, \tilde{b}_4 = \tfrac{2}{5} + \tilde{\Psi}(\mu, Y),$$

and hence,

$$\begin{aligned} Re\{b_4 + \alpha b_1 b_2\} &\leq Re\{\tilde{b}_4 + \alpha\tilde{b}_1\tilde{b}_2\} \\ &= \tfrac{2}{5} - (6-4\alpha)\mu^2 - \left\{\tfrac{2}{3}(1-\alpha)AY^2 + \tfrac{4}{27}(2-\alpha)BY^3\right\}\mu - \tfrac{2}{405}(7-5\alpha)ABY^5 \\ &= \tfrac{2}{5} - 2(3-2\alpha)\left[\mu + \left\{\frac{1-\alpha}{6(3-2\alpha)}AY^2 + \frac{2-\alpha}{27(3-2\alpha)}BY^3\right\}\right]^2 \\ &\quad + \frac{2(2-\alpha)^2}{729(3-2\alpha)}B^2Y^6 - \frac{2(11-14\alpha+5\alpha^2)}{405(3-2\alpha)}ABY^5 + \frac{(1-\alpha)^2}{18(3-2\alpha)}A^2Y^4. \end{aligned}$$

Here $\tilde{b}_1$, $\tilde{b}_2$ and $\tilde{b}_4$ are the coefficients of the function $\tilde{g}(z{:}X, Y)$ in Lemma 6, and

$$A = 3-4\cos^2\psi,\quad B = -\cos\psi\,(9-8\cos^2\psi).$$

Hence, under condition (2), we obtain the inequality

$$Re\left\{b_4 + \alpha b_1 b_2\right\} \leq \tfrac{2}{5} + \frac{3^6}{5\cdot 2^5}\left(\frac{\alpha-1}{\alpha-2}\right)^{10}\Lambda^2.$$

Equality occurs only for $g(z) = \tilde{g}(z{:}X_0, Y_0)$, where

$$Y_0 = \tfrac{9}{2}\left(\frac{\alpha-1}{\alpha-2}\right)^2\frac{A}{B},\quad \mu_0 = 16X_0^4 - 12X_0^2 + 1 = -\tfrac{27}{8}\left(\frac{\alpha-1}{\alpha-2}\right)^5\frac{A^3}{B^2}.$$

Case (2) $-0.6 \le b_1 \le -0.46$, $0 \le Re\, b_2 \le 0.33$. As in Case (1) we have

$$Re\,\{b_4 + \alpha b_1 b_2\} \le \frac{2}{5} + \frac{3^6}{5\cdot 2^5}\left(\frac{\alpha-1}{\alpha-2}\right)10\wedge^2.$$

Case (3) $-0.46 \le b_1 \le 0$, $Re\, b_2 \ge 0.425$. By Lemma 2 we have

$$4\,|\,b_4\,|^2 \le 1-2\times 0.425^2 < \frac{16}{25},$$

and hence

$$Re\,\{b_4 + \alpha b_1 b_2\} \le Re\, b_4 < \frac{2}{5}.$$

Case (4) $-0.6 \le b_1 \le -0.46$, $Re\, b_2 \ge 0.33$. By Lemma 2 we have

$$4\,|\,b_4\,|^2 \le 1-0.46^2 - 2\cdot 0.33^2 < \frac{16}{25},$$

and hence

$$Re\,\{b_4 + \alpha b_1 b_2\} \le Re\, b_4 < \frac{2}{5}.$$

Case (5) $b_1 \le -0.6$, $Re\, b_2 \ge 0$. By Lemma 2 we have

$$4\,|\,b_4\,|^2 \le 1 - 0.6^2 = \frac{16}{25},$$

and hence

$$Re\,\{b_4 + \alpha b_1 b_2\} \le Re\, b_4 \le \frac{2}{5}.$$

Case (6) $b_1 \le 0$, $Re\, b_2 \le 0$. In this case by Lemma 3 we have

$$Re\,\{b_4 + \alpha b_1 b_2\} \le \mathrm{Re}\,\{b_4 + b_1 b_2\} \le \frac{2}{5}.$$

Next we consider the case $b_1 \ge 0$. We also divide this case into several subcases.

Case (1) $0 \le b_1 \le 0.43$, $-0.49 \le \mathrm{Re}\, b_2 \le 0$, or $0.43 \le b_1 \le 0.6$, $-0.38 \le Re\, b_2 \le 0$. In this case by Lemma 5 there are real numbers X and Y, $Y \ge 0$, satisfying condition (1) such that

$$b_1 = -2\mu + 3Y^2,\quad Re\, b_2 = -2\mu + 2Y^3$$

$$\mu = 16X^4 - 12X^2 + 1.$$

Using Lemma 1 and Lemma 4 we have

$$Re\, b_4 \leq Re\, b_4^* = \frac{2}{5} + \Psi^*(\mu, Y),$$

and hence we obtain

$$Re\,\{b_4 + \alpha b_1 b_2\} \leq \frac{2}{5} - (6-4\alpha)\mu^2 + \{(8-4\alpha)Y^3 + (6-6\alpha)Y^2\}\mu - \left(\frac{42}{5} - 6\alpha\right)Y^5$$

$$= \frac{2}{5} - 2(3-2\alpha)\,[\mu - \left\{\frac{2-\alpha}{3-2\alpha}Y^3 + \frac{3(1-\alpha)}{2(3-2\alpha)}Y^2\right\}]^2$$

$$+ \frac{2(\alpha-2)^2}{3-2\alpha}Y^6 - \frac{30\alpha^2 - 84\alpha + 66}{5(3-2\alpha)}Y^5 + \frac{9(\alpha-1)^2}{2(3-2\alpha)}Y^4$$

$$\leq \frac{2}{5} + \frac{3^6}{5\cdot 2^5}\left(\frac{\alpha-1}{\alpha-2}\right)^{10}.$$

Here equality occurs for $g(z) = g^*(z{:}X_0, Y_0)$, where

$$Y_0 = \frac{3}{2}\left(\frac{\alpha-1}{\alpha-2}\right)^2, \quad \mu_0 = 16X_0^4 - 12X_0^2 + 1 = \frac{27}{8}\left(\frac{\alpha-1}{\alpha-2}\right)^5.$$

In the remaining cases: (2) $0 \leq b_1 \leq 0.43$, $Re\, b_2 \leq -0.49$, (3) $0.43 \leq b_1 \leq 0.6$, $Re\, b_2 \leq -0.38$, (4) $0.6 \leq b_1$, $Re\, b_2 \leq 0$, (5) $b_1 \geq 0$, $Re\, b_2 \geq 0$, we have, by using Lemma 2 and Lemma 3,

$$Re\,\{b_4 + \alpha b_1 b_2\} \leq \frac{2}{5}.$$

References

[1] Garabedian, P. R., and M. Schiffer, A coefficient inequality for schlicht functions. Ann. of Math. 61, pp. 116 - 136 (1955).

[2] Gronwell, T. H., Some remarks on conformal representation. Ann. of Math. 16, pp. 72 - 76 (1914-15).

[3] Jenkins, J. A., On certain coefficients of univalent functions. Analytic Functions. Princeton Univ. Press. (1960) pp. 159 - 194.

[4] Jenkins, J. A., An addendum to the general coefficient theorem. Trans. Amer.

Math. Soc. 107, pp. 125 - 128 (1963)

[5] Kubota, Y., On the fourth coefficient of meromorphic univalent functions. Kōdai Math. Sem. Rep. 26, pp. 267 - 288 (1974-75)

[6] Kubota, Y., A remark on the third coefficient of meromorphic univalent functions. Kōdai Math. Sem. Rep. 29, pp. 197 - 206 (1977).

[7] Leung, Y. J., and G. Schober, Low order coefficient estimates in the class Σ. Ann. Acad. Sci. Fenn. Ser. A I Math. 11, pp. 39 - 61 (1986).

[8] Schiffer, M., Sur un probléme d′ extrémum de la représentation conforme. Bull. Soc. Math. France 66, pp. 48 - 55 (1938).

[9] Schober,G., and J. K. Williams, On coefficient estimates and conjectures for the class Σ. Math. Z. 186, pp. 309 - 320 (1984).

[10] Schober, G., Some conjectures for the class Σ. Contemporary Math. 38, pp. 13 - 21 (1985).

Department of Mathematics
Tokyo Gakugei University

AN INCLUSION LATTICE OF AUTOMORPHISM GROUPS OF COMPACT RIEMANN SURFACES OF GENUS 5

Akikazu Kuribayashi and Hideyuki Kimura

The purpose of this paper is to study the canonical representation $\text{Aut}(X) \to GL(g,\mathfrak{C})$ with the space of holomorphic differentials on X as its representation module, where X is a compact Riemann surface of genus g=5. For a subgroup $\mathfrak{G}$ of $\text{Aut}(X)$ we denote its image by $R(X,\mathfrak{G})$.

It is interesting to determine the $GL(g,\mathfrak{C})$–conjugate class of $R(X,\mathfrak{G})$. Because it reveals an invariant of holomorphic family of Riemann surfaces which is defined by the subgroup of Teichmüller modular group corresponding to the pair $(X,\mathfrak{G})$. See [1, 2, 3, 6, 7, 8, 9].

In this paper by the result in [4] we consider an inclusion lattice of $R(X,\mathfrak{G})$ and construct the surjective homomorphisms ϕ for each top groups of the lattice. Roughly speaking the existence of ϕ means the existence of $R(X,\mathfrak{G})$. Hence, if we make sure the existence of ϕ for each top groups of the lattice, then we can verify obviously that all subgroups downward of the lattice come from automorphism groups of Riemann surfaces. In relation with the existence of ϕ, we shall give two exceptional groups which satisfy both the CY-condition and the RH-condition but do not satisfy the EX-condition. For these definitions see [4]. As a by-product of the inclusion lattice we can give an interesting example which seems to us that it is related to Rauch's conjecture [11].

Our programme of investigation is as follows:

§0. Preliminaries.

§1. Inclusion lattice.

§2. Surjective homomorphisms.

§3. Exceptional groups.

Notations.

$$A(a,b,c,d,e) = \begin{bmatrix} a &&&& \\ & b &&& \\ && c && \\ &&& d & \\ &&&& e \end{bmatrix}. \qquad B(a,b,c,d,e) = \begin{bmatrix} a &&&& \\ & b &&& \\ && c && \\ &&& 0 & d \\ &&& e & 0 \end{bmatrix}. \qquad C(a,b,c,d,e) = \begin{bmatrix} a &&&& \\ & 0 & b && \\ & c & 0 && \\ &&& d & \\ &&&& e \end{bmatrix}.$$

$$D(a,b,c,d,e) = \begin{bmatrix} 0 & a &&& \\ b & 0 &&& \\ && c && \\ &&& d & \\ &&&& e \end{bmatrix}. \qquad E(a,b,c,d,e) = \begin{bmatrix} a &&&& \\ & 0 & 0 & b & 0 \\ & 0 & c & 0 & 0 \\ & d & 0 & 0 & 0 \\ & 0 & 0 & 0 & e \end{bmatrix}. \qquad F(a,b,c,d,e) = \begin{bmatrix} a &&&& \\ & 0 & 0 & 0 & b \\ & 0 & 0 & c & 0 \\ & 0 & d & 0 & 0 \\ & 0 & e & 0 & 0 \end{bmatrix}.$$

$$G(a,b,c,d,e) = \begin{bmatrix} a &&&& \\ & 0 & b && \\ & c & 0 && \\ &&& 0 & d \\ &&& e & 0 \end{bmatrix}. \qquad H(a,b,c,d,e) = \begin{bmatrix} a &&&& \\ & 0 & 0 & b & 0 \\ & 0 & 0 & 0 & c \\ & d & 0 & 0 & 0 \\ & 0 & e & 0 & 0 \end{bmatrix}. \qquad K(a,b,c,d,e) = \begin{bmatrix} 0 & a &&& \\ b & 0 &&& \\ && 0 & 0 & c \\ && 0 & d & 0 \\ && e & 0 & 0 \end{bmatrix}.$$

$$L(a,b,c,d,e) = \begin{bmatrix} a &&&& \\ & 0 & 0 & 0 & b \\ & c & 0 & 0 & 0 \\ & 0 & d & 0 & 0 \\ & 0 & 0 & e & 0 \end{bmatrix}. \qquad M(a,b,c,d,e) = \begin{bmatrix} a &&&& \\ & 0 & 0 & 0 & b \\ & 0 & 0 & c & 0 \\ & d & 0 & 0 & 0 \\ & 0 & e & 0 & 0 \end{bmatrix}. \qquad N(a,b,c,d,e) = \begin{bmatrix} a &&&& \\ & b &&& \\ && 0 & 0 & c \\ && d & 0 & 0 \\ && 0 & e & 0 \end{bmatrix}.$$

$$S = \begin{bmatrix} -i &&&& \\ & \zeta^5/\sqrt{2} & 0 & \zeta^7/\sqrt{2} & 0 \\ & 0 & 0 & 0 & i \\ & \zeta^7/\sqrt{2} & 0 & \zeta^5/\sqrt{2} & 0 \\ & 0 & i & 0 & 0 \end{bmatrix}, \ \zeta = \zeta_8.$$

$$T=\begin{bmatrix} 0 & 0 & 0 & 0 & 1 \\ 0 & \zeta^5/\sqrt{2} & 0 & \zeta^5/\sqrt{2} & 0 \\ 1 & 0 & 0 & 0 & 0 \\ 0 & \zeta^7/\sqrt{2} & 0 & \zeta^3/\sqrt{2} & 0 \\ 0 & 0 & 1 & 0 & 0 \end{bmatrix}, \zeta = \zeta_8.$$

$$Q = \begin{bmatrix} i & & & & \\ & 0 & 0 & 1/\sqrt{2} & -i/\sqrt{2} \\ & 0 & 0 & i/\sqrt{2} & -1/\sqrt{2} \\ & -1/\sqrt{2} & -i/\sqrt{2} & 0 & 0 \\ & i/\sqrt{2} & 1/\sqrt{2} & 0 & 0 \end{bmatrix}.$$

$$U = \begin{bmatrix} 0 & 0 & 0 & 0 & 1 \\ 0 & -1 & 0 & 0 & 0 \\ 0 & 0 & 0 & -1 & 0 \\ 0 & 0 & -1 & 0 & 0 \\ 1 & 0 & 0 & 0 & 0 \end{bmatrix}.$$

$$V = \begin{bmatrix} 0 & 1 & 0 & 0 & 0 \\ 0 & 0 & 0 & 0 & 1 \\ 1 & 0 & 0 & 0 & 0 \\ 0 & 0 & 1 & 0 & 0 \\ 0 & 0 & 0 & 1 & 0 \end{bmatrix}.$$

$$W = \begin{bmatrix} a & b & b & b & b \\ b & c & d & e & f \\ b & d & f & c & e \\ b & e & c & f & d \\ b & f & e & d & c \end{bmatrix},$$

where $a = -1/5$, $b = \sqrt{6}/5$,

$c = (3 - \sqrt{5})\ /10$, $f = (3 + \sqrt{5})\ /\ 10$,

$d = -\ (1 + \sqrt{5})\ /5$, $e = -\ (1 - \sqrt{5})/\ 5$.

$I = A(1,1,1,1,1)$, $J = (-1)\ I$ and $\zeta_n = \exp\ (2\pi i/n)$.

§0. Preliminaries.

We shall list up all groups $R(X,\mathfrak{G})$ in [4].

(a) Cyclic groups.

1.1 $G_1\ (2) = <A(1, 1, 1, -1, -1)>$.

1.2 $G_2\ (2) = <A(1, 1, -1, -1, -1)>$.

1.3 $G_3\ (2) = <A(1, -1, -1, -1, -1)>$.

1.4 $G_4\ (2) = <J>$.

2.1 $G_1\ (3) = <A(1, \omega, \omega, \omega^2, \omega^2)>, \omega = \zeta_3$.

2.2 $G_2\ (3) = <A(\omega, \omega, \omega, \omega^2, \omega^2)>$

3.1 $CG_1\ (4) = <A(1, 1, -1, i, -i)>$

3.2 $CG_2\ (4) = <A(1, i, i, -i, -i)>$

3.3 $CG_3\ (4) = <A(1, -1, i, i, -i)>$

3.4 $CG_4\ (4) = <A(i, i, i, -i, -i)>$

3.5 $CG_5\ (4) = <A(-1, i, i, -i, -i)>$

3.6 $CG_6\ (4) = <A(-1, i, i, i, -i)>$.

4.1 $G(5) = < A(1, \zeta, \zeta^2, \zeta^3, \zeta^4) >, \zeta = \zeta_5$.

5.1 $CG_1\ (6) = < A(1, \zeta, \zeta^4, \zeta^2, \zeta^5) >, \zeta = \zeta_6$.

5.2 $CG_2\ (6) = < A(-1, \zeta, \zeta, \zeta^5, \zeta^5) >$.

5.3 $CG_3\ (6) = < A(\zeta, \zeta, \zeta^4, \zeta^2, \zeta^5) >$.

5.4 $CG_4\ (6) = <A(-1, \zeta, \zeta, \zeta^2, \zeta^5) >$.

5.5 $CG_5\ (6) = <A(-1, \zeta, \zeta^4, \zeta^2, \zeta^5) >$.

6.1 $CG_1\ (8) = < A(1, \zeta, \zeta^5, \zeta^3, \zeta^7) >, \zeta = \zeta_8$.

6.2 $CG_2\ (8) = < A(\zeta^2, \zeta, \zeta, \zeta^5, \zeta^3) >$.

6.3 $CG_3\ (8) = < A(\zeta^2, \zeta, \zeta^5, \zeta^3, \zeta^7) >.$

7.1 $CG\ (10) = < A(-1, \zeta. \zeta^7, \zeta^3, \zeta^9) >, \zeta = \zeta_{10}.$

8.1 $G_1\ (11) = < A(\zeta, \zeta^2, \zeta^3, \zeta^4, \zeta^6) >, \zeta = \zeta_{11}.$

8.2 $G_2\ (11) = < A(\zeta, \zeta^2, \zeta^3, \zeta^4, \zeta^5) > .$

9.1 $CG\ (12) = < A(\zeta^3, \zeta, \zeta^{10}, \zeta^8, \zeta^5) >, \zeta = \zeta_{12}.$

10.1 $CG\ (15) = < A(\zeta^{10}, \zeta, \zeta^2, \zeta^8, \zeta^4) >, \zeta = \zeta_{15}.$

11.1 $CG\ (20) = < A(\zeta^{15}, \zeta, \zeta^7, \zeta^3, \zeta^9) >, \zeta = \zeta_{20}.$

12.1 $CG\ (22) = < A(\zeta. \zeta^{13}, \zeta^3, \zeta^{15}, \zeta^5) >, \zeta = \zeta_{22}.$

(b) Non-cyclic groups of order 2^p $(2 \leq p \leq 6)$.

1.1 $AG_1\ (4) = < G_1\ (2), A(1, 1, -1, 1, -1) >$, [2, 4; ---].

1.2 $AG_2\ (4) = < G_1\ (2), G_3\ (2) >$, [1, 4; 2, 2, 2, 2].

1.3 $AG_3\ (4) = < G_1\ (2), A(1, -1, -1, -1, 1) >$, [1, 4; 2, 2, 2, 2].

1.4 $AG_4\ (4) = < G_1\ (2), A(-1, -1, -1, 1, 1) >$, [0, 4; 2, 2, 2, 2, 2, 2, 2, 2].

1.5 $AG_5\ (4) = < G_1\ (2), A(-1, -1, -1, 1, -1) >$, [0, 4, 2, 2, 2, 2, 2, 2, 2, 2].

1.6 $AG_6\ (4) = < G_2\ (2), A(-1, -1, 1, 1, -1) >$, [0, 4; 2, 2, 2, 2, 2, 2, 2, 2].

2.1.1 $AG_1\ (8) = < CG_1\ (4), A(1, -1, -1, 1, 1) >$, [1, 8; 2, 2].

2.1.2 $AG_2\ (8) = < CG_2\ (4), A(1, 1, -1, 1, -1) >$, [1, 8; 2, 2].

2.1.3 $AG_3\ (8) = < CG_2\ (4), A(-1, 1, 1, 1, -1) >$, [0, 8; 2, 2, 2, 4, 4].

2.1.4 $AG_4\ (8) = < CG_2\ (4), A(-1, 1, -1, 1, -1) >$, [0, 8; 2, 2, 2, 4, 4].

2.1.5 $AG_5\ (8) = < CG_3\ (4), A(-1, -1, 1, 1, 1) >$, [0, 8; 2, 2, 2, 4, 4].

2.1.6 $AG_6\ (8) = < CG_3\ (4), A(-1, -1, 1, -1, 1) >$, [0, 8; 2, 2, 2, 4, 4].

2.1.7 $AG_7\ (8) = < CG_4\ (4), A(1, 1, -1, 1, -1) >$, [0, 8; 2, 2, 2, 4, 4].

2.1.8 $AG_8\ (8) = < CG_5\ (4), A(1, 1, -1, 1, -1) >$, [0, 8; 4, 4, 4, 4].

2.1.9 $AG_9(8) = < CG_6(4), G_1(2) >$, [0, 8; 4, 4, 4, 4].

2.1.10 $AG_{10}(8) = < AG_1(4), A(1, -1, 1, 1, -1) >$, [1, 8; 2, 2].

2.1.11 $AG_{11}(8) = < AG_1(4), A(-1, -1, 1, 1, 1) >$, [0, 8; 2, 2, 2, 2, 2, 2].

2.1.12 $AG_{12}(8) = < AG_1(4), A(-1, -1, 1, 1, -1) >$, [0, 8; 2, 2, 2, 2, 2, 2].

2.1.13 $AG_{13}(8) = < AG_2(4), A(-1, 1, -1, 1, -1) >$, [0, 8; 2, 2, 2, 2, 2, 2].

2.2.1 $G_1(8) = < CG_1(4), B(1, -1, 1, 1, 1) >$, [1, 8; 2,2].

2.2.2 $G_2(8) = < CG_1(4), B(-1, -1, 1, 1, 1) >$, [0, 8; 2, 2, 2, 2, 2, 2].

2.2.3 $G_3(8) = < CG_2(4), F(-1, 1, 1, 1, 1) >$, [0, 8; 2, 2, 2, 2, 2, 2].

2.2.4 $G_4(8) = < CG_5(4), F(1, 1, 1, 1, 1) >$, [0, 8; 2, 2, 2, 4, 4].

2.2.5 $G_5(8) = < CG_2(4), F(-1, i, i, i, i) >$, [0, 8; 4, 4, 4, 4].

3.1.1 $AG_1(16) = < CG_2(8), A(1, -1, 1, -1, 1) >$, [0, 16; 4, 8, 8].

3.1.2 $AG_2(16) = < AG_2(8), A(-1, 1, 1, 1, -1) >$, [0, 16; 2, 2, 4, 4].

3.1.3 $AG_3(16) = < AG_{10}(8), A(-1, 1, 1, 1, -1) >$, [0, 16; 2, 2, 2, 2, 2].

3.2.1 $G_1(16) = < CG_1(8), G(1, 1, 1, 1, 1,) >$, [1, 16; 2].

3.2.2 $G_2(16) = < CG_1(8), F(-1, 1, 1, 1, 1) >$, [0, 16; 2, 2, 2, 2, 2].

3.2.3 $G_3(16) = < CG_1(8), H(-1, 1, 1, 1, 1) >$, [0, 16; 2, 2, 4, 4].

3.2.4 $G_4(16) = < CG_2(8), G(1, 1, 1, 1, 1) >$, [0, 16; 4, 8, 8].

3.2.5 $G_5(16) = < AG_1(8), B(-1, 1, 1, 1, 1,) >$, [0, 16; 2, 2, 2, 2, 2].

3.2.6 $G_6(16) = < AG_1(8), G(1, 1, 1, 1, 1) >$, [1, 16; 2].

3.2.7 $G_7(16) = < AG_1(8), C(-1, 1, 1, 1, 1) >$, [0, 16; 2, 2, 4, 4].

3.2.8 $G_8(16) = < AG_1(8), C(-1, 1, 1, 1, -1) >$, [0, 16; 2, 2, 4, 4].

3.2.9 $G_9(16) = < AG_1(8), G(1, 1, 1, i, i) >$, [1, 16; 2].

3.2.10 $G_{10}(16) = < AG_2(8), G(-1, 1, 1, 1, 1) >$, [0, 16; 2, 2, 4, 4].

3.2.11 $G_{11}(16) = < AG_2(8), H(-1, 1, 1, 1, 1) >$, [0, 16; 2, 2, 2, 2, 2].

3.2.12 $G_{12}(16) = < AG_8(8), H(1, 1, 1, 1, 1) >$, [0, 16; 2, 2, 4, 4].

3.2.13 $G_{13}(16) = < AG_6(8), K(1, 1, 1, 1, 1) >$, [0, 16; 2, 2, 4, 4].

3.2.14 $G_{14}(16) = < AG_5(8), D(1, 1, 1, -1, 1) >$, [0, 16; 2, 2, 4, 4].

4.1 $G_1(32) = < AG_1(16), E(-1, 1, -1, 1, -1) >$, [0, 32; 2, 8, 8].

4.2 $G_2(32) = < AG_2(16), E(-i, i, i, i, 1) >$, [0, 32; 4, 4, 4].

4.3 $G_3(32) = < AG_2(16), H(-1, 1, 1, 1, 1) >$, [0, 32; 2, 2, 2, 4].

4.4 $G_4(32) = < AG_3(16), G(-1, 1, 1, 1, 1) >$, [0, 32; 2, 2, 2, 4].

4.5 $G_5(32) = < G_1(16), H(-1, 1, 1, 1, 1) >$, [0, 32; 2, 2, 2, 4].

4.6 $G_6(32) = < G_{12}(16), L(i, 1, i, 1, -i) >$, [0, 32; 4, 4, 4].

4.7 $G_7(32) = < G_4(16), Q>$, [0, 32; 2, 8, 8].

5.1 $G_1(64) = < G_1(32), S>$, [0, 64; 2, 4, 8].

5.2 $G_2(64) = < G_4(32), M(i, 1, 1, 1, 1) >$, [0, 64; 2, 4, 8].

(c) Non-cyclic groups of order $2^p \times 3$ $(1 \leq p \leq 6)$.

1.1 $G_1(6) = < G_1(3), F(1, 1, 1, 1, 1) >$, [1, 6; 3, 3].

1.2 $G_2(6) = < G_1(3), F(-1, 1, 1, 1, 1) >$, [0, 6; 2, 2, 2, 2, 3, 3].

2.1.1 $AG_1(12) = < CG_1(6), A(-1, 1, -1, -1, 1) >$, [0, 12; 2, 2, 6, 6].

2.1.2 $AG_2(12) = < CG_1(6), A(-1, -1, -1, -1, 1) >$, [0, 12; 2, 2, 6, 6].

2.2.1 $G_1(12) = < CG_1(6), F(1, 1, 1, 1, 1) >$, [1, 12; 3].

2.2.2 $G_2(12) = < CG_1(6), F(-1, 1, 1, 1, 1) >$, [0, 12; 2, 2, 2, 2, 3].

2.2.3 $G_3(12) = < CG_2(6), F(1, 1, 1, 1, 1) >$, [0, 12; 2, 2, 2, 2, 3].

2.2.4 $G_4(12) = < CG_5(6), F(1, 1, 1, 1, 1) >$, [0, 12; 2, 2, 6, 6].

2.2.5 $G_5(12) = < CG_1(6), F(1, i, 1, 1, i) >$, [1, 12; 3].

2.2.6 G_6 (12) = $<$ CG_2 (6), F(i, i, i, i, i) $>$, [0, 12; 2, 3, 4, 4].

2.2.7 G_7 (12) = $<$ AG_1 (4), N(ω, ω^2, 1, 1, 1) $>$, [0, 12; 3, 3, 3, 3].

3.1.1 AG (24) = $<$ CG (12), A(1, 1, −1, −1, 1) $>$, [0, 24; 2, 12, 12].

3.2.1 G_1 (24) = $<$ AG_1 (12), F(1, 1, 1, 1, 1) $>$, [0, 24; 2, 2, 2, 6].

3.2.2 G_2 (24) = $<$ AG_1 (12), F(i, i, i, i, i) $>$, [0, 24; 4, 4, 6].

3.2.3 G_3 (24) = $<$ AG_2 (12), F(−1, 1, 1, 1, 1) $>$, [0, 24; 2, 2, 2, 6].

3.2.4 G_4 (24) = $<$ G_7 (12), A(1, −1, −1, 1, 1) $>$, [0, 24; 3, 6, 6].

3.2.5 G_5 (24) = $<$ G_7 (12), J $>$, [0, 24; 2, 2, 3, 3].

3.2.6 G_6 (24) = $<$ G_7 (12), A(−1, −1, 1, −1, −1) $>$, [0, 24; 2, 2, 3, 3].

3.2.7 G_7 (24) = $<$ G_7 (12), K(1, 1, 1, 1, −1) $>$, [0, 24; 2, 2, 3, 3].

4.1 G_1 (48) = $<$ AG (24), F(1, 1, 1, 1, 1) $>$, [0, 48; 2, 4, 12].

4.2 G_2 (48) = $<$ G_4 (24), A(−1, 1, −1, 1, 1) $>$, [0,48; 2, 6, 6].

4.3 G_3 (48) = $<$ G_5 (24), K(i, i, i, i, i) $>$, [0, 48; 3, 4, 4].

4.4 G_4 (48) = $<$ G_7 (24), A(−1, −1, 1, 1, 1) $>$, [0, 48; 2, 2, 2, 3].

4.5 G_5 (48) = $<$ G_6 (24), K(i, i, 1, 1, 1) $>$, [0, 48; 3, 4, 4].

5.1 G_1 (96) = $<$ G_2 (48) , K(1, 1, −1, −1, −1) $>$, [0, 96; 2, 4, 6].

5.2 G_2 (96) = $<$ G_2 (32), T $>$, [0, 96; 3, 3, 4].

6.1 G (192) = $<$ G_2 (96), H(−1, ζ, −1, ζ^7, −1) $>$, [0, 192; 2, 3, 8].

(d) Non-cyclic groups of order $2^p \times 5$ ($1 \leq p \leq 5$).

1.1 G_1 (10) = $<$ G (5), F(1, 1, 1, 1, 1) $>$, [1, 10; 5].

1.2 G_2 (10) = $<$ G (5), F(−1, 1, 1, 1, 1) $>$, [0, 10; 2, 2, 2, 2, 5].

2.1 G_1 (20) = $<$ CG (10), F(1, 1, 1, 1, 1) $>$, [0, 20; 2, 2, 2, 10].

2.2 G_2 (20) = $<$ CG (10), F(i, i, i, i, i) $>$, [0, 20; 4, 4, 10].

3.1 $G(40) = < CG(20), F(1, 1, 1, 1, 1) >$, $[0, 40; 2, 4, 20]$.

4.1 $G(80) = < AG_3(16), V >$, $[0, 80; 2, 5, 5]$.

5.1 $G(160) = < G(80), G(-1, 1, 1, 1, 1) >$, $[0, 160; 2, 4, 5]$.

(e) Non-cyclic groups of order $2^p \times 3 \times 5$ $(1 \le p \le 3)$.

1.1 $G(30) = < CG(15), F(-1, 1, 1, 1, 1) >$, $[0, 30; 2, 6, 15]$.

2.1 $G(60) = < G(5), W >$, $[0, 60; 3, 3, 5]$.

3.1 $G(120) = < G(60), J >$, $[0, 120; 2, 3 10]$.

§1. Inclusion lattice.

Proposition 1. Let $G \to H$ mean $H \subset G$ up to $GL(5, \mathbb{C})$ -conjugate. There exists the following inclusion lattice among the non-cyclic groups in §0.

(I) (1) $G(160) \to G(80)$

(2) $G(40) \to$
- $\to G_1(20) \to$
  - $\to G_1(10)$
  - $\to G_2(10)$
- $\to G_2(20)$

(3) $G(120) \to G(60)$

(4) $G(30)$

(II) (1) $G(192) \to$
- $\to G_1(64)$
- $\to G_2(96)$

(2) $G_1\ (64) \rightarrow$
- $\rightarrow G_2\ (32) \rightarrow AG_2\ (16) \rightarrow$
  - $\rightarrow AG_2\ (8)$
  - $\rightarrow AG_3\ (8) \rightarrow AG_5\ (4)$
  - $\rightarrow AG_9\ (8)$
- $\rightarrow G_1\ (32) \rightarrow AG_1\ (16)$
- $\rightarrow G_3\ (32) \rightarrow$
  - $\rightarrow G_9\ (16)$
  - $\rightarrow G_8\ (16) \rightarrow AG_{13}\ (8) \rightarrow AG_3\ (4)$
  - $\rightarrow G_{11}\ (16)$

(3) $G_2\ (64) \rightarrow$
- $\rightarrow G_4\ (32) \rightarrow$
  - $\rightarrow AG_3\ (16) \rightarrow$
    - $\rightarrow AG_{10}\ (8) \rightarrow AG_1\ (4)$
    - $\rightarrow AG_{11}\ (8)$
  - $\rightarrow G_6\ (16)$
  - $\rightarrow G_5\ (16) \rightarrow$
    - $\rightarrow AG_1\ (8) \rightarrow AG_2\ (4)$
    - $\rightarrow G_1\ (8)$
    - $\rightarrow G_2\ (8)$
  - $\rightarrow G_7\ (16)$
- $\rightarrow G_6\ (32) \rightarrow$
  - $\rightarrow G_{13}\ (16) \quad \rightarrow AG_6\ (8) \rightarrow AG_6\ (4)$
  - $\rightarrow G_{12}\ (16)$
    - $\rightarrow AG_8\ (8)$
    - $\rightarrow G_4\ (8)$
- $\rightarrow G_7\ (32) \rightarrow G_4\ (16)$

(4) $G_5\ (32) \rightarrow$
- $\rightarrow G_1\ (16)$
- $\rightarrow G_2\ (16) \rightarrow G_3\ (8)$
- $\rightarrow G_3\ (16) \rightarrow G_5\ (8)$
- $\rightarrow G_{10}\ (16) \rightarrow AG_4\ (8)$

(5) $^*G_{14}\ (16) \rightarrow$
- $\rightarrow AG_5\ (8) \rightarrow AG_4\ (4)$
- $\rightarrow AG_7\ (8)$
- $\rightarrow AG_{12}\ (8)$

. However $G_1(48) \rightarrow G_{14}\ (16)$.

(III) (1) G_1 (96) →

- → G_2 (48) →
  - → G_4 (24)
  - → G_6 (24)
- → G_5 (48) → G_5 (12)
- → G_4 (48) → G_7 (24) → G_7 (12)
- → G_3 (24) → AG_2 (12)

(2) G_1 (48) →

- → G_4 (12)
- → AG (24)
- → G_1 (24) →
  - → AG_1 (12)
  - → G_1 (12) → G_1 (6)
  - → G_2 (12) → G_2 (6)
  - → G_3 (12)
- → G_{14} (16)*
- → G_2 (24) → G_6 (12)

(3) G_3 (48) → G_5 (24)

Remark. Every non-cyclic group appears at least in one branch of the lattice. Obviously some of them can belong to other branches.

Proof. Almost all inclusions are trivial and so we shall give proof for some typical cases.

G_2 (20): If we consider the order of groups, there are five possible groups, i.e., G(40), G(80), G(160), G(60), and G(120) which contain G_2 (20). However, if we consider the Riemann-Hurwitz datum, then all groups except for G(40) are excluded. As for G(40), we see that

$$G_2\ (20) \sim \langle A(-1, \epsilon, \epsilon^7, \epsilon^3, \epsilon^9),\ F\ (\kappa^{15}, \kappa^9, \kappa^3, \kappa^7, \kappa)\rangle$$

where $\epsilon = \zeta_{10}$ and $\kappa = \zeta_{20}$. The right hand group is GL (5, $\mathfrak{C}$)–conjugate to a subgroup of G(40).

$\underline{G(30)}$: If we consider the order of groups, there are two possible groups, i.e., G(60) and G(120). However by the Riemann-Hurwitz datum both of them are excluded.

$\underline{G_1(64)}$: It contains trivially G_1 (32) and it is easy to see that

$$G_1(64) \sim \; < G_2\ (32), \begin{bmatrix} \zeta^2 & & & & \\ & \zeta & & & \\ & & 0 & 0 & \zeta \\ & & 0 & \zeta^3 & 0 \\ & & \zeta & 0 & 0 \end{bmatrix} >;$$

$$G_1\ (64) \sim \; <G_3\ (32), \begin{bmatrix} i & & & & \\ & \zeta & & & \\ & & 0 & 0 & 1 \\ & & 0 & \zeta^3 & 0 \\ & & 1 & 0 & 0 \end{bmatrix} >$$

where $\zeta = \zeta_8$.

$\underline{G_2\ (64)}$: It contains trivially G_4 (32) and G_6 (32). Further

$$G_7(32) \sim \; < \begin{bmatrix} i & & & & \\ & 0 & 0 & 1 & 0 \\ & 0 & 0 & 0 & -1 \\ & 0 & 1 & 0 & 0 \\ & 1 & 0 & 0 & 0 \end{bmatrix}, A\ (1, -1, -1, 1, 1), \begin{bmatrix} 1 & & & & \\ & 0 & 0 & -1 & 0 \\ & 0 & 0 & 0 & -1 \\ & 0 & -1 & 0 & 0 \\ & 1 & 0 & 0 & 0 \end{bmatrix} >$$

where the right hand group is a subgroup of G_2 (64).

$\underline{G_5(32)}$: Both G_1 (64) and G_2 (64) cannot contain G_5 (32), since in each case, we contradict to the CY-condition or the RH-condition.

$\underline{G_{14}(16)}$: Since this is hyperelliptic, groups of order 32, 64, 96 and 192 are excluded and G_1 (48) is possible. Further we see that

$G_1(48) \supset \langle A(-i, i, -1, 1, i), A(1, 1, -1, -1, 1), F(1, 1, 1, 1, 1)\rangle \sim G_{14}$ (16).

$\underline{G_5(48)}$: Since an element $K(i,i,1,1,1) \epsilon G_5(48)$ is GL (5, $\mathfrak{C}$)-conjugate to K(1, -1, 1, 1, 1) we see easily that G_1 (96) has a subgroup $\sim G_5$ (48).

$\underline{G_3(24)}$: We see that

G_1 (96) $\supset \langle N(\zeta^5, \zeta, 1, 1, 1), A(1, -1, -1, -1, -1), K(1, 1, -1, -1, -1) \rangle \sim G_3$ (24).

$\underline{G_3(48)}$: Since this is hyperelliptic, this is a top group.

$\underline{G_{10}(16)}$: Since $G_5(48)$ has no element of order 2 of trace -1, there is no inclusion relation for $G_5(48)$. However it is easily seen $G_{10}(16) \subset G_5(32)$ in our sense. Because $G_5(32)$ contains a group $\langle G(1, i, i, -i, -i,), G(1, 1, 1, 1, 1), F(-1, \zeta^7, \zeta^3, \zeta^5, \zeta)\rangle \sim G_{10}(16)$.

§2. Surjective homomorphisms.

Assume that G satisfies the CY-condition and the RH-condition. Let RH(G) be $[g_0, n; m_1, \ldots, m_r]$ and let Γ be a Fuchsian group

$$\langle \alpha_1, \beta_1, \ldots, \alpha_{g_0}, \beta_{g_0}, \gamma_1, \ldots, \gamma_r \rangle$$

with relations $\prod_j^r \gamma_j \prod_i^{g_0} [\alpha_i, \beta_i] = 1, \gamma_1^{m_1} = \ldots = \gamma_r^{m_r} = 1$. If we have a surjective homomorphism ϕ: $\Gamma \to G$ such that $\#\phi(\gamma_j) = m_j$ and $2-2\mathrm{Re}(\mathrm{Tr}(\phi(\gamma_j))) > 0$ $(1 \leq j \leq r)$, we say that G satisfies the EX-condition. If G satisfies the EX-condition there exist an X and a $\tilde{\mathfrak{C}}$ of Aut(X) isomorphic to G. Further in the case of g=5 it can be shown that there exists an R(X, $\mathfrak{G}$) which is GL(5, $\mathfrak{C}$)–conjugate to G by taking a suitable ϕ. See [8, 9, 10].

Proposition 2. There exists the surjective homomorphism ϕ for each top group of the inclusion lattice in Proposition 1.

Proof. G(192): We have, with $\zeta = \zeta_8$, $\phi(\gamma_1) = H(-1, \zeta, 1, \zeta^7, 1)$,

$$\phi(\gamma_2) = \begin{bmatrix} 0 & 0 & 0 & 0 & i \\ 0 & \zeta^3/\sqrt{2} & 0 & \zeta^7/\sqrt{2} & 0 \\ 1 & 0 & 0 & 0 & 0 \\ 0 & \zeta^5/\sqrt{2} & 0 & \zeta^5/\sqrt{2} & 0 \\ 0 & 0 & -i & 0 & 0 \end{bmatrix},$$

$$\phi(\gamma_3) = \begin{bmatrix} 0 & 0 & 0 & 0 & 1 \\ 0 & i/\sqrt{2} & 0 & -i/\sqrt{2} & 0 \\ 0 & 0 & i & 0 & 0 \\ 0 & i/\sqrt{2} & 0 & i/\sqrt{2} & 0 \\ i & 0 & 0 & 0 & 0 \end{bmatrix}.$$

Then we have $\#\phi(\gamma_1) = 2$, $\#\phi(\gamma_2) = 3$, $\#\phi(\gamma_3) = 8$ and $2 - 2\mathrm{Re}\left(\mathrm{Tr}(\phi(\gamma_j))\right) > 0$, $j = 1, 2, 3$. $\phi(\gamma_1)\,\phi(\gamma_2)\,\phi(\gamma_3) = I$ and $\phi: \Gamma \to G(192)$ is a surjective homomorphism.

G (160): We have $\phi(\gamma_1) = U$, $\phi(\gamma_2) = K(1, -1, -1, -1, 1)$ and $\phi(\gamma_3) = V$. Then we have $\#\phi(\gamma_1) = 2$, $\#\phi(\gamma_2) = 4$ $\#\phi(\gamma_3) = 5$ and $\phi(\gamma_1)\,\phi(\gamma_2)\,\phi(\gamma_3)\big) = I$. Further $\phi: \Gamma \to G(160)$ is a surjective homomorphism.

G (120): We have G (60) = $<\alpha, \beta,>$ with $\alpha^5 = \beta^2 = (\alpha\beta)^3 = I$. So we have $\phi(\gamma_1) = \alpha\beta$, $\phi(\gamma_2) = \beta\alpha$, $\phi(\gamma_3) = \alpha^3$ with $\phi(\gamma_1)\,\phi(\gamma_2)\,\phi(\gamma_3) = I$. As for G (120) we have G (120) = G (60) $\times$ $<J>$. Hence $\phi(\gamma_1) = J\beta$, $\phi(\gamma_2) = \alpha\beta$, $\phi(\gamma_3) = J\beta\alpha^4\beta$ with $\phi(\gamma_1)\,\phi(\gamma_2)\,\phi(\gamma_3) = I$. $\phi: \Gamma \to G(120)$ is a surjective homomorphism.

G_1 (96): We have $\phi(\gamma_1) = K(1, 1, -1, -1, -1\big)$, $\phi(\gamma_3) = N(\zeta^2, \zeta, 1, -1, 1)$

$$\phi(\gamma_2) = \begin{bmatrix} 0 & \zeta^5 & 0 & 0 & 0 \\ \zeta^4 & 0 & 0 & 0 & 0 \\ 0 & 0 & -1 & 0 & 0 \\ 0 & 0 & 0 & 0 & -1 \\ 0 & 0 & 0 & 1 & 0 \end{bmatrix},$$

with $\zeta = \zeta_6$. Then $\#\phi(\gamma_1) = 2$, $\#\phi(\gamma_2) = 4$, $\#\phi(\gamma_3) = 6$ and $\phi(\gamma_1)\,\phi(\gamma_2)\,\phi(\gamma_3) = I$. ϕ: $\Gamma \to G_1$ (96) is a surjective homomorphism.

$\underline{G_2(64)}$: We have $\phi(\gamma_1) = A(-1, -1, -1, -1, 1)$, $\phi(\gamma_2) = M\ (i, -1, -1, -1, -1)$,

$$\phi(\gamma_3) = \begin{bmatrix} i & & & & \\ & 0 & 0 & 1 & 0 \\ & 0 & 0 & 0 & -1 \\ & 0 & 1 & 0 & 0 \\ & 1 & 0 & 0 & 0 \end{bmatrix}$$

with $\phi(\gamma_1)\,\phi(\gamma_2)\,\phi(\gamma_3) = I$. We see that $\#\phi(\gamma_1) = 2$, $\#\phi(\gamma_2) = 4$, $\#\phi(\gamma_3) = 8$ and ϕ: $\Gamma \to G_2(64)$ is a surjective homomorphism.

$\underline{G_1(48)}$: We have $\phi(\gamma_1) = F(-1, -1, 1, 1, -1)$, $\phi(\gamma_2) = F(\zeta^3, \zeta^5, \zeta^8, \zeta^{10}, \zeta)$, and $\phi(\gamma_3) = A(\zeta^3, \zeta^5, \zeta^2, \zeta^4, \zeta)$ with $\zeta = \zeta_{12}$. We see that $\#\phi(\gamma_1) = 2$, $\#\phi(\gamma_2) = 4$, $\#\phi(\gamma_3) = 12$ and $\phi(\gamma_1)\,\phi(\gamma_2)\,\phi(\gamma_3) = I$. ϕ: $\Gamma \to G_1(48)$ is a surjective homomorphism.

$\underline{G_3(48)}$: We have $\phi(\gamma_3) = K(i, i, i, i, i)$ and

$$\phi\,(\gamma_1) = \begin{bmatrix} \omega^2 & & & & \\ & \omega & & & \\ & & 0 & -1 & 0 \\ & & 0 & 0 & 1 \\ & & -1 & 0 & 0 \end{bmatrix}, \qquad \phi(\gamma_2) = \begin{bmatrix} 0 & i\zeta^5 & & & \\ i\zeta & 0 & & & \\ & & i & 0 & 0 \\ & & 0 & 0 & i \\ & & 0 & -i & 0 \end{bmatrix}$$

with $\zeta = \zeta_6$. We see that ϕ: $\Gamma \to G_3(48)$ is a surjective homomorphism and $\phi(\gamma_1)\ \phi(\gamma_2)\ \phi(\gamma_3) = I$ with $\#\phi(\gamma_1) = 3$, $\#\phi(\gamma_2) = 4$, $\#\phi(\gamma_3) = 4$.

G(40): We have $\phi(\gamma_1) = F(-1, -1, -1, -1, -1)$, $\phi(\gamma_2) = F(\zeta^5, \zeta^3, \zeta, \zeta^9, \zeta^7)$ and $\phi(\gamma_3) = A(\zeta^5, \zeta^3, \zeta, \zeta^9, \zeta^7)$ with $\zeta = \zeta_{20}$. We see that ϕ: $\Gamma \to G(40)$ is a surjective homomorphism and $\phi(\gamma_1)\ \phi(\gamma_2)\ \phi(\gamma_3) = I$ with $\#\phi(\gamma_1) = 2$, $\#\phi(\gamma_2) = 4$, $\#\phi(\gamma_3) = 20$.

$G_5(32)$: We see that $\phi(\gamma_1) = H(-1, 1, 1, 1, 1)$, $\phi(\gamma_2) = F(-1, 1, 1, 1, 1)$, $\phi(\gamma_3) = F(-1, \zeta^3, \zeta^7, \zeta, \zeta^5)$ and $\phi(\gamma_4) = H(-1, \zeta^3, \zeta^7, \zeta, \zeta^5)$ with $\zeta = \zeta_8$. We see that ϕ: $\Gamma \to G_5(32)$ is a surjective homomorphism and $\phi(\gamma_1)\ \phi(\gamma_2)\ \phi(\gamma_3)\ \phi(\gamma_3)$=2,= I with $\#\phi(\gamma_1) = 2$, $\#\phi(\gamma_2) = 2$, $\phi(\gamma_3)$=2, $\#\phi(\gamma_4) = 4$.

G(30): We see that $\phi(\gamma_1) = F(-1, \zeta^{12}, \zeta^9, \zeta^6, \zeta^3)$, $\phi(\gamma_2) = F(-\zeta^{10}, \zeta^4, \zeta^8, \zeta^2, \zeta)$ and $\phi(\gamma_3) = A\ (\zeta^5, \zeta^2, \zeta^4, \zeta, \zeta^8)$ with $\zeta = \zeta_{15}$. We see that ϕ: $\Gamma \to G(30)$ is a surjective homomorphism and $\phi(\gamma_1)\ \phi(\gamma_2)\ \phi(\gamma_3) = I$ with $\#\phi(\gamma_1) = 2$, $\#\phi(\gamma_2) = 6$, $\#\phi(\gamma_3) = 15$.

$G_{10}(16)$: We see that $\phi(\gamma_1) = G(-1, -i, i, i, -i)$, $\phi(\gamma_2) = G(-1, 1, 1, 1, 1)$, $\phi(\gamma_3) = G(-1, i, i, -i, -i)$ and $\phi(\gamma_4) = G(-1, -1, 1, -1, 1)$. We see that ϕ: $\Gamma \to G_{10}(16)$ is a surjective homomorphism and $\phi(\gamma_1)\ \phi(\gamma_2)\ \phi(\gamma_3)\ \phi(\gamma_4) = I$ with $\#\phi(\gamma_1) = 2$, $\#\phi(\gamma_2) = 2$, $\#\phi(\gamma_3) = 4$, $\#\phi(\gamma_4) = 4$.

$G_{14}(16)$: We see that $\phi(\gamma_1) = D(1, 1, -1, 1, -1)$, $\phi(\gamma_2) = J$, $\phi(\gamma_3) = A(1, -1, i, i, -i)$ and $\phi(\gamma_4) = D(-1, 1, -i, i, i)$. We see that ϕ: $\Gamma \to G_{14}(16)$ is a surjective homomorphism and $\phi(\gamma_1)\ \phi(\gamma_2)\ \phi(\gamma_3)\ \phi(\gamma_4) = I$ with $\#\phi(\gamma_1) = 2$, $\#\phi(\gamma_2) = 2$, $\#\phi(\gamma_3) = 4$, $\#\phi(\gamma_4) = 4$.

Remark. In case of $G_2(64)$, if we take $\phi(\gamma_1) = A(-1, -1, -1, -1, 1)$,

$$\phi(\gamma_2) = \begin{bmatrix} -i & 0 & 0 & 0 & 0 \\ 0 & 0 & 0 & -1 & 0 \\ 0 & 0 & 0 & 0 & -1 \\ 0 & 0 & -1 & 0 & 0 \\ 0 & -1 & 0 & 0 & 0 \end{bmatrix},$$

$\phi(\gamma_3) = M(-i, -1, 1, 1, 1)$ then all conditions are satisfied. But $\mathrm{Tr}\phi(\gamma_2) = -i$ considering $\phi(\gamma_2)$ as an element of $G \subset GL(5, \mathbb{C})$. On the other hand by the formula in [8; Remark 1.2] Tr $\phi(\gamma_2) = i$ since $\mathrm{Tr}\phi(\gamma_2) = 1 + 2i/(1-i)$. This is a contradiction. Hence we must take a suitable ϕ: $\Gamma \to G$ which does not conflict with the formula in [8] in each case.

Important remark. $G_5(32)$ comes from Riemann surfaces as we have seen. Then, the family of Riemann surfaces from which $G_5(32)$ comes makes a one–dimensional space. Because the formula which gives the dimension, $3g_0 - 3 + r$, shows one in this case. However the groups $G_5(32)$ is not contained in any groups of higher order in the lattice. Because $G_5(32)$ contains an element of order 8, $A(1, \zeta, \zeta^5, \zeta^3, \zeta^7)$, $\zeta = \zeta_8$. Obviously this element has no fixed point. However the possible groups of higher order for $G_5(32)$ are $G(192)$, $G_1(96)$, $G_2(64)$ and $G(160)$. But each of them has no element of order 8 which has no fixed point. We can reinforce the proof by showing directly that any of them cannot contain a group isomorphic to $G_5(32)$. Thus we obtain a groups which makes a

one–dimensional space but is not contained in any groups of higher order. This phenomenon does not appear in lower genera. It seems to us that this is related to Rauch's conjecture [11, p. 22].

§3. Exceptional cases.

It is interesting to study what group G cannot be GL(5, $\mathfrak{C}$)–conjugate to R(X,$\mathfrak{G}$) for some X and some $\mathfrak{G}$ of Aut (X) even if G satisfies the CY–condition and the RH–condition. We know two groups $\mathfrak{D}$ and $\mathfrak{Q}$. See [10].

The groups $\mathfrak{D}$: = <A(1, i, i, –i, –i), F(1, 1, 1, 1, 1)> and $\mathfrak{Q}$: = <A(1, i, i, –i, –i), F(1, i, i, i, i)> actually satisfy the CY– and RH–conditions. Assume that these groups satisfy the EX–condition, i.e., there exists a surjective homomorphism ϕ: $\Gamma \to \mathfrak{D}$ or $\mathfrak{Q}$. Here $\Gamma = \langle \alpha, \beta, \gamma_1, \gamma_2; \alpha\beta\alpha^{-1}\beta^{-1}\gamma_1\gamma_2 = I, \gamma_1^2 = \gamma_2^2 = I\rangle$, Then we have $\phi(\gamma_1) = \phi(\gamma_2) = A(1, -1, -1, -1, -1)$ and $\phi(\alpha)\phi(\beta) = \phi(\beta)\phi(\alpha)$ since RH[$\mathfrak{D}$] or RH[$\mathfrak{Q}$]= [1, 8; 2, 2]. Since A(1, –1, –1, –1, –1) is an element of the center of $\mathfrak{D}$ or $\mathfrak{Q}$, we see that $\mathfrak{D}$ or $\mathfrak{Q}$ is abelian. This is absurd.

Remark. When we remove the condition 2– 2 Re $(\mathrm{Tr}(\phi(\gamma_j))) > 0$ in the definition of the EX-condition, if it is satisfied, there exist a Riemann surface X and a subgroup $\tilde{\mathfrak{G}}$ $\subset$ Aut (X) isomorphic to G. However in that case there does not always exist R(X, $\mathfrak{G}$) which is GL(5, $\mathfrak{C}$)–conjugate to G just as we have seen in the above. This is an important matter in our investigation.

$G_{10}(16)$: Since $G_5(48)$ has no element of order 2 of trace –1, there is no inclusion relation for $G_5(48)$.

Summarizing above sections we get the following theorem. See [4].

Theorem. Let G be a finite subgroup of GL(5, $\mathfrak{C}$). Then the following two conditions are equivalent except for $\mathfrak{D}$ and $\mathfrak{Q}$ and some groups including $\mathfrak{D}$ or $\mathfrak{Q}$:

(1) There is a compact Riemann surface X of genus 5 and an automorphism group $\mathfrak{G}$ of X such that R(X, $\mathfrak{G}$) is GL(5, $\mathfrak{C}$)–conjugate to G.

(2) G satisfies the CY–condition and the RH-condition.

Remark. Corresponding theorems hold also for genera 2, 3 and 4 without any exceptional groups. See [5, 7, 8].

References

[1] Gilman, J., On conjugacy classes in the Teichumüller modular group, Michigan Math. J., 23 (1976), pp. 53 - 63.

[2] Harvey, W. J., On branch loci in Teichumüller space, Trans. Amer. Math. Soc., 153 (1971), pp. 387 - 399.

[3] Guerrero, I., Holomorphic families of compact Riemann surfaces with automorphisms, Illinois J. of Math., 26 (1982), pp. 212 - 225.

[4] Kuribayashi, A., and H. Kimura, On automorphism groups of compact Riemann surfaces of genus five (to appear in J. of Algebra).

[5] Kuribayashi, I., and A. Kuribayashi, On automorphism groups of compact Riemann surfaces of genus 3 and 4 (to appear).

[6] Kuribayashi, I., On certain curves of genus three with many automorphisms, Tsukuba J. Math., 6 (1982), pp. 271 - 288.

[7] -------, On an algebraization of the Riemann-Hurwitz relation. Kodai Math. J., 7 (1984) pp. 222 - 237.

[8] ------, Classification of automorphism groups of compact Riemann surfaces of genus two (Preprint at Tsukuba 1986).

[9] ------, On automorphism groups of a curve as linear groups, J. Math. Soc. Japan, 29 (1987), pp. 51 - 77.

[10] Momose, F., Galois coverings of curves with given representations on differential forms (Preprint at Tsukuba 1986).

[11] Rauch, H. E., A transcendental view of algebraic Riemann surfaces, Bull. Am. Math. Soc. 71 (1965), pp. 1 - 39.

Akikazu Kuribayashi
Department of Mathematics
Chuo University

Hideyuki Kimura
Department of Mathematics
Tokyo Institute of Technology

VARIATIONAL FORMULAS UNDER QUASICONFORMAL DEFORMATION

Fumio Maitani

We will summarize variational formulas for function theoretic quantities on general open Riemann surfaces induced by quasiconformal deformations depending on a complex parameter. In the case of closed Riemann surfaces of genus > 1, it is well known that all elements of Riemann's period matrices are holomorphic functions on a corresponding Teichmüller space endowed with appropriate complex structure (Ahlfors [2], Bers [4]). Our study began with its generalization to the case of general open Riemann surfaces. The method was applicable to potential theoretic quantities, i.e., Green's functions, Robin's constants, period reproducers and some others. We show not only the first variational formulas of them but also the second variational formulas, where the "first" above suggests the first derivative. As the quasiconformal deformation we consider Riemann surfaces with conformal structures decided by certain Beltrami differentials depending holomorphically on a complex parameter. For the systematic study we need the notion of behavior spaces which come from Shiba's idea [27]. We treat meromorphic differentials whose boundary behaviors are restricted by behavior spaces. The meromorphic differentials with the boundary behavior continuously deform as the surface deforms. We estimate their distortions by the Beltrami differentials and obtain our variational formulas of Rauch type [22], [23]. We give some examples of quasiconformal deformations under our situation. One of them is a deformation by which branch points vary holomorphically on a covering surface and then all the elements of Riemann's period matrices with respect to normalized holomorphic differentials with normal behavior are holomorphic. Further this is applied to a Torelli type theorem on abelian Teichmüller disks due to Kra [7]. As the other application, the variational formula of Bergman kernel shows Suita's formula [28] which gives a relation between Robin's constant and Bergman kernel. The generalized form gives Lewittes' characterization [12] of the points at which Gaussian curvature with respect to Bergman metric vanishes on a general Riemann surface.

1. Quasiconformal deformations depending holomorphically on a complex parameter

Riemann's investigation, in his theory of abelian integrals, started from the physical conception of a steady flow of a fluid over a surface and described stationary states represented by meromorphic differentials on a compact surface in a well-rounded fullness. How does the flow change if the surface is altered? Hadamard gave the variational formula of the Green's function on planar domains. Schiffer systematically investigated variations of finite bordered surfaces and obtained the various variational formulas of the Green's function, the bilinear differentials and the differentials of the first kind [26]. Our purpose lies in the same line and our object is a general open Riemann surface which changes momentarily. We first draw the movement of the surface. We imagine the surface varing naturally, same as each point varies holomorphically in a domain of complex plane from the initial state. It is the movement which Mañé-Sad-Sullivan's λ-lemma maintains to be a quasiconformal deformation. The topological type of the surface is preserved but the conformal type is not preserved. We describe it as a change of conformal structure on a 2-dimensional differentiable manifold or by quasiconformal mappings from a starting surface to terminal surfaces.

Let S be an open Riemann surface and M(S) be the set of Beltrami differentials:

$$\left\{\mu = \tilde{\mu}\,\frac{d\bar{z}}{dz};\ \tilde{\mu} \text{ is measurable and } \|\,\mu\,\|_{\infty} = \operatorname{esssup}|\,\tilde{\mu}\,| < 1\right\}$$

on S. From $\mu \in$ M(S) we get another Riemann surface S_μ with the Riemannian metric

$ds = \lambda(z) \mid dz + \tilde{\mu}(z)d\bar{z} \mid$. We consider Beltrami differentials $\mu(z, t)$ with a complex parameter t varying in a domain about zero. We assume that

$$\begin{aligned} &\text{i. } \mu(z, t) \text{ is measurable, } \mu(z, 0) = 0 \text{ and } \operatorname*{esssup}_{z} \mid \mu(z, t) \mid < 1 \\ &\text{ii. } \operatorname*{esssup}_{z} \mid \mu(z, t + h) - \mu(z, t) \mid \leq M_t \mid h \mid \text{ for sufficiently small h,} \\ &\text{iii. } \mu(z, t) \text{ is holomorphic with respect to t.} \end{aligned} \tag{A}$$

Let $f_{\mu(z, t)}$ $(= f_t)$ be the quasiconformal mapping from S to S_μ $(= S_t)$ whose Beltrami coefficient is $\tilde{\mu} = (f_t)_{\bar{z}} / (f_t)_z$. We say f_t satisfies the condition (A). Such a family $\{S_t\}$ is treated as a quasiconformal deformation of S. We express f_t sometimes as $\zeta = f_t(z)$ in terms of respective generic local parameters z and ζ of S and S_t, as $\mu = \zeta_{\bar{z}}/\zeta_z$. We consider the quasiconformal deformation to be a dynamic flow of a surface. There are some specific kind of quasiconformal deformations satisfying the condition (A).

Example 1. Deformation due to branch points

Let S be a covering surface of a Riemann surface R and p be a branch point. There is a neighborhood V of p whose local coordinate w satisfies $w = z^n$, where z is a local coordinate in R about the projection point of p. We may assume that V is projected to $\{q \in R; \mid z(q) \mid < 1\}$ and contains no branch point except for p. Let S_t be the covering surface which has a branch point of order n on q_t $(z(q_t) = t)$ instead of p and has the same branch points as S except for p. Consider the quasiconformal mapping $\underline{f}_t$ which preserves the unit circle;

$$\underline{f}_t(z) = \frac{z + t}{1 - t\bar{z}}, \mid t \mid < 1, \mid z \mid < 1.$$

Let f_t be the lift of $\underline{f}_t(z)$ to V and be identity mapping on S − V. We can regard f_t as a quasiconformal mapping from S to S_t. The $\{S_t, f_t\}$ becomes a quasiconformal deformation which satisfies the condition (A). By using this we can treat branch points varied holomorphically on a covering surface. Remark that we can treat a countable number of branch points if we take disjoint parametric disks about the branch points.

Example 2. Boundary deformation

Let S be a Riemann surface with a compact border C and take an annular neighborhood V of C which is conformal $\{z\ ;\ a < |z| \le 1\}$, where C corresponds to the unit circle. Consider a function F (z, t) such that F(z, t) is analytic with respect to z and t on a neighborhood of $\{z: |z| = 1\} \times \{0\}$, is injective on $\{z; |z| = 1\}$ for a fixed t and $F(z, 0) = z$. Set

$$f(z, t) = \frac{a(1 - |z|)z}{(1 - a)\ |z|} + \frac{|z| - a}{1 - a}F\left(\frac{z}{|z|}, t\right) \text{ on V.}$$

For a sufficiently small t, it gives a quasiconformal mapping from V to an annular region V(t). We regard $(S - V) \cup V(t)$ as a Riemann surface S_t and

$$f_t = \begin{cases} f(z, t) & \text{on } V \\ \text{identity mapping on } S - V \end{cases}$$

as a quasiconformal mapping from S to S_t which satisfies the condition (A). This represents a movement which each boundary point varies holomorphically.

Example 3. Trivial deformation

We consider a quasiconformal self mapping on S which represents a small displacement of a parametric disk. Take a point p on R and a parametric disk V = $\{z; |z| < 2\}$ about p. Consider a function

$$f_t = \begin{cases} z + t & |z| < 1/2 \\ 2t(1 - |z|) + z & 1/2 < |z| < 1 \\ \text{identity} & \text{the other points} \end{cases}$$

where $|t| < 1/2$, the local parameter also denotes the corresponding point on S. The f_t satisfies the condition (A) and is used to variation of a certain meromorphic differential whose poles vary holomorphically on S.

Example 4. Pinching deformation

At last we refer to the one in Taniguchi's treatment [29], [30], [31] of a Riemann surface with nodes, which represents a pinching deformation along a curve. Let C be a simple closed analytic curve on S and V be an annular region containing C such that it is conformally equivalent to $\{z;\ a < |z| < a^{-1}\}$ by a conformal mapping h and $h(C) = \{z;\ |z| = 1\}$. Take two local variables of V;

$$z_1 = ah(p) \text{ and } z_2 = a/h(p).$$

We construct a Riemann surface S_t by defining;

$S_t = (S - \{p;\ ar < |h(p)| < a^{-1}r\}) \cup \{\zeta_1;\ |a+t|^2 < |\zeta_1| < 1\} \cup \{\zeta_2;\ |a+t|^2 < |\zeta_2| < 1\}$, where $|a+t| < r < 1$), ζ_1 (resp. ζ_2) is identified with p if $\zeta_1 = (a+t)\,h(p)$, $|\zeta_1| > r$ (resp. $\zeta_2 = (a+t)/h(p)$, $|\zeta_2| > r$), and ζ_2 is identified with ζ_1 if $\zeta_1\,\zeta_2 = (a+t)^2$. A quasiconformal mapping f_t from S to S_t is defined by

$$\zeta_i = \begin{cases} \dfrac{a+t}{a}\, z_i & a \le |z_i| \le d \\[2ex] z_i \left(\dfrac{r}{|z_i|}\right)^{A(t)} & d < |z_i| < r \end{cases}$$

and identity on the other points,

where $|a+t|d/a < r$ and $A(t) = \log((a+t)/a)/(\log r/d)$. The f_t satisfies the condition (A) and $\{S_t, f_t\}$ is called a pinching deformation of S along C.

We can use these examples being combinated if the supports of their Beltrami coefficients are in disjoint regions.

2. Behavior spaces

For the extension of classical theory of abelian integrals to open Riemann surfaces, Nevanlinna restricted concerned functions and differentials to those which have bounded Dirichlet integrals near the ideal boundary, Ahlfors [3] limited their boundary behaviors by distinguished differentials and Kusunoki [8] by canonical differentials. Sario's principal functions [25] also give prescribed modes of behavior near the ideal boundary. The notion of behavior spaces come from these studies on open Riemann surfaces. Inheriting them, Yosida [34] give the extended form and Shiba [27] introduced the notion of behavior spaces. We use the part of a simplified condition by Matsui [19] as our behavior space. If concerned differentials are limited to those which are like to elements of a behavior space near the ideal boundary, they well behave as those of compact surface and have some extremal properties. If we use some specific kind of behavior spaces, we can systematically give the variational formulas of typical functions and differentials such as the Green's function, Neumann's function, slit mapping and various reproducing differentials. A behavior space is a certain subspace of square integrable differentials.

Let Λ be the real Hilbert space of square integrable complex differentials whose inner product is given by

$$< \sigma, \omega > = \text{Real part of} \iint \sigma \wedge * \overline{\omega} = \mathrm{Re}\,(\sigma, \omega),$$

where $*\omega$ denotes the harmonic conjugate differential of ω and $\overline{\omega}$ denotes the complex conjugate of ω. The following subspaces of Λ will be used

$\Lambda_h = \{\omega \in \Lambda\colon\ \omega$ is a complex harmonic differential$\}$,

$\Lambda_{e0} = \{\omega \in \Lambda : \omega$ is a closed differential which is orthogonal to $\Lambda_h\}$,

$\Gamma_h = \{\omega \in \Lambda_h\colon \omega$ is a real differential$\}$,

$\Gamma_{hse} = \{\omega \in \Gamma_h\colon \int_c \omega = 0$ for every dividing closed analytic curve c$\}$,

$\Gamma_{he} = \{\omega \in \Gamma_h : \int_c \omega = 0$ for every closed analytic curve c$\}$,

$$\Gamma_{h0} = \{\omega \in \Gamma_h : <\omega, \sigma> = 0 \text{ for every } \sigma \in \Gamma_{he}\},$$

$$\Gamma_{hm} = \{\omega \in \Gamma_h : <\omega, \sigma> = 0 \text{ for every } \sigma \in \Gamma_{hse}\} \quad (\text{cf. [3]}).$$

We are concerned with a subspace which is represented by $\wedge_x = \Gamma_x + i*\Gamma_x^{\perp}$, where Γ_x is a subspace of Γ_h and $*\Gamma_x^{\perp} = \{\omega \in \Gamma_h : <\sigma, \omega> = 0 \text{ for any } \sigma \in *\Gamma_x\}$. We call such a space $\wedge_x$ a behavior space. If $\wedge_x$ is a behavior space, $i\wedge_x$ is also a behavior space. As useful examples we have behavior spaces $\wedge_0 = i\,\Gamma_h$, $\wedge_{hm} = \Gamma_{hm} + i\,\Gamma_{hse}$, $\wedge_{h0} = \Gamma_{h0} + i\,\Gamma_{he}$. When $\Gamma_x = *\,\Gamma_x^{\perp}$, we can regard the $\wedge_x = \Gamma_x + i*\Gamma_x^{\perp}$ as a Hilbert space over the complex number field. It becomes an important behavior space. If S is compact, then $\wedge_{h0}$ satisfies this condition. On an arbitrary Riemann surface there exists such a behavior space [13]. For a behavior space $\wedge_x$, we call a meromorphic differential ϕ to have $\wedge_x$-behavior if ϕ equals to an element of $\wedge_x + \wedge_{e0}$ outside of a compact set. We can construct the first, second and third kind abelian differentials with $\wedge_x$-behavior by the method of orthogonal projection. That is to say, there exist the following meromorphic differentials with $\wedge_x$-behavior [15]:

$\psi_{c,x}$; $\psi_{c,x} = \tau + i * \tau$, where $* \tau \in * \Gamma_x^{\perp}$ is a period reproducing differential of $*\Gamma_x^{\perp}$ for a closed analytic simple curve c, i.e. $<\sigma, * \tau> = \int_c \sigma$ for $\sigma \in * \Gamma_x^{\perp}$,

$\psi_{n,p,x}$; this coinsides with an element in $\wedge_x + \wedge_{e0}$ on S – V and has singularity

$\quad dz/\{z - z(p)\}^{n+1}$ at only p, $(n \geq 1)$,

$\psi_{p,q,x}$; this coinsides with an element in $\wedge_x + \wedge_{e0}$ on S – V and has singularities

$\quad dz/(z - z(p))$ at p and $- dz/(z - z(q))$ at q,

where V is a parametric disk, z is the local parameter and points p, q are contained in V. We call these the fundamental differentials with $\wedge_x$-behavior. Arbitrary meromorphic differential with $\wedge_x$-behavior can be represented by a finite number of linear sum of fundamental differentials with Λ_x-behavior. We can give a correspondence between differentials on two Riemann surfaces through the pull back by a quasiconformal mapping. For a differential $\omega = a d\zeta + b d\bar{\zeta}$ on S_t, denote the pull back by f_t

$$\omega o f_t = \{(a o f_t)\, \zeta_z + (b o f_t)\, \bar{\zeta}_z\}\, dz + \{(a o f_t)\, \zeta_{\bar{z}} + (b o f_t)\, \bar{\zeta}_{\bar{z}}\}\, d\bar{z},$$

where the derivatives are taken in the sense of distribution. The pull back gives an isomorphism between $\Lambda(S_t)$ and $\Lambda(S)$, because it preserves the finiteness of Dirichlet norm. Further it preserves the periods of closed differentials [18], [20] and gives an isomorphism between $\Lambda_{e0}(S_t)$ and $\Lambda_{e0}(S)$. Direct calculation shows the following relation between the inner products on S_t and S [14];

$$(*(\omega_1 o f_t), (*\omega_2) o f_t)_S = (\omega_1, \omega_2)_{S_t}. \tag{1}$$

For a behavior space $\Lambda_x(S) = \Gamma_x(S) + i*\Gamma_x{}^{\perp}(S)$, set

$$\Gamma_x(S_t, f_t)^{\perp} = \{\sigma \in \Gamma_h(S_t);\ <\sigma, \omega o f_t{}^{-1}> = 0 \text{ for } \omega \in \Gamma_x(S)\}$$

and we have a behavior space $\Lambda_x(S_t, f_t) = \Gamma_x(S_t, f_t) + i*\Gamma_x{}^{\perp}(S_t, f_t)$ on S_t. From (1) we obtain that $\Lambda_0(S_t, f_t) = \Lambda_0(S_t)$, $\Lambda_{hm}(S_t, f_t) = \Lambda_{hm}(S_t)$, $\Lambda_{he}(S_t, f_t) = \Lambda_{he}(S_t)$ and $\Lambda_x(S_t, f_t) = i\, \Lambda_x(S_t, f_t)$ if $\Lambda_x(S) = i\, \Lambda_x(S)$. Further we remark, by use of (1), that $\psi^t o f_t - \psi^0 \in \Lambda_x + \Lambda_{e0}$, where ψ^0 is a fundamental differential with $\Lambda_x(S)$-behavior on S, ψ^t is the corresponding fundamental differential with $\Lambda_x(S_t, f_t)$ -behavior which has the same singularity with ψ^0, i.e. $\psi^t o f_t - \psi^0$ is regular at the singular points of ψ^0. A meromorphic differential ϕ^t on S_t is uniquely determined if $\phi^t o\, f_t - \phi^0 \in \Lambda_x(S) + \Lambda_{e0}(S)$. Let write $\phi^t o f_t - \phi^0 = d\Phi^t$ on a parametric disk which contains p and q. We obtain the following representations for the period and the expansion coefficient with respect to the local variable.

$$\int_c \phi^t \circ f_t - \phi^0 = - <(i\, \phi^t) \circ f_t - i\, \phi^0, \bar{\psi}_{c, x'}> - i<\phi^t \circ f_t - \phi^0, \bar{\psi}_{c, x}>, \tag{2}$$

$$\Phi^t(p) - \Phi^t(q) = -\frac{1}{2\pi}\{ <\phi^t \circ f_t - \phi^0, \overline{\psi}_{p,q,x}> - i <(i\,\phi^t)\circ f_t - i\,\phi^0, \overline{\psi}_{p,q,x'}>\}, \quad (3)$$

$$\frac{d^n}{dz^n}\Phi^t(p) = -\frac{1}{2\pi}\{ < \phi^t \circ f_t - \phi^0, \overline{\psi}_{n,p,x}> - i <(i\,\phi^t)\circ f_t - i\,\phi^0, \overline{\psi}_{n,p,x'}>\}, \quad (4)$$

where x' denotes $i\Lambda_x$-behavior and we use the cauchy's principal integral for singular integral. Remark that f_t need not be conformal at p, q in (3) but it is assumed that f_t is conformal at p in (4). These are fundamental quantities on S_t. Further we refer to the other important functions which closely relate to specific behavior spaces.

2.1 Neumann's functions and the behavior space Λ_{he}

Let $N^t_{p,\,q}$ be the Newmann function on S_t with poles at $f_t(p)$ and $f_t(q)$. Then the differential $dN^t_{p,\,q} + i{*}dN^t_{p,\,q}$ has Λ_{he}-behavior and it coinsides with the third kind fundamental differential $\psi^t_{p,q,he}$.

2.2. Slit mappings and the behavior space Λ_{hm}

Let M^t_x be a meromorphic function on S_t whose differential dM^t_x has Λ_x-behavior. If f_t is conformal at the poles of M^0_x and $M^t_x \circ f_t$ has the same poles as M^0_x, then $dM^t_x \circ f_t - dM^0_x \in \Lambda_x + i\Lambda_{e0}$. In particular M^t_{hm} (resp. M^t_{hse}) is a vertical (resp. horizontal) slit-mapping if the genus of S_t is finite.

2.3 Extremal length and the behavior space Λ_0

Let $\lambda\,(C^t)$ be the extremal length of the family of curves homologous to $f_t(C)$. It is well known that $\lambda\,(C^t) = \|\,\psi^t_{c,0}\|^2/2$.

2.4. Normal solutions of Dirichlet problems and the behavior space Λ_0

Let S^*_t be the Royden's compactification [5] of S_t, $h(\zeta)$ be a real valued continuous function on S^*_0 whose restriction to S_0 is a Diriclet function. f_t has an extension to a homomorphism $\tilde{f}_t$ from S^*_0 to S^*_t [25] and set $h^t = h \circ \tilde{f}_t^{\,-1}$. Let H_{h^t} be the solution of the Dirichlet problem on S^*_t with the boundary value h^t [5] and $\psi^t_h = dH_{h^t} + i{*}dH_{h^t}$.

Since $H_{h^t} \circ f_t - H_{h^0}$ is a Dirichlet potential, $\psi_h^t \circ f_t - \psi_h^0 \in \Lambda_0 + \Lambda_{e0}$. Then

$$H_{h^t}(f_t(p)) - H_{h^0}(p) = \frac{1}{2\pi} < \psi_h^t \circ f_t - \psi_h^0, \overline{\psi}_p^0 >,$$

$$\int_{f_t(\gamma)} *dH_{h^t} = \frac{1}{2} < \psi_h^t, \psi_{\gamma,0}^t >. \tag{6}$$

When h is a characteristic function of a boundary component, then H_{h^t} is called a harmonic measure.

2.5 Green's functions and the behavior space Λ_0

Let G_p^t be the Green's function on S_t with pole at $f_t(p)$ and $\psi_p^t = dG_p^t + i{*}dG_p^t = 2\frac{\partial}{\partial\zeta}G_p^t d\zeta$. If f_t is conformal at p, then $\psi_p^t \circ f_t - \psi_p^0 \in \Lambda_0\,(S) + \Lambda_{e0}(S)$. Similarly as (3) we have

$$G_p^t(f_t(q)) - G_p^t(q) = \frac{1}{2\pi} < \psi_p^t \circ f_t - \psi_p^0, \overline{\psi}_q^0 > \quad \text{for } q \neq p. \tag{7}$$

2.6 Robin's constants and the behavior space Λ_0

A Robin's constant $\gamma^t(p)$ at $f_t(p)$ is defined by

$$\gamma^t(p) = \frac{1}{2\pi i} \int_{|z| = \epsilon} G_p^t\,(f_t(z))\frac{dz}{z},$$

where z is a local variable about p (identifying the corresponding point), for which the singularity of ψ_p^t is written as $- dz/z$. As for this Robin's constant

$$\gamma^t(p) - \gamma^0(p) = \frac{1}{2\pi} < \psi_p^t \circ f_t - \psi_p^0, \overline{\psi}_p^0 >. \tag{8}$$

3. Variational formulas

We started in our study to show Rauch's variational formulas [22], [23] by Ahlfors' method [1]. Let state only the results [15]. Take a meromorphic differential ϕ^t on S_t such that $\phi^t \circ f_t - \phi^0 \in \Lambda_x + \Lambda_{e0}$. Then

$$\frac{\phi^{t+u} \circ f_{t+u} \circ f_t^{-1} - \phi^t}{u} \quad \left(\text{resp. } \frac{\phi^{t+iv} \circ f_{t+iv} \circ f_t^{-1} - \phi^t}{v}\right)$$

converges to an element ϕ_u^t (resp. ϕ_v^t) in $\Lambda_x + \Lambda_{eo}$ in the norm sense as real u (resp. v) tends to zero. They satisfy that

$$\begin{aligned} \phi_u^t - i^*\phi_u^t &= -i\,(\phi_v^t - i\,{}^*\phi_v^t) \\ &= \phi^t \frac{2\tilde{\mu}_t\,(z(\zeta),\,t)}{1 - |\,\mu\,(z(\zeta),\,t)\,|^2} \frac{\zeta_z}{\overline{\zeta}_{\bar{z}}} \frac{d\overline{\zeta}}{d\zeta},\ (\tilde{\mu}_t = \frac{\partial\tilde{\mu}}{\partial t}). \end{aligned} \tag{9}$$

Set $\phi_{\bar{t}}^t = (\phi_u^t + i\,\phi_v^t)/2$ and $\phi_t^t = (\phi_u^t - i\,\phi_v^t)/2$. The $\phi_{\bar{t}}^t$ becomes a holomorphic differential. This fact plays a role of simplification of variational formulas. We have the following variational formulas.

Let meromorphic differentials ϕ^t, ψ^t satisfy that the poles of ϕ^0 and ψ^0 do not meet the support of μ and $\phi^t \circ f_t - \phi^0$, $\psi^t \circ f_t - \psi^0 \in \Lambda_x(S) + \Lambda_{e0}(S)$. Then

$$\frac{\partial}{\partial t}(\phi^t \circ f_t - \phi^0, \overline{\psi}^0) = \frac{1}{2}(\phi_t^t, \overline{\psi}^t)$$

$$= \frac{i}{2}\iint \underline{\phi}^t\, \underline{\psi}^t\, \tilde{\mu}_t\, \zeta_z{}^2 dz d\bar{z} \quad \left(\phi^t = \underline{\phi}^t d\zeta,\ \psi^t = \underline{\psi}^t d\zeta\right). \tag{10}$$

$$\frac{\partial^2}{\partial\bar{t}\partial t}\langle\phi^t \circ f_t - \phi^0, \overline{\psi}^0\rangle = \frac{1}{2}\left\{(\phi_{\bar{t}}^t, \overline{\psi}_t^t) + (\psi_{\bar{t}}^t, \overline{\phi}_t^t)\right\} \tag{11}$$

$$= \frac{i}{2}\iint(\underline{\phi}_{\bar{t}}^t\, \underline{\psi}^t + \underline{\psi}_{\bar{t}}^t\, \underline{\phi}^t)\, \tilde{\mu}_t \zeta_z{}^2 dz d\bar{z}\ (\phi_{\bar{t}}^t = \underline{\phi}_{\bar{t}}^t d\zeta,\ \psi_{\bar{t}}^t = \underline{\psi}_{\bar{t}}^t d\zeta).$$

Further, when $\Lambda_x = \Lambda_0$,

$$\frac{\partial}{\partial t}\langle\phi^t \circ f_t - \phi^0, \overline{\psi}^0\rangle = -\frac{1}{2}\left(\psi^t, \phi_{\bar{t}}^t\right),\ (\text{if } \psi^t \text{ is holomorphic}), \tag{12}$$

$$\frac{\partial^2}{\partial\bar{t}\,\partial t}\langle\phi^t \circ f_t - \phi^0, \overline{\psi}^0\rangle = -\langle\phi_{\bar{t}}^t, \psi_{\bar{t}}^t\rangle. \tag{13}$$

In particular, when ϕ^t, ψ^t are holomorphic and $\Lambda_x = \Lambda_0$,

$$\frac{\partial}{\partial t} <\phi^t, \psi^t> = \frac{1}{2}\left\{(\phi^t, \psi^t_{\bar{t}})\right) + (\psi^t, \phi^t_{\bar{t}})\Big\}, \tag{14}$$

$$\frac{\partial^2}{\partial \bar{t} \partial t} <\phi^t, \psi^t> = 2<\phi^t_{\bar{t}}, \psi^t_{\bar{t}}>, \tag{15}$$

$$\frac{\partial}{\partial t} \log<\phi^t, \phi^t> = \frac{1}{<\phi^t, \phi^t>} <\phi^t, \phi^t_{\bar{t}}>, \tag{16}$$

$$\frac{\partial^2}{\partial \bar{t}\, \partial t} \log<\phi^t, \phi^t> = \frac{1}{<\phi^t, \phi^t>^2}\Big\{2<\phi^t_{\bar{t}}, \phi^t_{\bar{t}}> <\phi^t, \phi^t> - | <\phi^t, \phi^{\,t}_{\bar{t}}> |^2\Big\}$$

$$\geq \frac{<\phi^t_{\bar{t}}, \phi^t_{\bar{t}}>}{<\phi^t, \phi^t>} \geq 0. \tag{17}$$

These formulas are, of course, applied to the above mentioned deformation, fundamental quantities and some important functions. Although we do not restate, we have many variational formulas due to their various combinations. Here, we give some remarks.

Remark 1.

When $\Gamma_x = *\Gamma_x{}^{\perp}$, $\Lambda_x = i\ \Lambda_x$ and $\psi_{c,x'} = \psi_{c,x'}$, $\psi_{p,q,x'} = \psi_{p,q,x'}$, $\psi_{n,p,x'} = \psi_{n,p,x'}$. For $\phi^{\,t} \circ f_t - \phi^0 \in \Lambda_x + \Lambda_{e0}$, by the formula (10),

$$\frac{\partial}{\partial \bar{t}} \int_c \phi^t \circ f_t - \phi^0 = -\frac{1}{2}\Big\{(\overline{\psi}^{\,t}_{c,x}\ ,\ i\ \phi^t_{\bar{t}}) + i(\overline{\psi}^{\,t}_{c,x}\ ,\ \phi^{\,t}_{\bar{t}})\Big\} = 0.$$

This shows the period of ϕ^t along $f_t(C)$ is holomorphic with respect to t. Similary $\Phi^t(p) - \Phi^t(q)$ and $\frac{d^n}{dz^n}\, \Phi^t\,(p)$ are holomorphic with respect to t. These are satisfied not only on a compact Riemann surface but also on an arbitrary open Riemann surface.

Remark 2.

Under the pinching deformation, A. Yamada [32] gave precise variational formulas on a compact Riemann surface and M. Taniguchi [29], [30], [31] also gave various variational formulas on an arbitrary Riemann surface with nodes. We also show a variational formula from our view point. Let $\phi^t \circ f_t - \phi^0, \psi^t \circ f_t - \psi^0 \in \Lambda_x + \Lambda_{e0}$. By Example 4

$$\tilde{\mu} = -\frac{z_i}{\overline{z}_i}\frac{A(t)}{2 - A(t)} \quad \text{on } d < | z_i | < r.$$

From (10) we obtain

$$\frac{\partial}{\partial t} <\phi^t \circ f_t - \phi^0, \overline{\psi}^0>$$

$$= \frac{i}{2}\iint_{S_t} \underline{\phi}^t\, \underline{\psi}^t \frac{\tilde{\mu}_t}{1-|\mu|^2} \frac{\zeta_z}{\overline{\zeta}_z}\, d\zeta\, d\overline{\zeta}$$

$$= \frac{-\, i\, A'(t)}{| 2 - A(t) |^2 - | A(t) |^2} \sum_{i=1}^{2} \iint_{d|a + t|/a<|\zeta_i|<r} \underline{\phi}^t \underline{\psi}^t \frac{\zeta_i}{\overline{\zeta}_i} d\zeta_i\, d\overline{\zeta}_i$$

Let $\phi^t = \Sigma\, a_n^i(t)\, \zeta_i^{\,n} d\zeta_i$ and $\psi^t = \Sigma\, b_n^i(t)\, \zeta_i^{\,n}\, d\zeta_i$. From the invariance of form we obtain

$$a^1_{-n-2} = -a_n^2(t)\, (a + t)^{2(n + 1)} \tag{18}$$

and the same equation for b_n^i. Thus

$$\frac{\partial}{\partial t}<\phi^t \circ f_t - \phi^0, \overline{\psi}^0>$$
$$= \pi\Sigma_n\left\{a_n^1(t)\, b_n^2(t) + a_n^2(t) b_n^1(t)\right\} (a + t)^{2n + 1}. \tag{19}$$

If we use a parameter $\tau = (a + t)^2$,

$$\frac{\partial}{\partial \tau}<\phi^t \circ f_t - \phi^0, \overline{\psi}^0>$$
$$= \frac{\pi}{2}\Sigma_n\left\{a_n^1(t)\, b_n^2(t) + a_n^2(t) b_n^1(t)\right\}\, \tau^n \tag{19'}$$

When ϕ^t and ψ^t are the fundamental differentials with Λ_x-behavior, we know, by the construction due to the methods of orthogonal projection, that $a_n^i(t)$ and $b_n^i(t)$ are uniformly bounded and by (18)

$$\lim_{t\to -a} a^i_{-n-2}(t) = 0 \qquad \text{for } n \geq 0.$$

If $\int_{|\zeta_i|=1}\phi^t = c_i(t)$ and $\lim_{t\to -a} c_i(t) = c_i$, then $\lim_{t\to -a} a^i_{-1}(t) = c_i$, $c_2 = -c_1$. The $a^i_n(t)$ and $b^i_n(t)$ are, by (4), (11), differentiable. Further if $\Lambda_x = i\,\Lambda_x$, they are holomorphic and $t = -a$ is a removable singularity. The formula (19') is given by differentiated form and the quantity $<\phi^t \circ f_t - \phi^0, \overline{\psi}^0>$ has a logarithmic singularity if $c_1 \neq 0$.

Remark 3.

As for the case of Λ_0, we have a remarkable fact [33], [15]. That is, by (15),

$\|\psi^t_{c,0}\|^2$, $\lambda\,(C^t)$, $\|dH_{h^t}\|^2$ are subharmonic,

and by (13)

$\gamma^t(p)$, $-\mathrm{Re}\left\{\Phi^t_{p,q,0}(p) - \Phi^t_{p,q,0}(q)\right\}$, $-\mathrm{Re}\frac{d^n}{dz^n}\Phi^t_{n,p,0}\,(p)$,

$-\,\mathrm{Im}\left\{\Phi^t_{p,q,0}(p) - \Phi^t_{p,q,0}(q)\right\}$, $-\,\mathrm{Im}\frac{d^n}{dz^n}\,\Phi^t_{n,p,0}(p)$ are superharmonic.

5. Applications

5.1. Let K^t_n be a Bergman kernel such that

$$(\omega,\, K^t_n) = \frac{d^n}{dz^n}w(t) \text{ for every holomorphic differential } \omega = dw \in \Lambda.$$

The K^t_n has a representation;

$$K^t_n = \frac{n!}{4\pi}\left(\psi_{n,t,0} - \psi_{n,t,h}\right), \tag{20}$$

where $\psi_{n,\,t,\,h}$ is the second kind fundamental differential with $i\Lambda_0$-behavior. We apply example 3 to our variational formula and obtain

$$\frac{\partial}{\partial t}K_n(t) = \left(K^t_n,\ \left(K^t_n\right)_{\overline{t}}\right), \text{ where } k_n(t)=(\,K^t_n\,,\,K^t_n). \tag{21}$$

Further we have

$$\frac{\partial^2}{\partial \overline{t}\,\partial t}K_n(t) = \left(\left(K^t_n\right)_{\overline{t}},\ \left(K^t_n\right)_{\overline{t}}\right) = K_{n+1}(t), \tag{22}$$

$$\frac{\partial^2}{\partial \overline{t}\,\partial t}\log K_n(t) = \frac{1}{K_n(t)^2}\left\{K_{n+1}(t)\,K_n(t) - \left|\left(K^t_{n+1},\ K^t_n\right)\right|^2\right\} \geq 0. \tag{23}$$

where $(K_n^t)_{\bar{t}} = \frac{n!}{4\pi}\left((\psi_{n,t,0})_{\bar{t}} - (\psi_{n,t,h})_{\bar{t}}\right) = K_{n+1}^t$. The $-2\pi K_0(t)$ is regarded as

Robin's constant at t, hence (22) is the one of Suita's formula [28], [17].

5.2. Since the Gaussian curvature of the Bergman metric is represented by

$\frac{-2}{K_1(t)} \frac{\partial^2}{\partial \bar{t} \partial t} \log K_1(t)$, by using (23) and Riemann-Roch theorem [24] we obtain Lewittes

theorem [12], [17]. Let S be a non-planar Riemann surface. If the Gaussian curvature of

the Bergman metric vanishes, then S is an (ultra) hyper-elliptic Riemann surface of

parabolic type. Conversely, if S is an (ultra) hyper-elliptic Riemann surface of parabolic

type, the branch points coinside with the zeros of the Gaussian curvature of the

Bergmann metric.

5.3. The formula of the same form shows a kind of Yamaguchi's theorem [33], [16].

Let S be a compact bordered Riemann surface of genus n and m boundary components

and $\{C_i\}_{1 \leq i \leq 2n+m-1}$ be canonical homology basis of S. If $\log \|\psi_{c_i,0}^t\|$ is harmonic

for every C_i, by (17) $\left(\psi_{c_i,0}^t\right)_{\bar{t}} = 0$ and by (14) $\|a\,\psi_{c_i,0}^t + b\,\psi_{c_j,0}^t\|$ is constant. Hence

$\mathrm{Im} \int_{f_t(c_j)} \psi_{c_i,0}^t = \frac{1}{2}<\psi_{c_i,0}^t, \psi_{c_j,0}^t>$ is constant. From Torelli's theorem on the double of

S_t it follows that $\{S_t\}$ are the same Riemann surfaces.

5.4. Let $\{A_j, B_j\}$ be a canonical homology basis modulo boundary on S and

$f_t(A_j) = A_j^t$, $f_t(B_j) = B_j^t$. Suppose $\Gamma_{hm} \subset \Gamma_x \subset \Gamma_{hse}$, $\wedge_x = i * \wedge_x$ and $\int_{A_j} \omega = 0$ for

every A_j and every $\omega \in \wedge_x$. There are normalized holomorphic differentials $\{\psi_i^t\}$ so that

ψ_i^t has a $\wedge_x$-behavior and $\int_{A_j^t} \psi_i^t = \delta_{ij}$. Further suppose $\mu(z,t)$ is given as $t\bar{\phi}_k/\phi_k$,

where ϕ_k is a holomorphic differential with $\wedge_x$-behavior. Set

$$\pi_{ij}\left(t\bar{\phi}_k/\phi_k\right) = \int_{B_j^t} \psi_i^t.$$

We have a Torelli type Theorem on abelian Teichmüller disks due to Kra [7], [16]. If $\pi_{ij}\left(t_1\overline{\phi}_1/\phi_1\right) = \pi_{ij}\left(t_2\overline{\phi}_2/\phi_2\right)$ for every i, j, then $t_1\overline{\phi}_1/\phi_1 = t_2\overline{\phi}_2/\phi_2$.

References

[1] Ahlfors, L., The complex analytic structure of the space of the closed Riemann surfaces, Analytic function, Princeton, (1960), pp. 45 - 66.

[2] Ahlfors, L., Lectures on quasiconformal mappings, Van Nostrand, (1966), 146pp.

[3] Ahlfors, L., & L. Sario, Riemann surfaces, Princeton Univ., Press (1960), 382pp.

[4] Bers, L., Holomorphic differentials as functions of moduli, Bull. Amer. Math. Soc., 67 (1961), pp. 206 - 210.

[5] Constantinescu, C., & A. Cornea, Ideale Ränder Riemannscher Flächen, Springer-Verlag, (1963), 244pp.

[6] Guerrero, I., Quasiconformal variation of the Green's function, Michigan Math. J., 26 (1979), pp. 351 - 360.

[7] Kra, I., Abelian Teichmüller disks, J. Analyse Math., 40 (1981), pp. 129 - 143.

[8] Kusunoki, Y., Theory of Abelian integrals and its applications to conformal mappings, Mem. Col. Sci. Univ. Kyoto Ser. A. Math., 32 (1959), pp. 235 - 258.

[9] Kusunoki, Y., Differentiability of period matrices in Teichmüller spaces, Hokkaido Math. J., 10 (1981), Special Issue pp. 435 - 446.

[10] Kusunoki, Y., & F. Maitani, Variations of Abelian differentials under quasiconformal deformations, Math. Z., 181 (1982) pp. 435 - 450.

[11] Kusunoki, Y., & M. Taniguchi, A continuity property of holomorphic differentials under quasiconformal deformations, Ann. Acad. Sci. Fenn. Ser. A. I. Math., 5 (1980), pp. 207 - 226.

[12] Lewittes, J., Differentials and metrics on Riemann surfaces, Trans, Amer. Math. Soc., 139 (1969), pp. 311 - 318.

[13] Maitani, F., The method of linear operators for square integrable differentials on open Riemann surfaces and its applications, J. Math. Kyoto Univ., 20 (1980), pp. 661 - 689.

[14] Maitani, F., Remarks on the isomorphism of certain spaces of harmonic differentials induced from quasiconformal homomorphism, Proc. Japan Acad., 56 (1980), pp. 311 - 314.

[15] Maitani, F., Variations of meromorphic differentials under quasiconformal deformations, J. Math. Kyoto Univ., 24 (1984), pp. 49 - 66.

[16] Maitani, F., Abelian Teichmüller disks restricted by behavior, Memo. F. Ind. Arts, Kyoto Inst. Tech. Sci. Tech., 35 (1986), pp. 1 - 11.

[17] Maitani, F., Covering properties of extremal vertical slit mappings, Kodai Math. J., 11(1988), pp361-371.

[18] Marden, A., The weakly reproducing differentials on open Riemann surfaces, Ann. Acad. Sci. Fenn. Ser. A. I., 359 (1965).

[19] Matsui, K., Convergence theorem of Abelian differentials with applications to conformal mapping I, J. Math. Kyoto Univ., 15 (1975), pp. 73 - 100; II. Ibid.,17 (1977), pp. 345 - 374.

[20] Minda, C. D., Square integrable differentials on open Riemann surfaces and quasiconformal mappings, Trans. Amer. Math. Soc., 195 (1974), pp. 365 - 381.

[21] Patt, C., Variations of Teichmüller and Torelli surfaces, J. d'Analyse Math., 11 (1963), pp. 221 - 247.

[22] Rauch, H. E., On the transcendental moduli of algebraic Riemann surfaces, Proc. National Acad. Sci., 41 (1955), pp. 42 - 49.

[23] Rauch, H. E., Weierstrass points, branch points, and the moduli of Riemann surfaces, Comm. pure and App. Math., 12 (1959), pp. 543 - 560.

[24] Rodin, B., & L. Sario, Principal functions, Van Nostrand, Princeton, (1968), 347pp.

[25] Sario, L., & M. Nakai, Classification Theory of Riemann Surfaces, Springer-Verlag, (1970), 446pp.

[26] Schiffer, M., & D. C. Spencer, Functionals of finite Riemann surfaces, Princeton Univ., Press (1954), 451pp.

[27] Shiba, M., On the Riemann-Roch theorem on open Riemann surfaces, J. Math. Kyoto Univ., 11 (1971), pp. 495 - 525.

[28] Suita, N., On capacities and kernels of Riemann surfaces, Arch. Rat. Mech. Anal., 46 (1972), pp. 212 - 217.

[29] Taniguchi, M., Variational formulas on arbitrary Riemann surfaces under pinching deformation, J. Math. Kyoto Univ., 27 (1987), pp. 507 - 530.

[30] Taniguchi, M., Supplements to my previous paper; a refinement and application, J. Math Kyoto Univ., 28 (1988), pp. 81 - 86.

[31] Taniguchi, M., Abelian differentials with normal behavior and complex pinching deformation, to appear.

[32] Yamada, A., Precise variational formulas for abelian differentials, Kodai Math., J., 3 (1980), pp. 114 - 143.

[33] Yamaguchi, H., Calcul des variations analytiques, Japanese J. Math., 7 (1981), pp. 319- 377.

[34] Yoshida, M., The method of orthogonal decomposition for differentials on open Riemann surfaces, J. Sci. Hiroshima Univ., Ser. A - I., 8 (1968), pp. 181 - 210.

Kyoto Institute of Technology

MEROMORPHIC FUNCTIONS WITH A PERFECT SET AS THE SET OF SINGULARITIES

Kikuji Matsumoto and Toshiko Kurokawa

§ 1. Introduction

For a totally disconnected compact set E in the extended z – plane $\hat{\mathbb{C}}$, we denote by M_E the totality of meromorphic functions each of which is defined in the domain complementary to E and has E as the set of transcendental singularities. A meromorphic function f(z) of M_E is said to be exceptionally ramified at a singularity $\zeta \in E$, if there exist values w_k, $1 \leq k \leq q$, and positive integers ν_k, $1 \leq k \leq q$, with

$$\sum_{k=1}^{q} \left(1 - \frac{1}{\nu_k} \right) > 2,$$

such that, in some neighborhood of ζ, the multiplicity of any w_k – point of f(z) is not less than ν_k. In [2], Kurokawa gave Cantor sets E for which each function of M_E cannot be exceptionally ramified at any singularity $\zeta \in E$. The theorem given there is the following

<u>Theorem A</u>. Let E be a Cantor set with successive ratios $\{\xi_n\}$ satisfying the condition

$$\xi_{n+1} = o(\xi_n^5),$$

then the domain complementary to E admits no exceptionally ramified meromorphic functions with E as the set of transcendental singularities.

The purpose of this paper is to show that we have the same conclusion for Cantor sets with much weaker condition. We shall prove the following

<u>Theorem</u>. Let E be a Cantor set with successive ratios $\{\xi_n\}$ satisfying the condition

$$\xi_{n+1} = o(\xi_n^2),$$

then the domain complementary to E admits no exceptionally ramified meromorphic functions with E as the set of transcendental singularities.

§ 2. Preliminaries

2.1. Let f be exceptionally ramified meromorphic in a domain G in the z-plane having three totally ramified values $\{w_j\}_{j=1,2,3}$ with $\{\nu_j\}_{j=1,2,3}$, positive integers explained in § 1, and let Δ be a triply connected subdomain of G with $\overline{\Delta} \subset G$ which is bounded by analytic curves $\{\Gamma_j\}_{j=1,2,3}$. We assume that they satisfy the following three conditions (1), (2) and (3):

(1) There exist mutually disjoint simply connected sectionally analytic domains $\{D_j\}_{j=1,..,\alpha}$ $(1 \leq \alpha \leq 3)$ with

$$|D_j| < \frac{1}{2} \lim_{k \neq m} \chi(w_k, w_m)$$

and the images $\{f(\Gamma_i)\}_{i=1,2,3}$ are covered with $\{D_j\}_{j=1,..,\alpha}$, each D_j containing $f(\Gamma_i)$ for at least one i, where $\chi(w_k, w_m)$ denotes the chordal distance between w_k and w_m and $|D_j|$ denotes the diameter of D_j.

(2) The number n of roots of the equation f(z) = w in Δ is constant and ≥ 1 for $w \in \hat{\mathbb{C}} - \bigcup_{j=1}^{\alpha} \overline{D}_j$.

(3) f has no ramified values on each boundary ∂D_j.

We remove from Δ all relatively noncompact components of $\{f^{-1}(\overline{D}_j)\}_{j=1,...,\alpha}$ with respect to Δ. Then there remains an open set, each component of which must be triply connected because of Lemma 2 in [2]. Hence the open set is a triply connected subdomain Δ' of Δ, whose boundary curves Γ_j' are homotopic to Γ_j (j = 1, 2, 3). The following 1), 2), 3) and 4) hold (see Lemma 3 in [2]).

1) The Riemannian image of Δ' under f belongs to one of the 25 classes listed in Table 1.

2) f has no ramified values other than $\{w_i\}_{i=1,2,3}$ in Δ'.

3) Each component of $\Delta - \Delta'$ is doubly connected and its image is contained in one of $\{D_j\}$ $j = 1, 2, \ldots, \alpha$.

4) Each D_j contains one of the totally ramified values $\{w_i\}$ $i = 1, 2, 3$.

Table 1

class	ν_1	ν_2	ν_3	m_1 $\ell_{1,j}$	m_2 $\ell_{2,j}$	m_3 $\ell_{3,j}$	n	σ_1	σ_2	σ_3
1	2	4	5	3 $\ell_{1,j} = 2$	1 $\ell_{2,1} = 4$	1 $\ell_{3,1} = 5$	6	0	2 $\{1, 1\}$	1 $\{1\}$
2	2	4	5	4 $\ell_{1,j} = 2$	2 $\ell_{2,j} = 4$	1 $\ell_{3,1} = 5$	8	0	0	3 $\{1,1,1\}$
3	2	3	7	4 $\ell_{1,j} = 2$	2 $\ell_{2,j} = 3$	1 $\ell_{3,1} = 7$	8	0	2 $\{1, 1\}$	1 $\{1\}$
4	2	3	7	4 $\ell_{1,j} = 2$	3 $\ell_{2,j} = 3$	1 $\ell_{3,1} = 7$	9	1 $\{1\}$	0	2 $\{1.1\}$
5	2	3	7	5 $\ell_{1,j} = 2$	3 $\ell_{2,j} = 3$	1 $\ell_{3,1} = 7$	10	0	1 $\{1\}$	2 $\{1,2\}$
6	2	3	7	5 $\ell_{1,j} = 2$	3 $\ell_{2,j} = 3$	1 $\ell_{3,1} = 8$	10	0	1 $\{1\}$	2 $\{1,1\}$
7	2	3	7	5 $\ell_{1,j} = 2$	3 $\{\ell_{2,1}, \ell_{2,2}, \ell_{2,3}\} = \{3, 3, 4\}$	1 $\ell_{3,1} = 7$	10	0	0	3 $\{1,1,1\}$

8	2	3	7	6 $\ell_{1,j} = 2$	4 $\ell_{2,j} = 3$	1 $\ell_{3,1} = 7$	12	0	0	3 {1,1,3}
9	2	3	7	6 $\ell_{1,j} = 2$	4 $\ell_{2,j} = 3$	1 $\ell_{3,1} = 7$	12	0	0	3 {1,2,2}
10	2	3	7	6 $\ell_{1,j} = 2$	4 $\ell_{2,j} = 3$	1 $\ell_{3,1} = 8$	12	0	0	3 {1,1,2}
11	2	3	7	6 $\ell_{1,j} = 2$	4 $\ell_{2,j} = 3$	1 $\ell_{3,1} = 9$	12	0	0	3 {1,1,1}
12	2	3	7	8 $\ell_{1,j} = 2$	5 $\ell_{2,j} = 3$	2 $\ell_{3,j} = 7$	16	0	1 {1}	2 {1,1}
13	2	3	7	9 $\ell_{1,j} = 2$	6 $\ell_{2,j} = 3$	2 $\ell_{3,j} = 7$	18	0	0	3 {1,1,2}
14	2	3	7	9 $\ell_{1,j} = 2$	6 $\ell_{2,j} = 3$	2 $\{\ell_{3,1}, \ell_{3,2}\}$ $= \{7, 8\}$	18	0	0	3 {1,1,1}
15	2	3	7	12 $\ell_{1,j} = 2$	8 $\ell_{2,j} = 3$	3 $\ell_{3,j} = 7$	24	0	0	3 {1,1,1}
16	3	3	4	1 $\ell_{1,1} = 3$	1 $\ell_{2,1} = 3$	0	3	0	0	3 {1,1,1}
17	2	4	5	2 $\ell_{1,j} = 2$	1 $\ell_{2,1} = 4$	0	4	0	0	3 {1,1,2}
18	2	3	7	2 $\ell_{1,j} = 2$	1 $\ell_{2,1} = 3$	0	4	0	1 {1}	2 {1, 3}

19	2	3	7	2 $\ell_{1,\,j} = 2$	1 $\ell_{2,\,1} = 3$	0	4	0	1 {1}	2 {2, 2}
20	2	3	7	1 $\ell_{1,\,1} = 2$	1 $\ell_{2,\,1} = 3$	0	3	1 {1}	0	2 {1, 2}
21	2	3	7	3 $\ell_{1,\,j} = 2$	2 $\ell_{2,\,j} = 3$	0	6	0	0	3 {1,1,4}
22	2	3	7	3 $\ell_{1,\,j} = 2$	2 $\ell_{2,\,j} = 3$	0	6	0	0	3 {1,2,3}
23	2	3	7	3 $\ell_{1,\,j} = 2$	2 $\ell_{2,\,j} = 3$	0	6	0	0	3 {2,2,2}
24	2 2 2 2	3 7 4 5	7 3 5 4	1 $\ell_{1,\,1} = 2$	0	0	2	0	1 {2}	2 {1, 1}
25	2 2 3	3 4 3	7 5 4	0	0	0	1	1 {1}	1 {1}	1 {1}

<u>Notations</u>.

m_i: the number of w_i – points of f(z) in Δ' (i = 1, 2, 3).

$\{\ell_{i,\,j}\}_{\,j\,=\,1,\,\ldots,\,m_i}$: the multiplicities of w_i – points.

σ_i: the number of Γ'_j in $\{\Gamma'_j\}_{\,j\,=\,1,\,2,\,3}$ with $f(\Gamma'_j) = \partial D_k$, $w_i \in D_k$, where $\sigma_i = 0$ means that none of $\{D_j\}_{j=1,\ldots,\alpha}$ contains w_i.

$\overset{\sigma_3}{\underset{\{1,2\}}{2}}$ means that two of $\{\Gamma_j'\}_{j=1,2,3}$ are mapped onto ∂D_k, $w_3 \in D_k$, and one of them has an image curve winding once around w_3, while the other has an image curve winding twice.

Now we shall show that we can delete the classes (8), (9), (19) and (22) from Table 1.

First we deal with the classes (8) and (9). From Table 1, we see that $\sigma_1 = \sigma_2 = 0$ and $\sigma_3 = 3$, and so $\alpha = 1$. Hence three boundary curves $\{\Gamma_i'\}_{i=1,2,3}$ are the inverse image of ∂D_1 under f, $w_3 \in D_1$ and $w_1, w_2 \in \hat{\mathbb{C}} - \overline{D}_1$. We join w_1 to w_3 with a simple analytic curve Λ which does not pass through any ramified values of f (see Figure 1).

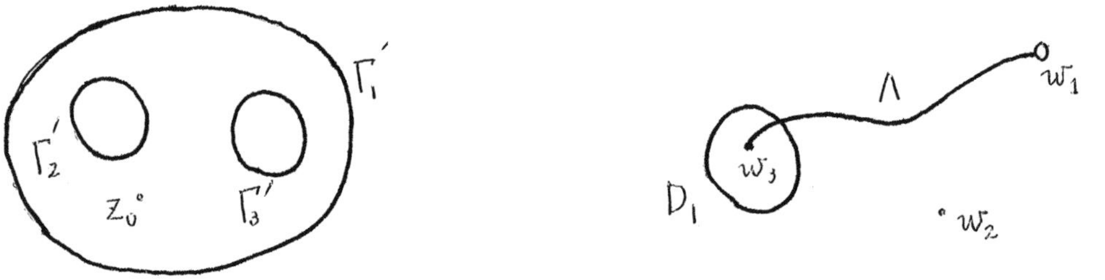

z – plane   w – plane

Figure 1

Since $m_3 = 1$, there is only one w_3 - point in Δ', which we denote by z_0. Since $m_1 = 6$ and $\ell_{1,j} = 2$, j = 1, ..., 6, there are six w_1–points with multiplicity 2. Therefore, for each w_1–point, the inverse image $f^{-1}(\Lambda)$ contains two curves starting there to the opposite two directions and ending at z_0 or $\{\Gamma_i'\}_{i=1,2,3}$. We connect these two curves at the w_1 – point and have six simple curves $\{\beta_i\}_{i=1,\ldots,6}$ in Δ' joining z_0 or $\{\Gamma_i'\}_{i=1,2,3}$ to z_0 or $\{\Gamma_i'\}_{i=1,2,3}$ and passing through a w_1– point.

We consider the Riemannian image S of Δ' under f and remove from S all curves over Λ. Then S is divided into four parts, each of which has a branch point with multiplicity 3 over w_2, because S has no branch points over $\hat{\mathbb{C}} - \Lambda$ other than $m_2 = 4$ branch points over w_2. This means that the domain Δ' is divided into four subdomains $\{\Delta_i'\}_{i=1,2,3,4}$ by $\{\beta_i\}_{i=1,...,6}$, which contain four w_2-points one by one.

We shall deal with the matter topologically for a while and consider that points $\{P_i\}_{i=1,2,3}$ represent $\{\Gamma_i'\}_{i=1,2,3}$. We set $z_o = P_o$. Then, such a subdomain Δ_i' must be represented topologically by one of the following four domains (A), (B), (C) and (D) (see Figure 2).

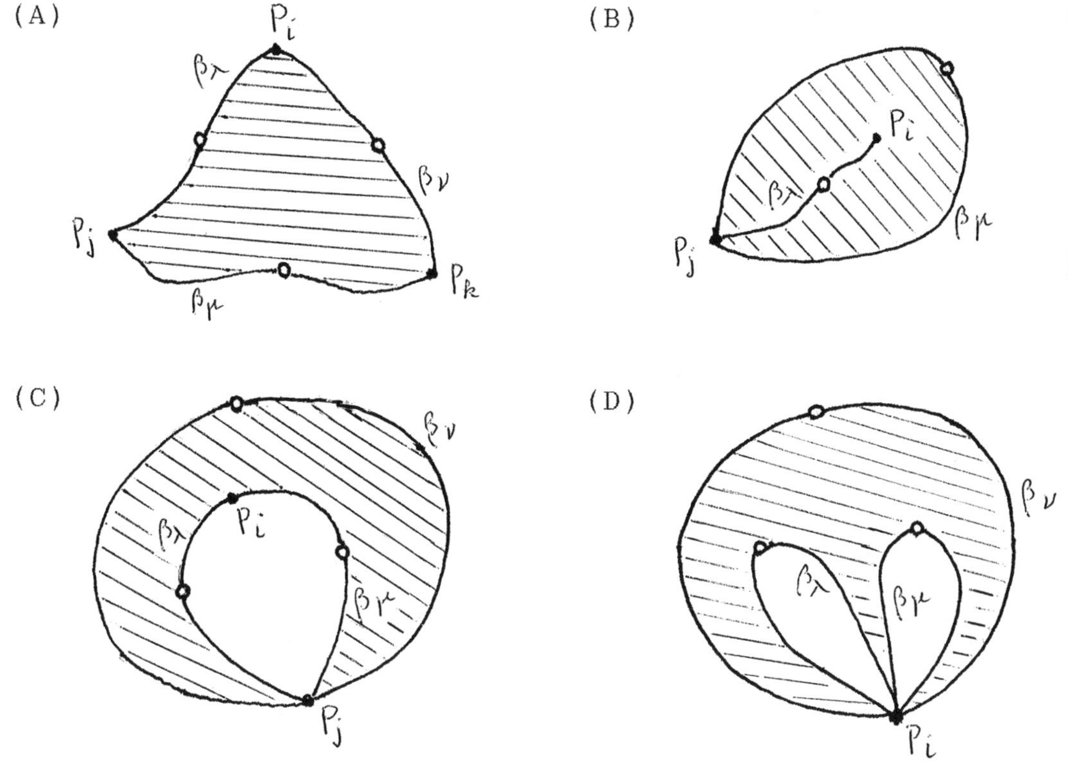

Figure 2

Here a mark o denotes a w_1-point of f.

Our problem is how to fill up $\hat{\mathbb{C}} - \{P_i\}_{i = 0, 1, 2, 3}$ univalently with four domains being topological to one of (A), (B), (C) and (D), where we are permitted to use domains of the same type repeatedly. Since the multiplicity of the w_1-point z_0 is not less than 7, we see that using (C) or (D) is necessary and there are only two possibilities (i) and (ii) (see Figure 3). Now (i) and (ii) correspond to the classes (10) and (11) of Table 1 respectively and hence the classes (8) and (9) do not occur.

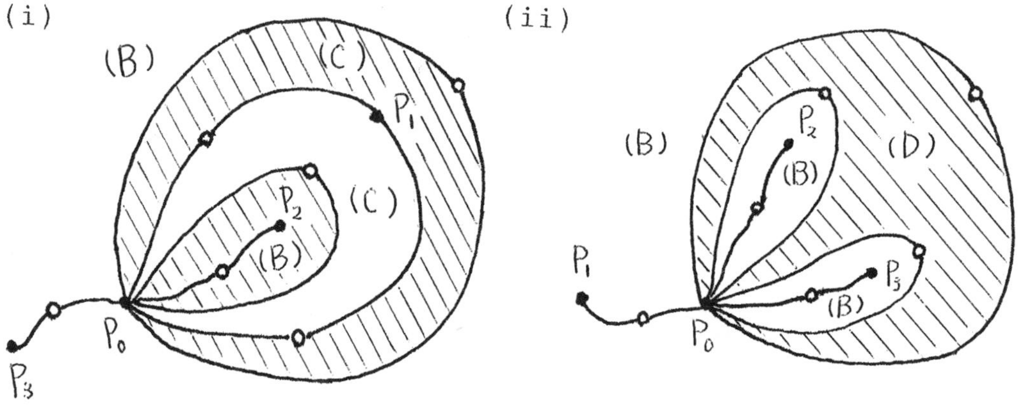

Figure 3

Next we deal with the class (19). From Table 1, we see that $\sigma_1 = 0$, $\sigma_2 = 1$ and $\sigma_3 = 2$, so that $\alpha = 2$. We may assume that $w_3 \in D_1$, $w_2 \in D_2$, $w_1 \in \hat{\mathbb{C}} - \overline{D}_1 \cup \overline{D}_2$, $f^{-1}(\partial D_2) = \Gamma'_2$ and $f^{-1}(\partial D_1) = \Gamma'_1 \cup \Gamma'_3$. We join w_1 to w_3 with a simple analytic curve Λ in $\hat{\mathbb{C}} - \overline{D}_2$ which does not pass through any ramified values of f (see Figure 4).

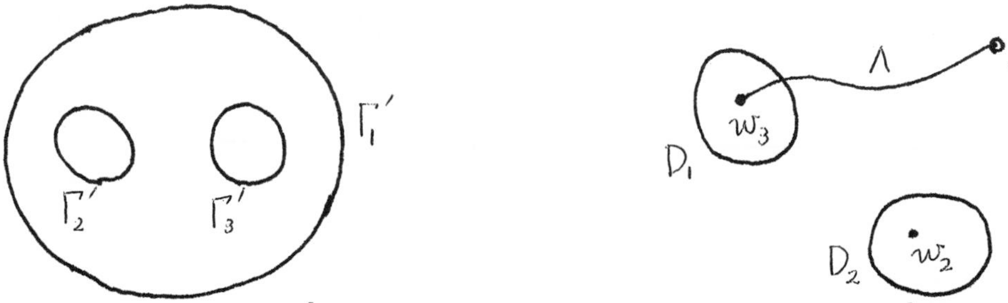

z – plane w – plane

Figure 4

Since $m_1 = 2$, and $\ell_{1,1} = \ell_{1,2} = 2$, we see similarly as above that the inverse image $f^{-1}(\Lambda)$ consists of two curves $\{\beta_i\}_{i = 1, 2}$ in Δ' joining $\{\Gamma'_i\}_{i = 1, 3}$ to $\{\Gamma'_i\}_{i = 1, 3}$ and passing through a w_1-point. We remove from S all curves over Λ. Then S is divided into two parts such that one of them has a branch point with multiplicity 3 over w_2 and the other covers $\hat{C} - \Lambda \cup \bar{D}_1 \cup \bar{D}_2$ univalently, because $n = 4$, only Γ'_2 in $\{\Gamma'_i\}_{i = 1, 2, 3}$ is mapped onto ∂D_2, it has the image curve winding once around w_2 and S has no branch points over $\hat{C} - \Lambda$ other than $m_2 = 1$ branch point over w_2. This means that the domain Δ' is divided into two subdomains by $\{\beta_j\}_{j = 1, 2}$ such that one of them is topologically equivalent to one of (A), (B), (C) and (D) and the other is a doubly connected domain bounded by one of $\{\beta_j\}_{j = 1, 2}$ plus a part of Γ'_1 and Γ'_2. Thus only (iii) is possible (see Figure 5). Since (iii) corresponds to the class (18), there are no covering surfaces S belonging to the class (19).

(iii)

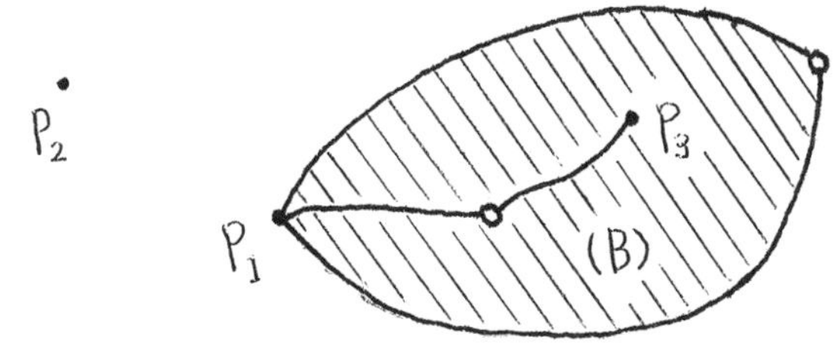

Figure 5

For the class (22), $n = 6$, $m_1 = 3$, $\ell_{1,1} = \ell_{1,2} = \ell_{1,3} = 2$ and S has $m_2 = 2$ branch points with multiplicity 3 over w_2. Thus we have three simple curves $\{\beta_i\}_{i = 1, 2, 3}$ dividing Δ' into two subdomains, each of which is topologically equivalent to one of (A), (B), (C) and (D). Hence we have only two possibilities (iv) and (v) (see Figure 6).

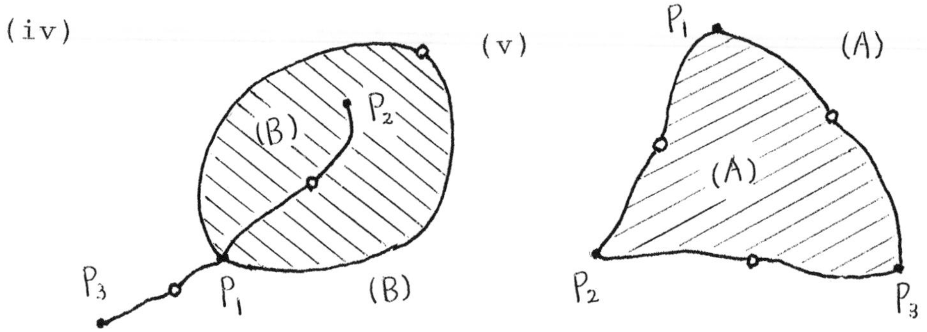

Figure 6

Since (iv) and (v) correspond to the classes (21) and (23), respectively, there are no covering surfaces S belonging to the class (22).

2.2. We form a Cantor set in the usual manner. Let $\{\xi_n\}$ be a sequence of positive numbers satisfying $0 < \xi_n < \frac{2}{3}$, n = 1, 2, 3, ... We remove first an open interval of length $(1 - \xi_1)$ from the interval $I_{0,\,1}$: $[-\frac{1}{2}, \frac{1}{2}]$, so that on both sides there remains a closed interval of length $\frac{\xi_1}{2} \equiv \eta_1$. The remaining intervals are denoted by $I_{1,\,1}$ and $I_{1,\,2}$. Inductively we remove an open interval of length $(1 - \xi_n) \prod_{p=1}^{n-1} \eta_p$, with $\eta_p = \frac{1}{2}\xi_p$ (p = 1, 2, ...), from each interval $I_{n-1,\,k}$ of length $\prod_{p=1}^{n-1} \eta_p$, $k = 1, 2, \ldots, 2^{n-1}$, so that on both sides there remains a closed interval of length $\prod_{p=1}^{n} \eta_p$. The remaining intervals are denoted by $I_{n,2k-1}$ and $I_{n,2k}$. By repeating this procedure endlessly, we obtain an infinite sequence of closed intervals $\{I_{n,k}\}$ $n = 1, 2, \ldots, k = 1, 2, \ldots, 2^n$. The set given by

$$E = \bigcap_{n=1}^{\infty} \bigcup_{k=1}^{2^n} I_{n,k}$$

is said to be the Cantor set in the interval $I_{0,1}$ with successive ratios $\{\xi_n\}$.

Set

$$R_{n,k} = \left\{ z\, ;\ \prod_{p=1}^{n} \eta_p < |\, z - z_{n,k}\,| < \frac{1}{3} \prod_{p=1}^{n-1} \eta_p \right\}$$

and

$$\Gamma_{n,k} = \left\{ z; \mid z - z_{n,k} \mid = \prod_{p=1}^{n-1} \eta_p \sqrt{\frac{\eta_n}{3}} \right\},$$

where $z_{n,k}$ is the midpoint of $I_{n,k}$. Denoting by $\mu_n = \mu(R_{n,k})$ the harmonic modulus of $R_{n,k}$, we have

$$\mu_n = \log \frac{1}{3\eta_n} = \log \frac{2}{3\xi_n}.$$

We assume that $\lim_{n\to\infty} \xi_n = 0$ and have

Lemma 1 (Lemma 4 in [2]). Let f be an exceptionally ramified meromorphic function in the domain $G = \hat{\mathbb{C}} - E$. Then, for sufficiently large n, we have

$$\mid f(\Gamma_{n,k}) \mid < M \exp(- \mu_n / 2) ,$$

where $\mid f(\Gamma_{n,k}) \mid$ denotes the diameter of $f(\Gamma_{n,k})$ with respect to the chordal distance and M is a positive constant depending only on E and f.

By our previous result ([3]), f in Lemma 1 has just three totally ramified values $\{w_i\}_{i = 1, 2, 3}$. Since $\mid f(\Gamma_{n,k}) \mid < \delta_n = M \exp(- \mu_n / 2) = M \sqrt{3\xi_n/2}$ by Lemma 1, we can take a spherical disc $D_{n,k}$ of radius δ_n containing $f(\Gamma_{n,k})$. We denote by $\Delta_{n,k}$ the triply connected domain bounded by $\Gamma_{n,k}$, $\Gamma_{n+1,2k-1}$ and $\Gamma_{n+1,2k}$. Taking n so large that $\delta_n < \frac{1}{12} \min_{i \neq j} \chi(w_i , w_j)$, we consider the union $D = \overline{D}_{n,k} \cup \overline{D}_{n+1,2k-1} \cup \overline{D}_{n+1,2k}$, which consists of at most three, say α, components.

If $\alpha = 1$, that is, D is connected, it is possible that D is doubly connected, and we take a disc $\tilde{D}_1$ of radius at most $\delta_n + 2\delta_{n+1}$ containing D. If $\alpha = 2$ or 3, we denote the components of D by $\{\tilde{D}_j\}_{j = 1, \ldots, \alpha}$, which are simply connected.

When $\alpha = 1$ and f takes in $\Delta_{n,k}$ no values outside $\tilde{D}_1$, $f(\overline{\Delta}_{n,k}) \subset \tilde{D}_1$, we say that $\Delta_{n,k}$ is degenerate (f). When $\alpha = 1$ and f takes in $\Delta_{n,k}$ values outside $\tilde{D}_1$ or when $\alpha =$ 2 or 3, we say that $\Delta_{n,k}$ is non-degenerate (f). Then f, $\Delta_{n,k}$ and $\{\tilde{D}_j\}_{j = 1, \ldots, \alpha}$ satisfy three conditions (1), (2) and (3) stated at the beginning of 2.1, so that by 4) stated there, each $\tilde{D}_j$ contains one w_j^* of the totally ramified values $\{w_i\}_{i = 1, 2, 3}$ and the union $\bigcup_{j=1}^{\alpha} \tilde{D}_j \supset D$ is contained in $\bigcup_{i=1}^{3} D(w_i, 2(\delta_n + 2\delta_{n+1}))$, where we denote by $D(w, \delta)$ the spherical disc of radius δ and with center at w. We assume $2\delta_{n+1} < \delta_n$ and set $\tilde{D}'_j = D(w_j^*, 4\delta_n)$, $j = 1, ..., \alpha$. Then f, $\Delta_{n,k}$ and $\{\tilde{D}'_j\}_{j = 1, ..., \alpha}$ again satisfy three conditions (1), (2) and (3), so that there exists a triply connected subdomain $\Delta'_{n,k}$ of $\Delta_{n,k}$ such that 1), 2), 3) and 4) stated there hold. The Riemannian image $S_{n,k}$ of $\Delta'_{n,k}$ under f belongs to one of the classes of Table 1 other than the classes (8), (9), (19) and (22). The boundary curves of $\Delta'_{n,k}$ are denoted by $\check{\gamma}_{n,k}$, $\hat{\gamma}_{n+1,2k-1}$ and $\hat{\gamma}_{n+1,2k}$ being homotopic to $\Gamma_{n,k}$, $\Gamma_{n+1,2k-1}$ and $\Gamma_{n+1,2k}$, respectively. Each γ of them has an image curve winding around some w^* of w_1, w_2 and w_3, and we denote its winding number by $s(\gamma)$. The value w^* corresponds to one $\tilde{w}$ of three totally ramified values for the class in Table 1 to which $S_{n,k}$ belongs, and we can read the ν-value, the minimum of the multiplicities of $\tilde{w}$-points, in Table 1, which we denote by $\nu(\gamma)$.

Suppose now that $S_{n,k}$ belongs to a class other than (23). Reading Table 1, we see that the image curves of at least two of $\check{\gamma}_{n,k}$, $\hat{\gamma}_{n+1,2k-1}$ and $\hat{\gamma}_{n+1,2k}$ have the winding number 1. Hence $s(\hat{\gamma}_{n+1,2k-1}) = 1$ or $s(\hat{\gamma}_{n+1,2k}) = 1$, say $s(\hat{\gamma}_{n+1,2k}) = 1$, where we assume $\nu(\hat{\gamma}_{n+1,2k-1}) \leq \nu(\hat{\gamma}_{n+1,2k})$ if $s(\hat{\gamma}_{n+1,2k-1}) = s(\hat{\gamma}_{n+1,2k}) = 1$. The adjacent $\Delta_{n+1,2k}$ is degenerate (f) or non-degenerate (f). Suppose that $\Delta_{n+1,2k}$ is non-degenerate (f). Then $\hat{\gamma}_{n+1,2k}$ and $\check{\gamma}_{n+1,2k}$ wind around the same totally ramified value w^* and bound a doubly connected domain where f takes the value w^*. Hence we have

$$s(\hat{\gamma}_{n+1,2k}) + s(\check{\gamma}_{n+1,2k}) \geq \max \{ \nu(\hat{\gamma}_{n+1,2k}), \ \nu(\check{\gamma}_{n+1,2k}) \},$$

that is,

$$s(\check{\gamma}_{n+1,2k}) \geq \max \{ \nu(\hat{\gamma}_{n+1,2k}), \ \nu(\check{\gamma}_{n+1,2k}) \} - 1,$$

because $s(\hat{\gamma}_{n+1,2k}) = 1$. From Table 1, we see that only the pairs $\{\Delta_{n,k}, \Delta_{n+1,2k}\}$ listed below satisfy this inequality.

Table 2

$\Delta_{n,k}$			$\Delta_{n+1,2k}$		
class	$\nu(\hat{\gamma}_{n+1,2k})$	$s(\hat{\gamma}_{n+1,2k})$	class	$\nu(\check{\gamma}_{n+1,2k})$	$s(\check{\gamma}_{n+1,2k})$
			(4)	2	1
(20)	2	1	(20)	2	1
			(25)	2	1
(3)	3	1			
(5)	3	1			
(18)	3	1	(24)	3	2
(24)	3	1			
(25)	3	1			

Remark. The pair of $\Delta_{n,k}$ ((20), 2, 1) and $\Delta_{n+1,2k}$ ((24), 3, 2) satisfies the inequality, but, under the assumption that f is exceptionally ramified, we can omit it, because $S_{n,k}$ and $S_{n+1,2k}$ have branch points of multiplicity 2 over distinct totally ramified values.

From Table 1, we see that, if $\Delta_{n+1,2k}$ of the right side of Table 2 is of class (4), (20) or (24), one of $\hat{\gamma}_{n+2,4k-1}$ and $\hat{\gamma}_{n+2,4k}$, say $\hat{\gamma}_{n+2,4k}$, satisfies $s(\hat{\gamma}_{n+2,4k}) = 1$ and

$\nu(\hat{\gamma}_{n+2,4k}) = 7$, and if it is of class (25), $s(\hat{\gamma}_{n+2,4k}) = 1$ and $\nu(\hat{\gamma}_{n+2,4k}) \geq 5$. Therefore $\Delta_{n+2,4k}$ must be degenerate (f). Thus we have

Lemma 2. If $\Delta_{n,k}$ is non-degenerate (f) and belongs to a class other than (23), then for at least one of $\hat{\gamma}_{n+1,2k-1}$ and $\hat{\gamma}_{n+1,2k}$, say $\hat{\gamma}_{n+1,2k}$, $s(\hat{\gamma}_{n+1,2k}) = 1$. If the adjacent $\Delta_{n+1,2k}$ is non-degenerate (f), then for at least one of $\hat{\gamma}_{n+2,4k-1}$ and $\hat{\gamma}_{n+2,4k}$, say $\hat{\gamma}_{n+2,4k}$, $s(\hat{\gamma}_{n+2,4k}) = 1$ and the adjacent $\Delta_{n+2,4k}$ is degenerate (f).

Concluding § 2, we state a theorem due to Teichmüller for the moduli of ring domains as a lemma, which we shall often use later.

Lemma 3. If a ring domain R in $\mathbb{C}$ separates two points 0 and $r_1 e^{i\theta_1}$ from two points $r_2 e^{i\theta_2}$ and ∞ ($r_1 > 0$, $r_2 > 0$), then

$$\text{har. mod. } R \leq \log\left(16 \frac{r_2}{r_1} + 8 \right)$$

(cf. Lehto and Virtanen [4], pp. 54 - 62).

§ 3. Proof of Theorem

3.1. Contrary suppose that a function f of M_E is exceptionally ramified at a singularity $\zeta_o \in E$. By our previous result in [3], f has just three totally ramified values $\{w_i\}_{i = 1, 2, 3}$ near ζ_o with $\{\nu_i\}_{i = 1, 2, 3}$, ν_i being the minimum of the multiplicities of w_i - points, satisfying

$$\sum_{i=1}^{3} \left(1 - \frac{1}{\nu_i} \right) > 2 ,$$

where we may assume without any loss of generality that $w_1 = \infty$, $w_2 = 1$ and $w_3 = 0$.

From our assumption $\xi_{n+1} = o\ (\xi_n^2)$, we can take n_o so large that $\delta_n = M \sqrt{3\xi_n/2} < \sqrt{2} / 24$ and $\delta_{n+1} < (1/2)\delta_n$ for $n \geq n_o$. Here we may assume that Γ_{n_o,k_o} surrounds ζ_o and f is exceptionally ramified in the part G_o of $G = \hat{\mathbb{C}} - E$ surrounded with Γ_{n_o,k_o}. Then if $\Delta_{n,k}$ in G_o is degenerate (f), $f(\overline{\Delta}_{n,k})$ is contained in a disc $\tilde{D}_{n,k}$ of radius at most $\delta_n + 2\,\delta_{n+1} < 2\,\delta_n$.

Now suppose that all $\Delta_{n,k}$ in G_o are degenerate (f). The image $f(\overline{\Delta}_{n_o,k_o})$ is contained in $\tilde{D}_{n_o,k_o}$. Since $\tilde{D}_{n_o,k_o} \cap \tilde{D}_{n_o+1,2k_o-1} \neq \phi$ and $\tilde{D}_{n_o,k_o} \cap \tilde{D}_{n_o+1,2k_o} \neq \phi$, $f(\overline{\Delta}_{n_o,k_o} \cup \overline{\Delta}_{n_o+1,2k_o-1} \cup \overline{\Delta}_{n_o+1,2k_o})$ is contained in a disc D_2 of radius at most $2\delta_{n_o} + 4\delta_{n_o+1} < 2\,\delta_{n_o}\,(1 + 2^0)$ and with the same center w_o as $\tilde{D}_{n_o,k_o}$. If $f(\overline{\Delta}_{n_o,k_o} \cup (\bigcup_{p=1}^{m} (\bigcup_k{}' \overline{\Delta}_{n_o+p,k})))$ is contained in a disc D_m of radius at most $2\delta_{n_o}(1 + \sum_{p=1}^{m} \frac{1}{2^{p-1}})$ and with center at w_o, then $f(\overline{\Delta}_{n_o,k_o} \cup (\bigcup_{p=1}^{m+1} (\bigcup_k{}' \overline{\Delta}_{n_o+p,k})))$ is contained in a disc D_{m+1} of radius at most $2\delta_{n_o}\,(1 + \sum_{p=1}^{m} \frac{1}{2^{p-1}}) + 4\delta_{n_o+m+1} < 2\delta_{n_o}\,(1 + \sum_{p=1}^{m+1} \frac{1}{2^{p-1}})$ and with center at w_o, because $D_m \cap \tilde{D}_{n_o+m+1,\,k} \neq \phi$ for each $\Delta_{n_o+m+1,k}$ in G_o, where $\bigcup_k{}' \overline{\Delta}_{n_o+p,k}$ means the union taken over all the $\Delta_{n_o+p,k}$'s in G_o. By induction, we conclude that $f(G_o)$ is contained in a disc of radius at most $2\delta_n\,(1 + \sum_{p=1}^{\infty} \frac{1}{2^{p-1}}) = 6\delta_{n_o} < \sqrt{2}\,/\,4$. This means that f is bounded in G_o. Since E is of linear measure zero, each point of $E \cap G_o$ must be a removable singularity for f (cf. Besicovitch [1]), which contradicts our assumption that f has E as the set of transcendental singularities. Thus we see that there are infinitely many $\Delta_{n,k}$ in G_o being non-degenerate (f).

We take such a domain $\Delta_{n,k}$. If $\Delta_{n,k}$ belongs to a class other than (23), we may assume from Lemma 2 that $s(\hat{\gamma}_{n+1,2k}) = 1$ and the adjacent $\Delta_{n+1,2k}$ is degenerate (f). We shall show that $f(\Gamma_{n+2,4k-1}) \cup f(\Gamma_{n+2,4k}) \subset D(w_i, 8\delta_{n+2})$ for some $w_i \in \{w_i\}$ $i = 1, 2, 3$.

As we mentioned after we had stated Lemma 1, for $\Delta_{m,\ell}$ being non-degenerate (f), the union $\overline{D}_{m,\ell} \cup \overline{D}_{m+1,2\ell-1} \cup \overline{D}_{m+1,2\ell}$ is contained in $\bigcup_{i=1}^{3} D(w_i, 2(\delta_m + 2\delta_{m+1})) \subset \bigcup_{i=1}^{3} D(w_i, 4\delta_m)$. Therefore, if $f(\Gamma_{m,\ell}) \not\subset \bigcup_{i=1}^{3} D(w_i, 8\delta_m)$, then $\Delta_{m,\ell}$ is degenerate (f) and $f(\overline{\Delta}_{m,\ell})$ is contained in a disc $\tilde{D}_{m,\ell}$ of radius at most $2\delta_m$. We have

$\tilde{D}_{m,\ell} \cap \bigcup_{i=1}^{3} D(w_i, 4\delta_m) = \phi$. Since $2\delta_{m+1} < \delta_m$, we see that $f(\Gamma_{m+1,2\ell-1}) \not\subset \bigcup_{i=1}^{3} D(w_i, 8\delta_{m+1})$ and $f(\Gamma_{m+1,2\ell}) \not\subset \bigcup_{i=1}^{3} D(w_i, 8\delta_{m+1})$ so that $\Delta_{m+1,2\ell-1}$ and $\Delta_{m+1,2\ell}$ both are degenerate (f).

Suppose now $f(\Gamma_{n+2,4k-1}) \not\subset \bigcup_{i=1}^{3} D(w_i, 8\delta_{n+2})$. Then, by induction, we see that all $\Delta_{m,\ell}$ in the part of G surrounded with $\Gamma_{n+2,4k-1}$ are degenerate (f). However this is impossible as we saw above. Thus $f(\Gamma_{n+2,4k-1}) \subset \bigcup_{i=1}^{3} D\left(w_i, 8\delta_{n+2}\right)$. Quite similarly we have $f(\Gamma_{n+2,\,4k}) \subset \bigcup_{i=1}^{3} D(w_i, 8\delta_{n+2})$. Since $\Delta_{n+1,2k}$ is degenerate (f), $f(\Gamma_{n+2,4k-1})$ and $f(\Gamma_{n+2,\,4k})$ must be contained in the same disc, say $D(w_i, 8\delta_{n+2})$, $w_i \in \{w_i\}$ $i = 1, 2, 3$.

We consider the part of the Riemannian image of $\Delta_{n,k} \cup \Gamma_{n+1,2k} \cup \Delta_{n+1,2k}$ under f over the annulus $R = \{w;\ 8\delta_{n+2} < \chi(w, w_i) < \sqrt{2}/2\}$. Since $s(\hat{\gamma}_{n+1,2k}) = 1$ and $\Delta_{n+1,2k}$ is degenerate (f), its component $\tilde{R}$ containing $f(\hat{\gamma}_{n+1,2k})$ covers R univalently, so that $\tilde{R}$ is also an annulus and its harmonic modulus is equal to that of R. The inverse image $f^{-1}(\tilde{R})$ is a ring domain separating $\Gamma_{n,k} \cup \Gamma_{n+1,2k-1}$ from $\Gamma_{n+2,4k-1} \cup \Gamma_{n+2,4k}$. By Lemma 3, we have

$$\log\left(16\,\frac{Y_n(1-\xi_{n+1})}{Y_{n+1}} + 8\right) \geq \text{har. mod. } R$$

$$= \log\left(\frac{1/2}{8\delta_{n+2} / \sqrt{1-(8\delta_{n+2})^2}}\right)$$

and hence

$$\log\left(32 / \xi_{n+1}\right) \geq \log\left(1 / 32\, M \sqrt{3\,\xi_{n+2} / 2}\right),$$

so that

$$\xi_{n+2} \geq \frac{1}{3 \cdot 2^{19} \cdot M^2}\, \xi_{n+1}^2,$$

where $Y_n = \prod_{p=1}^{n} \eta_p$. Thus there are only finitely many $\Delta_{n,k}$ in G_o being non-degenerate (f) which belong to classes other than the class (23), for otherwise, the inequality holds for infinitely many n contradicting our assumption $\xi_{n+1} = o(\ \xi_n^2\ )$. Now we may assume that all $\Delta_{n,k}$ in G_o being non–degenerate (f) are of class (23).

3.2. Let $\Delta_{n,k}$ be non–degenerate (f) and belong to the class (23). Then the image $f(\partial\Delta_{n,k})$ of the boundary of $\Delta_{n,k}$ is contained in one of $\{D(w_i, 4\delta_n)\}_{i\,=\,1,\,2,\,3}$, say $D(w_3, 4\delta_n)$, $w_3 = 0$. Both of adjacent $\Delta_{n+1,2k-1}$ and $\Delta_{n+1,2k}$ are degenerate (f). In fact, if $\Delta_{n+1,2k-1}$ is non–degenerate (f), $s(\hat{\gamma}_{n+1,2k-1}) = s(\check{\gamma}_{n+1,2k-1}) = 2$, because $\Delta_{n+1,2k-1}$ is also of class (23), and hence $s(\hat{\gamma}_{n+1,2k-1}) + s(\check{\gamma}_{n+1,2k-1}) < \max \{\nu(\hat{\gamma}_{n+1,2k-1}), \nu(\check{\gamma}_{n+1,2k-1})\} = 7$, which contradicts the inequality given in 2.2 just before we gave Table 2. Hence $\Delta_{n+1,2k-1}$ is degenerate (f). Similarly we see that $\Delta_{n+1,2k}$ is also degenerate (f). Now at least one of $\Delta_{n+2,4k-1}$ and $\Delta_{n+2,4k}$, say $\Delta_{n+2,4k}$, is degenerate (f), because, if both of them are non-degenerate (f), then they are of class (23) and f has a zero in the domain bounded by $\hat{\gamma}_{n+1,2k}$, $\check{\gamma}_{n+2,4k-1}$ and $\check{\gamma}_{n+2,4k}$, so that

$$s(\hat{\gamma}_{n+1,2k}) + s(\check{\gamma}_{n+2,4k-1}) + s(\check{\gamma}_{n+2,4k}) \geq 7,$$

while $s(\hat{\gamma}_{n+1,2k}) = s(\check{\gamma}_{n+2,4k-1}) = s(\check{\gamma}_{n+2,4k}) = 2$, a contradiction. The other $\Delta_{n+2,4k-1}$ is degenerate (f) or is non-degenerate (f), is of class (23) and $f(\partial\Delta_{n+2,4k-1}) \subset D(0, 4\delta_{n+2})$. Here we give a lemma.

Lemma 4. Let $\Delta_{m,\ell}$ and adjacent $\Delta_{m+1,2\ell-1}$ satisfy the following two conditions 1) and 2).

1) $\Delta_{m,\ell}$ is degenerate (f).

2) $\Delta_{m+1,2\ell-1}$ is non-degenerate (f) and of class (23), and

$$f(\partial\Delta_{m+1,2\ell-1}) \subset D(0, 4\delta_{m+1}).$$

Then there exists a circle $\tilde{\Gamma}_{m+1,2\ell-1}$ in $R_{m+1,2\ell-1} = \{z;\ Y_{m+1} < |z - z_{m+1,2\ell-1}| < Y_m / 3\}$ concentric with the circle $\Gamma_{m+1,2\ell-1}$ such that $|f(z)| = o(\xi_m^2)$ for $z \in \tilde{\Gamma}_{m+1,2\ell-1}$.

Proof. For small $d > 0$, we denote by S_d the covering surface of class (23) over $\hat{C} - \bar{D}(0, d)$. When $d = 4\delta_{m+1}$, S_d is the Riemannian image $S_{m+1,2\ell-1}$ of the subdomain $\Delta'_{m+1,2\ell-1}$ of $\Delta_{m+1,2\ell-1}$ under f. As the limit surface as $d \to 0$, we have a six – sheeted covering surface of $\hat{C} - \{0\}$ having three pinholes over 0. We stop up these holes and obtain a six – sheeted covering surface Φ of $\hat{C}$, which is planar and has three branch points of multiplicity 2 over $w_1 = \infty$, two branch points of multiplicity 3 over $w_2 = 1$ and three branch points of multiplicity 2 over $w_3 = 0$. Let $w = \varphi(\omega)$ be a conformal mapping of the extended ω– plane onto Φ with $\varphi(0) = \varphi(1) = \varphi(\infty) = 0$. Consider S_d, $d = 4\delta_{m+1}$, as a subdomain of Φ. Its inverse image $\varphi^{-1}(S_d)$ is a triply connected domain $\hat{C} - \bigcup_{i=1}^{3} B_i$, where $\partial B_1 = \varphi^{-1} \circ f(\hat{\gamma}_{m+2,4\ell-3})$, $\partial B_2 = \varphi^{-1} \circ f(\hat{\gamma}_{m+2,4\ell-2})$ and $\partial B_3 = \varphi^{-1} \circ f(\check{\gamma}_{m+1,2\ell-1})$. We may assume that $\omega_1 = 0 \in B_i$, $\omega_2 = 1 \in B_2$ and $\omega_3 = \infty \in B_3$. If m is sufficiently large, that is, d is sufficiently small, for each i, ∂B_i is nearly a circle of chordal radius $\alpha_i\sqrt{d}$ and with center at ω_i, where $\{\alpha_i\}$ $i = 1, 2, 3$ are positive constants not depending on d and hence on m. The annulus $R = \{\omega;\ 2\alpha_3\sqrt{d} < \chi(\omega, \infty) < 1 / \sqrt{5}\}$ separates $B_1 \cup B_2$ from B_3, so that its image $f^{-1} \circ \varphi(R)$ is a ring domain in $\Delta'_{m+1,2\ell-1} \subset \Delta_{m+1,2\ell-1}$ separating $\Gamma_{m+2,4\ell-3} \cup \Gamma_{m+2,4\ell-2}$ from $\Gamma_{m+1,2\ell-1}$ and has the same harmonic modulus as R. We set

$$r_1 = \max\{\ |z - z_{m+2,4\ell-3}|\ ;\ z \in \hat{\gamma}_{m+2,4\ell-3}\},$$

$$r_2 = \max\{\ |z - z_{m+2,4\ell-2}|\ ;\ z \in \hat{\gamma}_{m+2\ 4\ell-2}\}$$

and

$$r = \min \{ | z - z_{m+1,2\ell-1} | ; z \in \check{\gamma}_{m+1,2\ell-1} \}.$$

By Lemma 3, we have

$$\log (16 (\frac{r}{Y_{m+1}/ 2}) + 8) \geq \text{har. mod. } R = \log \frac{\sqrt{1 - (2\alpha_3\sqrt{d})^2 / 2\alpha_3\sqrt{d}}}{2}.$$

Hence

$$64\, r / Y_{m+1} \geq 1 / 8\, \alpha_3 \sqrt{d} ,$$

so that we have

$$r \geq Y_{m+1} / 2^9\, \alpha_3 \sqrt{d} = K\, Y_m\, \xi^{3/4}_{m+1}.$$

Similarly we have

$$r_1 \leq K_1 Y_{m+1}\, \xi^{1/4}_{m+2} \quad \text{and} \quad r_2 \leq K_2 Y_{m+1}\, \xi^{1/4}_{m+2}$$

with K_1 and K_2 not depending on m. Now we denote by $\gamma_{m+1,2\ell-1}$ the circle $| z - z_{m+1,2\ell-1}| = K\, Y_m\, \xi^{3/4}_{m+1}$. Since the image of the ring domain $\{ z;\ 2\, Y_{m+1} < | z - z_{m+1,2\ell-1}| < K\, Y_m\, \xi^{3/4}_{m+1}\}$ under $\varphi^{-1} o\, f$ separates $B_1 \cup B_2$ from B_3, we have again by Lemma 3,

$$32 \min \{|\omega|;\ \omega \in \varphi^{-1} o\, f(\gamma_{m+1,2\ell-1})\} \geq K / 2\, \xi^{1/4}_{m+1} = K' / \sqrt{d}.$$

This means that $| f (z) | \leq \alpha\, d = 4\, \alpha\, \delta_{m+1}$ in the domain bounded by $\gamma_{m+1,2\ell-1}$ and $\varGamma_{m+1,2\ell-1}$, where α does not depend on m.

(a) Let f have no zeros in the annulus $R_{m+1,2\ell-1}$. Then the image curve of any closed curve in $R_{m+1,2\ell-1}$ homotopic to $\varGamma_{m\,+\,1,2\ell-1}$ under f winds twice around 0, because $s(\check{\gamma}_{m+1,2\ell-1}) = 2$. Therefore $f^{1/2}$ is single - valued there. Since $\Delta_{m,\ell}$ is degenerate (f) , $| f(\Delta_{m,\ell}) | < 4\delta_m$, while $f(\varGamma_{m+1,2\ell-1}) \subset D(0,\ 4\delta_{m+1})$, so that $f(\Delta_{m,\ \ell}) \subset D(0,\ 4(\delta_m + \delta_{m+1}))$. Thus f, consequently $f^{1/2}$, is bounded in $\{z;\ K\, Y_m\, \xi^{3/4}_{m+1} < | z - z_{m+1,2\ell-1} | < Y_m/3\}$. Apply the Cauchy integral formula to $f^{1/2}$ there. We have

$$\frac{df^{1/2}}{dz}(z) = \frac{1}{2\pi i}\left\{\int_{|\zeta - z_{m+1,2\ell-1}| = \frac{1}{3}Y_m} - \int_{\gamma_{m+1,2\ell-1}}\right\}\frac{f^{1/2}(\zeta)}{(\zeta - z)^2}\, d\zeta.$$

We take $z \in \Gamma_{m+1,2\ell-1}$. Noting that $f(\Delta_{m,\ell}) \subset D(0, 4(\delta_m + \delta_{m+1}))$ and $|f(z)| \le 4\alpha\delta_{m+1}$ on $\gamma_{m+1,2\ell-1}$, we have

$$\left|\frac{df^{1/2}}{dz}(z)\right| \le \frac{1}{2\pi}\left\{\frac{\sqrt{12\,\delta_m}}{(Y_m/3 - Y_m\sqrt{\xi_{m+1}/6})^2}\cdot 2\pi\cdot\frac{Y_m}{3}\right.$$

$$\left. + \frac{\sqrt{8\,\alpha\,\delta_{m+1}}}{(Y_m\sqrt{\xi_{m+1}/6} - KY_m\,\xi_{m+1}^{3/4})^2}\cdot 2\pi\cdot K\,Y_m\,\xi_{m+1}^{3/4}\right\} = O(1/Y_m).$$

Thus the length of the curve $f^{1/2}(\Gamma_{m+1,2\ell-1})$ is dominated by

$$\int_{\Gamma_{m+1,\, 2\ell-1}}\left|\frac{df^{1/2}}{dz}(\zeta)\right||d\zeta| = O(\xi_{m+1}^{1/2}).$$

Since the curve $f^{1/2}(\Gamma_{m+1,2\ell-1})$ winds once around 0, we see that $|f^{1/2}(z)| \le O(\xi_{m+1}^{1/2})$ on $\Gamma_{m+1,2\ell-1}$. Now we have $|f(z)| = O(\xi_{m+1}) = o(\xi_m^{2})$ on $\Gamma_{m+1,2\ell-1}$, so that $\tilde{\Gamma}_{m+1,\, 2\ell-1} = \Gamma_{m+1,2\ell-1}$ satisfies all conditions of the lemma.

(b) Let f have zeros in the annulus $R_{m+1,2\ell-1}$. We set

$$R' = \left\{z;\ \left(\frac{Y_m}{2^{2/5}\cdot 3^{3/5}}\right)\xi_{m+1}^{2/5} < |z - z_{m+1,2\ell-1}| < Y_m/3\right\}$$

and

$$R'' = \left\{z;\ Y_{m+1} < |z - z_{m+1,2\ell-1}| < \left(\frac{Y_m}{2^{2/5}\cdot 3^{3/5}}\right)\xi_{m+1}^{2/5}\right\}.$$

(b_1) The case that f has no zeros in R'. Since $R' \subset \Delta_{m,\ell}$, f is bounded in R' and has at least one zero in the domain bounded by the circle $\{z;\ |z - z_{m+1,2\ell-1}| = (\frac{Y_m}{2^{2/5}\cdot 3^{3/5}})\,\xi_{m+1}^{2/5}\}$ and $\check{\gamma}_{m+1,2\ell-1}$, whose multiplicity is not less than 7. Since

$s(\check{\gamma}_{m+1,2\ell-1}) = 2$, the image curve of any closed curve in R' homotopic to $\Gamma_{m+1,2\ell-1}$ under f winds at least 5, say p, times around 0. Thus the function $f^{1/p}$ is single-valued and bounded in R'. We take as $\tilde{\Gamma}_{m+1,2\ell-1}$ the circle dividing R' into two annuli whose harmonic moduli both are one half of that of R', that is, $\tilde{\Gamma}_{m+1,2\ell-1} = \left\{z;\ |z - z_{m+1,2\ell-1}| = (\frac{Y_m}{2^{1/5}\cdot 3^{4/5}})\ \xi_{m+1}^{1/5}\right\}$. Applying the Cauchy integral formula to $f^{1/p}$ in R', we see similarly as above that the length of the curve $f^{1/p}(\tilde{\Gamma}_{m+1,2\ell-1})$ is dominated by $O(\delta_m^{1/p}\ \xi_{m+1}^{1/5})$. Thus $|f(z)| = O(\delta_m\ \xi_{m+1}^{p/5}) = o(\xi_m^2)$ for $z \in \tilde{\Gamma}_{m+1,2\ell-1}$.

(b_2) The case that f has zeros in R'. Let z_o be a point in R' with $f(z_o) = 0$ and $\tilde{\Gamma}_{m+1,2\ell-1}$ be the circle with center at $z_{m+1,2\ell-1}$ passing through the point z_o, that is, the circle $\{z;\ |z - z_{m+1,2\ell-1}| = |z_o - z_{m+1,2\ell-1}|\}$, where $|z_o - z_{m+1,2\ell-1}| = \frac{Y_m}{3}(\frac{3}{2}\ \xi_{m+1})^{\lambda}$, $\lambda \in [0, \frac{2}{5}]$. We have seen above that if $f(\Gamma_{m+1,2\ell-1}) \subset D(0, d)$, then $|f(z)| \le \alpha d$ in the domain bounded by the circle $\gamma = \{z;\ |z - z_{m+1,2\ell-1}| = Y_{m+1}/\ 2^9\alpha_3\sqrt{d}\}$ and $\Gamma_{m+1,2\ell-1}$. Applying the Cauchy integral formula to f in the domain bounded by $\Gamma_{m,\ell}$, $\Gamma_{m+1,2\ell-1}$ and γ, we have

$$f'(z) = \frac{1}{2\pi i}\left\{\int_{\Gamma_{m,\ell}} - \int_{\Gamma_{m+1,2\ell}} - \int_{\gamma}\right\} \frac{f(\zeta)}{(\zeta - z)^2}\, d\zeta.$$

Since $f(\Delta_{m,\ell}) \subset D(0, 6\delta_m)$ and $f(\Gamma_{m+1,2\ell-1}) \subset D(0, 4\delta_{m+1})$ so that we can take $d = 4\delta_{m+1} = 4M\sqrt{3\xi_{m+1}/2}$, we have

$$| f'(z) | \le \frac{1}{2\pi}\left\{\frac{12\,\delta_m}{\left(Y_{m-1}\sqrt{\frac{\xi_m}{6}}\right)^2}\cdot 2\pi\cdot Y_{m-1}\sqrt{\frac{\xi_m}{6}} + \frac{12\delta_m}{\left(\frac{Y_m}{2}\right)^2}\cdot 2\pi\cdot Y_m\sqrt{\frac{\xi_{m+1}}{6}}\right.$$

$$\left. + \frac{8\,\alpha\,\delta_{m+1}}{\left(Y_m\sqrt{\frac{\xi_{m+1}}{6}}\right)^2}\cdot 2\pi\cdot\frac{Y_{m+1}}{2^9\,\alpha_3\sqrt{4\delta_{m+1}}}\right\} = \epsilon_m\,\frac{\xi_m^{1/2}}{Y_m}\,,\ z\in\Gamma_{m+1,2\ell-1},$$

where ϵ_m is positive and tends to 0 as $m \to \infty$. Hence the length of the curve $f(\Gamma_{m+1,2\ell-1})$ is dominated by $\epsilon_m \frac{\xi_m^{1/2}}{Y_m}\cdot 2\pi\cdot Y_m\sqrt{\frac{\xi_{m+1}}{6}} = \frac{2\pi}{\sqrt{6}}\epsilon_m\,\xi_m^{1/2}\,\xi_{m+1}^{1/2}$, while $f(\Gamma_{m+1,2\ell-1})$ winds around 0, so that $f(\Gamma_{m+1,2\ell-1}) \subset D(0, 3\epsilon_m\,\xi_m^{1/2}\,\xi_{m+1}^{1/2})$. Thus we can take $d = \xi_m^{1/2}\,\xi_{m+1}^{1/2}$. Repeating the above argument, we have $f(\Gamma_{m+1,2\ell-1}) \subset D(0, \epsilon'_m\,\xi_m^{3/4}\,\xi_{m+1}^{1/2})$ with $\epsilon'_m \to 0$ $(m \to \infty)$. Now we can take $d = \xi_m^{3/4}\,\xi_{m+1}^{1/2}$ and conclude that $| f(z) | \le \alpha\,\xi_m^{3/4}\,\xi_{m+1}^{1/2}$ in the domain bounded by the circle $\gamma = \{z;\ | z - z_{m+1,2\ell-1} | = (Y_m / 2^{10}\alpha_3)\,\xi_m^{-3/8}\,\xi_{m+1}^{3/4}\}$ and $\Gamma_{m+1,2\ell-1}$. Apply again the Cauchy integral formula to f in the domain bounded by $\Gamma_{m,\ell}$, $\Gamma_{m+1,2\ell}$ and γ. We have

$$f^{(4)}(z) = \frac{1}{2\pi i}\left\{\int_{\Gamma_{m,\ell}} - \int_{\Gamma_{m+1,2\ell}} - \int_{\gamma}\right\}\frac{f(\zeta)}{(\zeta - z)^5}\,d\zeta$$

and for $z \in \tilde{\Gamma}_{m+1,2\ell-1}$,

$$| f^{(4)}(z) | \le \frac{1}{2\pi}\left\{\frac{12\,\delta_m}{\left(Y_{m-1}\sqrt{\frac{\xi_m}{6}}\right)^5}\cdot 2\pi\cdot Y_{m-1}\sqrt{\frac{\xi_m}{6}}\right.$$

$$+ \frac{12\,\delta_m}{\left(\frac{Y_m}{2}\right)^5}\cdot 2\pi\cdot Y_m\sqrt{\frac{\xi_{m+1}}{6}}$$

$$\left. + \frac{2\alpha\,\xi_m^{3/4}\,\xi_{m+1}^{1/2}}{\left(\frac{Y_m}{3}\left(\frac{3}{2}\,\xi_{m+1}\right)^{\lambda}\right)^5}\cdot 2\pi\cdot\frac{Y_m}{2^{10}\alpha_3}\,\xi_m^{-3/8}\,\xi_{m+1}^{3/4}\right\}$$

$$= \frac{1}{Y_m^4}\left(\lambda_1\,\xi_m^{5/2} + \lambda_2\xi_m^{1/2}\,\xi_{m+1}^{1/2} + \lambda_3\xi_m^{3/8}\,\xi_{m+1}^{5/4-5\lambda}\right) = \frac{1}{Y_m^4}\,\Lambda.$$

Since $f'''(z_o) = f''(z_o) = f'(z_o) = f(z_o) = 0$, $| f'''(z) | = | f'''(z) - f'''(z_o) | =$ $| \int_{z_o}^{z} f^{(4)}(\zeta)\, d\zeta | \leq \frac{\Lambda}{Y_m^4} \cdot 2\pi \cdot \frac{Y_m}{3} \left(\frac{3}{2}\, \xi_{m+1}\right)^{\lambda}$ for $z \in \tilde{\Gamma}_{m+1,\, 2\ell-1}$. Repeating this procedure, we have, for $z \in \tilde{\Gamma}_{m+1,\, 2\ell-1}$,

$$| f(z) | \leq \frac{\Lambda}{Y_m^4} \left(\cdot 2\pi \cdot \frac{Y_m}{3} \left(\frac{3}{2}\, \xi_{m+1}\right)^{\lambda}\right)^4 = \left(\frac{2}{3}\right)^{4-4\lambda} \pi^4 \Lambda \xi_{m+1}^{4\lambda}$$

$$< \left(\frac{2\pi}{3}\right)^4 \left(\lambda_1\, \xi_m^{5/2}\, \xi_{m+1}^{4\lambda} + \lambda_2 \xi_m^{1/2}\, \xi_{m+1}^{1/2+4\lambda} + \lambda_3 \xi_m^{3/8}\, \xi_{m+1}^{5/4-\lambda}\right)$$

$$\leq \left(\frac{2\pi}{3}\right)^4 \left(\lambda_1\, \xi_m^{5/2} + \lambda_2 \xi_m^{1/2}\, \xi_{m+1}^{1/2} + \lambda_3 \xi_m^{3/8}\, \xi_{m+1}^{17/20}\right) = o\left(\xi_m^2\right),$$

because $0 \leq \lambda \leq 2/5$. Our proof of the lemma is now complete.

3.3. We shall proceed with our proof of Theorem.

We denote by Ω the sextuply connected domain bounded by $\Gamma_{n,k}$, $\Gamma_{n+1,2k-1}$ and $\{\Gamma_{n+3,8k-j}\}_{j = 0, 1, 2, 3}$ if $\Delta_{n+2,4k-1}$ and $\Delta_{n+2,4k}$ are degenerate (f) (Case (i)), or the quintuply connected domain bounded by $\Gamma_{n,k}$, $\Gamma_{n+1,2k-1}$, $\tilde{\Gamma}_{n+2,4k-1}$ and $\{\Gamma_{n+3,8k-j}\}_{j = 0, 1}$ if $\Delta_{n+2,4k-1}$ is non-degenerate (f) and $\Delta_{n+2,4k}$ is degenerate (f), where $\tilde{\Gamma}_{n+2,4k-1}$ is the circle of Lemma 4 for $\Delta_{n+2,4k-1}$ satisfying $f(\tilde{\Gamma}_{n+2,4k-1}) \subset D(0, o(\xi_{n+1}^2))$ (Case (ii)). We showed at the beginning of our proof of Theorem that $f(\Gamma_{m,\ell}) \subset D(w_i, 8\delta_m)$ for some $w_i \in \{w_i\}_{i = 1, 2, 3}$. Therefore $f(\Gamma_{n+3,8k-j}) \subset D(0, 8\delta_{n+3})$, $j = 0, 1, 2, 3$. Thus we see that the image $f(\partial\Omega - \Gamma_{n,k} \cup \Gamma_{n+1,2k-1})$ is contained in $D(0, \tilde{\epsilon}_{n+1}\xi_{n+1}^2)$ with $\tilde{\epsilon}_{n+1} \to 0$ $(n \to \infty)$, because $\delta_{n+3} = M \sqrt{3\xi_{n+3}/2} = o(\xi_{n+1}^2)$ from the assumption $\xi_{n+1} = o\left(\xi_n^2\right)$. We consider the part of the Riemannian image S of Ω under f over the annulus $R = \{w;\ \tilde{\epsilon}_{n+1}\xi_{n+1}^2 < | w | < 1/2 \}$. Since $s(\hat{\gamma}_{n+1,2k}) = 2$ and $\Delta_{n+1,2k}$ and $\Delta_{n+2,4k}$ are degenerate (f), its component $\tilde{R}$ containing $f(\hat{\gamma}_{n+1,2k})$ covers R divalently and the boundary of $\tilde{R}$ over $|w| = 1/2$ is the curve over $|w| = 1/2$ in $S_{n,k} \subset S$ so that it consists of a single curve. Hence $\tilde{R}$ is doubly or triply connected. Suppose that $\tilde{R}$ is doubly connected. The inverse image $f^{-1}(\tilde{R})$ is a ring domain separating $\{z_{n+2,4k-1}, z_{n+2,4k}\}$ from $\{z_{n+1,2k-1}, \infty\}$ and its harmonic modulus is equal to one half of that of R, that is, $\frac{1}{2}\log(1/2\,\tilde{\epsilon}_{n+1}\xi_{n+1}^2)$. By Lemma 3, we have

$$\log\left(16\frac{Y_n(1-\eta_{n+1})}{Y_{n+1}(1-\eta_{n+2})}+8\right) \geq \frac{1}{2}\log(1/2\,\tilde{\epsilon}_{n+1}\xi_{n+1}^2),$$

so that $\tilde{\epsilon}_{n+1} \geq 1/2^{13}$. This is impossible for sufficiently large n. Hence we may assume that $\tilde{R}$ is triply connected. Then by the Hurwitz formula, we see that $\tilde{R}$ has just one branch point with multiplicity 2, whose projection we denote by w^*. Since the part of $\tilde{R}$ over $\{w;\ |w^*| < |w| < 1/2\}$ is doubly connected, we have by Lemma 3

$$\log\left(16\frac{Y_n(1-\eta_{n+1})}{Y_{n+1}(1-\eta_{n+2})}+8\right) \geq \frac{1}{2}\log(1/2|w^*|),$$

that is

$$|w^*| > \frac{1}{2^{13}}\xi_{n+1}^2.$$

The inverse image of the curve in $\tilde{R}$ over the circle $\{w;\ |w| = |w^*|\}$ is an eight-shaped closed curve crossing at the point z^* with $f(z^*) = w^*$, that is,it consists of two simple closed curves C_1 and C_2 with $C_1 \cap C_2 = \{z^*\}$. The part of $\tilde{R}$ over the annulus $R' = \{w;\ \tilde{\epsilon}_{n+1}\xi_{n+1}^2 < |w| < |w^*|\}$ consists of two annuli $\tilde{R}_1'$ and $\tilde{R}_2'$ each of which covers R' univalently and has the harmonic modulus being equal to that of $R' \geq \log 1/(2^{13}\tilde{\epsilon}_{n+1}) \to \infty\ (n \to \infty)$. The inverse image $\{f^{-1}(\tilde{R}_1')\}_{i=1,2}$ are ring domains separating in the same manner at least one of $\{\Gamma_{n+3,8k-j}\}_{j=0,1,2,3}$ in Case (i) or at least one of $\Gamma_{n+1,2k-1}$ and $\{\Gamma_{n+3,8k-j}\}_{j=0,1}$ in Case (ii) from the remainder not being empty. Suppose that they separate $\Gamma_{n+3,8k-j_o}$ with $j_o \in \{0, 1\}$ from

$\{\Gamma_{n+3,8k-j}\}$, $j \neq j_o$ and $j \epsilon \{0, 1, 2, 3\}$, or $\Gamma_{n+3,8k-j_o} \cup \Gamma_{n+3,8k-j_1}$ with $j_o \epsilon \{0, 1\}$ and $j_1 \epsilon \{2, 3\}$ from $\Gamma_{n+3,8k-j_2} \cup \Gamma_{n+3,8k-j_3}$, $\{j_2, j_3\} = \{0, 1, 2, 3\} - \{j_o, j_1\}$.

Then one of $f^{-1}(\tilde{R}_1')$ and $f^{-1}(\tilde{R}_2)$, say $f^{-1}(\tilde{R}_1')$, separates $\{ z_{n+3,8k-j_o}, \infty \}$ from $\{z_{m+3,8k-j}\}$, $j \neq j_o$ and $j \epsilon \{0, 1, 2, 3\}$, so that we have by Lemma 3

$$\log\left(16 \frac{Y_{m+2}(1 - \eta_{n+3})}{Y_{m+1}(1 - \eta_{n+2})} + 8 \right) \geq \text{har. mod. } f^{-1}(\tilde{R}_1) \geq \log 1/2^{13} \tilde{\epsilon}_{n+1},$$

and hence

$$\tilde{\epsilon}_{n+1} \geq 1 / 2^{16} \cdot 3,$$

because $| z_{n+3,8k-1} - z_{n+3,8k} | = Y_{m+2}(1 - \eta_{n+3})$ and $\min \{ | z_{n+3,8k-j} - z_{n+3,8k-\ell}| ; j \epsilon \{0, 1\}, \ell \epsilon \{2, 3\} \} = | z_{n+3,8k-2} - z_{n+3,8k-1}| \geq Y_{m+1}(1 - \xi_{n+2})$. This is impossible for sufficiently large n. We may assume now that in Case (i), C_1 surrounds $\Gamma_{n+3,8k-3}$ and $\Gamma_{n+3,8k-2}$ and C_2 surrounds $\Gamma_{n+3,8k-1}$ and $\Gamma_{n+3,8k}$. Similarly we see in Case (ii) that C_1 surrounds $\Gamma_{n+2,4k-1}$ and C_2 surrounds $\Gamma_{n+3,8k-1}$ and $\Gamma_{n+3,8k}$. One of $\Delta_{n+3,8k-1}$ and $\Delta_{n+3,8k}$ is degenerate (f), for otherwise, C_2, $\check{\gamma}_{n+3,8k-1}$ and $\check{\gamma}_{n+3,8k}$ bound a domain where f takes the value 0, so that $s(C_2) + s(\check{\gamma}_{n+3,8k-1}) + s(\check{\gamma}_{n+3,8k}) \geq 7$, while $s(C_2) = 1$ and $s(\check{\gamma}_{n+3,8k-1}) = s(\check{\gamma}_{n+3,8k}) = 2$. We denote by Ω' the quintuply connected domain bounded by C_2 and $\{\Gamma_{n+4,16k-j}\}$ $j = 0, 1, 2, 3$ if $\Delta_{n+3,8k-1}$ and $\Delta_{n+3,8k}$ both degenerate (f), or the quadruply connected domain bounded by C_2, $\tilde{\Gamma}_{n+3,8k-1}$ and $\{\Gamma_{n+4,8k-j}\}$ $j = 0, 1$ if one of them, say $\Delta_{n+3,8k-1}$, is non-degenerate (f), where $\tilde{\Gamma}_{n+3,8k-1}$ is the circle of Lemma 4 for $\Delta_{n+3,8k-1}$ satisfying $f(\tilde{\Gamma}_{n+3,8k-1}) \subset D(0, o(\xi_{n+2}^2))$. Since $f(\Gamma_{n+4, 16k-j}) \subset D\left(0, 8\delta_{n+4}\right)$ for $j = 0, 1, 2, 3$ and $\delta_{n+4} = M\sqrt{3\xi_{n+4}/2}$, we see from the assumption $\xi_{n+1} = o(\xi_n^2)$ that the image $f(\partial\Omega' - C_2)$ is contained in $D(0, \tilde{\epsilon}'_{n+1} \xi_{n+1}^2 \xi_{n+2})$ with $\tilde{\epsilon}'_{n+1} \to 0$ as $n \to \infty$. We consider the part of the

Riemannian image of Ω' over the annulus $R'' = \left\{w;\ 2\tilde{\epsilon}'_{n+1}\xi^2_{n+1}\ \xi_{n+2} < |w| < |w^*|\right\}$.

Since $s(C_2) = 1$, the component $\tilde{R}''$ having $f(C_2)$ as a boundary curve covers R'' univalently, so that its harmonic modulus is equal to that of R''. The inverse image $f^{-1}(\tilde{R}'')$ is a ring domain separating $\{\Gamma_{n+3,8k-j}\}_{j = 2,\ 3}$ from $\{\Gamma_{n+4,16k-j}\}\ j = 0,\ 1,\ 2,\ 3$ and has the same harmonic modulus as $\tilde{R}''$. By Lemma 3, we have

$$\log\left(16\,\frac{Y_{n+1}}{Y_{n+2}(1 - \xi_{n+3})} + 8\right) \geq \log\frac{|\ w^*\ |}{2\tilde{\epsilon}'_{n+1}\xi^2_{n+1}\xi_{n+2}},$$

so that $64/\xi_{n+2} \geq |\ w^*\ |\ /2\tilde{\epsilon}_{n+1}\xi^2_{n+1}\xi_{n+2}$, that is, $\tilde{\epsilon}_{n+1} \geq 1/2^{24}$ because $|\ w^*| > \xi^2_{n+1}/2^{13}$ as we have seen just above. It is impossible for sufficiently large n. Our proof is now complete.

References

[1] Besicovitch, A. S., On sufficient conditions for a function to be analytic, and on behaviour of analytic functions in the neighbourhood of non-isolated singular points, Proc. Lond. Math. Soc., Ser. 2, 32(1930), pp. 1 - 9.

[2] Kurokawa, T., Exceptionally ramified meromorphic functions with a non-enumberable set of essential singularities, Nagoya Math. J., 88 (1982), pp. 133 -154.

[3] Kurokawa, T., and K. Matsumoto, On totally ramified values, Tôhoku Math. J., 28 (1976), pp. 245 - 256.

[4] Lehto, O., and K. I. Virtanen, Quasiconformal mappings in the plane, Springer - Verlag, 1973.

K. Matsumoto
Department of Mathematics
College of General Education
Nagoya University

T. Kurokawa
Department of Mathematics
Faculty of Education
Mie University

GAP SERIES

Takafumi Murai

1. Introduction

According to Weierstraß [116], Riemann said to his student that $\sum_{n=1}^{\infty} (\sin n^2 t)/n^2$ is nowhere differentiable. This tale is sometimes quoted as the beginning of the study of gap series. As is well known, this problem was solved by Gerver [35] (cf. Hardy [42]) in the negative, and at present various differentiability properties are established. The theme of gap series is to deduce the behaviour of power series

$$f(z) = \sum_{n=0}^{\infty} c_n z^n \tag{1}$$

and Fourier series

$$P(t) = \sum_{n=-\infty}^{\infty} c_n^* e^{int} \tag{2}$$

from their spectra $S(f) = \{n \geq 0; c_n \neq 0\}$ and $S^*(P) = \{n; c_n^* \neq 0\}$. We write by $n_1(f)$, $n_2(f)$, $\cdots$ the elements of $S(f)-\{0\}$ according to the order of growth. We write simply $n_k = n_k(f)$ $(k \geq 1)$, $n_0 = 0$. Then (1) is expressed in the following form:

$$f(z) = \sum_{k=0}^{\infty} a_k z^{n_k}. \tag{3}$$

In this note, we shall mainly show the value-distribution theory of gap series. There are many topics about gap series. Prior to stating the value-distribution theory, we begin by showing some theorems about gap series in the other areas.

1.1 Natural boundaries

Hadamard [41] (1982) shows that if f(z) of form (3) satisfies

$$\inf_{k \geq 1} n_{k+1}/n_k > 1 \text{ (Hadamard's gap condition)}, \tag{4}$$

then the natural boundary of f(z) is identical with the circle $\{|z|=r\}$ (r = the radius of convergence of f(z)). Fabry [28] (1894) shows that this assertion holds with (4) replaced by

$$\lim_{k\to\infty} k/n_k = 0 \quad \text{(Fabry's gap condition).} \tag{5}$$

Pólya [97] (1942) shows that (5) is best possible.

1.2 Sidon sets

Let $L^s(0, 2\pi)$ $(1\leq s\leq\infty)$ denote the L^s space on $[0, 2\pi)$ with respect to the 1-dimensional Lebesgue measure $|\cdot|$. Sidon [107] (1927) shows that if f(z) of form (3) satisfies (4), then

$$\sum_{k=0}^{\infty} |a_k| \leq C\|f(e^{i\cdot})\|_{L^\infty(0,2\pi)},$$

where C is a constant depending only on S(f). A set E of integers is called a Sidon set if, for any P(t) of form (2) satisfying $S^*(P) \subset E$,

$$\sum_{n=-\infty}^{\infty} |c_n^*| \leq C\|P\|_{L^\infty(0,2\pi)},$$

where C is a constant depending only on E. Stečkin [111] shows that a finite union of Hadamard gap series is a Sidon set. Drury [21] shows that a finite union of Sidon sets is a Sidon set. For s>1, a set E of integers is called a $\Lambda(s)$ set if, for any P(t) satisfying $S^*(P) \subset E$,

$$\|P\|_{L^s(0,2\pi)} \leq C\|P\|_{L^1(0,2\pi)},$$

where C is a constant depending only on E and s. Rudin [100] shows that {Sidon sets} $\subset \cap_{s>1}\{\Lambda(s)$ sets}. Various relations between Sidon sets and $\Lambda(s)$ sets are known (cf. Kahane [51], Pisier [95]). Sidon sets on discrete groups are also studied (cf. López-Ross [69], Hewitt-Ross [49]).

1.3 Probability theory

Salem-Zygmund [103] shows the following theorem. For a real Fourier series

$$P(t) \sim \sum_{k=0}^{\infty} (\alpha_k \cos n_k t + \beta_k \sin n_k t) \quad (n_0 = 0),$$

we put

$$B_k = \{\tfrac{1}{2} \sum_{j=0}^{k} (\alpha_j^2 + \beta_j^2)\}^{1/2}, \; P_k(t) = \sum_{j=0}^{k} (\alpha_j \cos n_j t + \beta_j \sin n_j t) \quad (k \geq 0).$$

If $\lim_{k\to\infty}(\alpha_k^2 + \beta_k^2)/B_k^2 = 0$ and $(n_k)_{k=1}^{\infty}$ satisfies (4), then, for any real numbers α, β $(\alpha < \beta)$ and any interval $I \subset [0,2\pi]$,

$$\lim_{k\to\infty} |\{t \in I;\ \alpha < P_k(t)/B_k < \beta\}| = \frac{|I|}{\sqrt{2\pi}} \int_{\alpha}^{\beta} \exp(-x^2/2)\, dx$$

(the central limit theorem). M. Weiss [118] shows that if $\alpha_k^2 + \beta_k^2 = o(B_k^2 \log\log B_k)$ $(k\to\infty)$ and $(n_k)_{k=1}^{\infty}$ satisfies (4), then

$$\limsup_{k\to\infty} P_k(t)/\sqrt{2B_k^2 \log\log B_k} = 1 \text{ a.e.}$$

(the law of the iterated logarithm). This is an improvement of results of Salem-Zygmund [104] and Erdös-Gál [25]. There are many articles about the central limit theorem and the law of the iterated logarithm (cf. Bingham [13], Kahane [54], Kawata [59], Takahashi [114]). Erdös [24] shows that if $\alpha_k^2 + \beta_k^2 = 1$, $n_{k+1}/n_k \geq 1 + d_k/\sqrt{k}$ $(k \geq 1)$, $\lim_{k\to\infty} d_k = \infty$, then P(t) obeys the central limit theorem. This gap condition is best possible in a sense. It is also known that $\sum_{k=0}^{\infty} \cos([e^{\sqrt{k}}]t)$ obeys the central limit theorem, where $[e^{\sqrt{k}}]$ is the integral part of $e^{\sqrt{k}}$ ([81]). The almost sure invariance property is studied by Philipp-Stout [94] and the recurrence property is studied in [43] and [80]. Applying the law of the iterated logarithm of gap series, Makarov [73] gives an estimate of harmonic measure. There are many applications of probabilistic properties of gap series to function theory.

1.4 The completeness problem and the uniqueness problem

For a sequence E of non-negative integers, the spanning radius of E is defined by

$$D_{com}(E) = \inf\{D > 0; < e^{int}>_{n \epsilon E} \neq L^2(-\pi D,\pi D)\},$$

where $<e^{int}>_{n\epsilon E}$ is the subspace of $L^2(-\pi D,\pi D)$ spanned by $(e^{int})_{n\epsilon E}$. Levinson [68] and Beurling-Malliavin [10] investigate $D_{com}(E)$. A sequence E of non-negative integers is called a U-set if, for any $f(e^{it})$ with $S(f) \subset E$,

$$\text{"}f(e^{it}) = 0 \text{ on some interval"} \Rightarrow f \equiv 0.$$

Applying the results of [10], Kahane [52] characterizes U-sets. He [53] also discusses U-sets replacing intervals by Cantor sets (see Levinson [68], Kahane-Salem [57], Korevaar [61]). The set $S^*(P)$ is the support of $(\hat{P}(n))_{n=-\infty}^{\infty}$ where $\hat{P}(n)=(1/2\pi)\int_0^{2\pi} P(t)e^{-int}\,dt$. Hence the theme of gap series is nothing else to deduce the behaviour of functions from the information of the supports of Fourier transforms. Recall, for example, a famous theorem in this direction: $f \epsilon L^2(-\pi,\pi)$ is of subexponential type A if and only if supp(f) $\subset [-A,A]$ (Paley-Wiener's theorem). As a summary of the introduction, we here quote Kahane's description [54]: Even so, we wish to give the feeling that the field of lacunary series, though very tiny, is a very fertile one. Therefore it may be better if a few people go on and plow it and are able from time to time to dig out new curious plants that grow therein.

This is a translation of my Japanese survey [87] with some recent results. I would like to express my thanks to Professor W. K. Hayman who advised to translate [87]. I wish to express my thanks to Professor W. H. J. Fuchs for his encouragement. I am indebted to Professor J. M. Anderson, Ch. Pommerenke, L. R. Sons, J. Hawkes, D. Gnuschke-Hauschild, R. M. Gethner for their valuable suggestions and conversations at the conference in honor of Professor W. K. Hayman.

2. Entire functions (Approximation theory)

2.1 Müntz-Szász' approximation

As is well known, $<t^n>_{n=0}^{\infty} = C(0,1)$ (the Weierstraß approximation theorem). Müntz [77] (1914) and Szász [113] (1916) improve this theorem as follows: $<t^{n_k}>_{k=0}^{\infty} = C(0,1)$ if and only if

$$\sum_{k=1}^{\infty} 1/n_k = \infty. \tag{6}$$

Considering a transform $x \to e^{-x}$, we immediately see that $<e^{-n_k t}>_{k=0}^{\infty} = C_0(0,\infty)$ if and only if $(n_k)_{k=1}^{\infty}$ satisfies (6), where $C_0\ (0,\infty)$ is the space of continuous functions on $[0,\infty)$ vanishing at infinity. Taking account of this fact, Macintyre [71] shows that if a non-constant entire function f(z) of form (3) satisfies

$$\sum_{k=1}^{\infty} 1/n_k < \infty \text{ (Fejér's gap condition)}, \tag{7}$$

then f(z) is unbounded on $(0,\infty)$, and shows that, for any $(n_k)_{k=1}^{\infty}$ satisfying (6), there exists a non-constant entire function f(z) of form (3) such that $S(f) = (n_k)_{k=1}^{\infty}$ and f(z) is bounded on $(0,\infty)$. From this fact, he conjectures that a non-constant entire function of form (3) satisfying (7) has no finite asymptotic value (Macintyre's conjecture). Korevaar [60] shows the following. Let $\gamma = \gamma(x)$ be an analytic curve such that $\gamma(0) = 0$ and Re $\gamma(x) \uparrow \infty$ $(x\uparrow\infty)$. Then $<e^{-n_k z}>_{k=0}^{\infty} = C_0(\gamma)$ if and only if $(n_k)_{k=1}^{\infty}$ satisfies (6). Here are some partial answers to Macintyre's conjecture. His conjecture is valid with (7) replaced by "$n_k \geq k(\log k)^{1+\epsilon}$ $(k\geq1)$ for some $\epsilon>0$" (Kövari [62]). His conjecture is valid with (7) replaced by "$\sum_{k=1}^{\infty}(\log n_k)^{1+\epsilon}/n_k < \infty$ for some $\epsilon>0$" (Anderson-Binmore [7]). His conjecture is valid under the additional condition that $(n_k/k)_{k=1}^{\infty}$ is increasing (Pavlov [92]).

[Method 1] Korevaar's method is typical in approximation theory. The proof of the "only if" part is as follows. Suppose that $(n_k)_{k=1}^{\infty}$ satisfies (7). It is sufficient to construct

$\phi \in C_0(\gamma)$ so that

$$\int_\gamma \phi\, dz \neq 0,\ \int_\gamma e^{-n_k z} \phi\, dz = 0\ \ (k \geq 1),\ \int_\gamma |\phi|\, |dz| < \infty.$$

Let $-\gamma = \{-z;\ z \in \gamma\}$. Then there exists $\psi \in C_0(-\gamma)$ such that

$$\int_{-\gamma} \psi\, dz \neq 0,\ D^{(m)}\psi(0) = 0\ (m \geq 0),\ \int_{-\gamma} |\prod_{k=1}^{\infty}(1-(D/n_k))\psi|\, |dz| < \infty,$$

where $D\psi$ is the derivative of ψ along $-\gamma$ (cf. Carleman [17]). Then $\phi(z) =$ $\{\Pi_{k=1}^{\infty}(1-(D/n_k))\ \psi\}(-z)$ is the required function.

[Method 2] Here is another fundamental tool. For an entire function f(z) and for a continuous arc γ with endpoints 0 and r (r>0),

$$\max_{z \in \gamma} |f(z)| \geq \left\{ \max_{0 \leq x \leq r} |f(x)| \right\}^2 / M(r,f),$$

where $M(r,f) = \max_{|z| \leq r} |f(z)|$ (Pólya's reflection principle). The proof is as follows. Let $L = \max_{z \in \gamma} |f(z)|$, $F(z) = f(z)\overline{f(\overline{z})}$ and let R be the compact set bounded by $\gamma \cup \{z;\ \overline{z} \in \gamma\}$. Then, on the boundary of R, $|F(z)| \leq L\, M(r,f)$. Hence the maximum modulus principle shows that $|f(x)|^2 = |F(x)| \leq L\, M(r,f)$ $(0 \leq x \leq r)$, which yields the required inequality.

These two tools are very important in approximation by gap series (cf. Anderson [4], Korevaar-Dixon [62]).

2.2 The "$\rho D<1/2$" theorems

Pólya [96] shows the following. If an entire function f(z) satisfies $\rho^*(f)D_P(f) < 1/2$, then f(z) is unbounded on $(0,\infty)$, where $\rho^*(f) = \limsup_{r\to\infty} \{\log \log M(r,f)\}/\log r$ (the order of f), $D_P(f) = \limsup_{\xi\to 1} \limsup_{r\to\infty} \{\omega(r,f) - \omega(r\xi,f)\}/(r-r\xi)$ (Pólya's upper density) and $\omega(r,f) =$ (the number of elements of $S(f) - \{0\}$ less than r) . Edrei [22] improves this theorem as follows: If $\rho^*(f)D_{\log}(f) < 1/2$, then f(z) is unbounded on $(0,\infty)$, where $D_{\log}(f) = \liminf_{r\to\infty}\{\Sigma_{n_k < r} 1/n_k\}/\log r$ (the lower logarithmic density). On the other hand, for any sequence $(n_k)_{k=1}^{\infty}$ with positive lower logarithmic density d, there exists an entire function f(z) such that $S(f) = (n_k)_{k=1}^{\infty}$, $\rho^*(f)d = 1/2$ and f(z) is bounded on $(0,\infty)$. Kövari [67] investigates the order on $(0,\infty)$: If $\rho^*(f)D_P(f) < 1/2$, then $\limsup_{r\to\infty}\log|f(r)|/\log M(r,f) \geq \cos\{\pi\rho^*(f)D_P(f)\}$. He [66] tries to replace $(0,\infty)$ by curves also: If $\rho^*(f)D_{gap}(f) < 1/\pi^2$, then f(z) is unbounded on any curve tending to infinity, where $D_{gap}(f) = \sup_{k\geq 1}(n_{k+1} - n_k)^{-1}$. It is already known that these facts hold with $\rho^*(f)$ replaced by $\rho_*(f) = \liminf_{r\to\infty}\{\log\log M(r,f)\}/\log r$ (the lower order of f).

[Method 3] It is very important to know the reason why the behaviour changes at $\rho D = 1/2$. Here is Carleman's theorem: For an analytic function $\phi(z)$ in $V = \{\mathrm{Re}\ z > 0\}$ having zeros at $\{m_k\}_{k=1}^{\infty}$, $0 < m_1 < m_2 < \cdots$,

$$\sum_{m_k<r}\{(1/m_k) - (m_k/r^2)\} \leq \frac{1}{2\pi}\int_1^r\{\log|\phi(ix)\phi(-ix)|\}/x^2\,dx$$

$$+\frac{1}{\pi r^2}\int_{-\pi}^{\pi}\log|\phi(re^{it})|\,dt + O(1)\quad (r\to\infty).\ (8)$$

This theorem is closely related to our question. The proof of the above Edrei's theorem is as follows. Supposing that f(z) is bounded on $(0,\infty)$, we show that $\rho D \geq 1/2$ ($\rho = \rho^*(f)$, $D = D_{\log}(f)$). We may assume that $f(0) = 0$. Then $g(z) = \int_1^{\infty} f(x)x^{-z-2}\,dx$ takes poles of order 1 on S(f). Hence $\{z \in V;\ \phi(z) = 0\} = \{m\geq 0;\ m \notin S(f)\}$, where $\phi(z) = g(z)\sin\pi z$. Applying Carleman's theorem to $\phi(z)$, we obtain (8). It is easily seen that (the second term in the left-hand side of (8)) = O(1) and (the second term in the right-hand side of (8)) = O(1). Thus

$$\sum_{m_k<r} 1/m_k \leq \frac{1}{2\pi}\int_1^r\{\log|\phi(ix)\phi(-ix)|\}/x^2\,dx + O(1).$$

Let $\epsilon > 0$. Since $\limsup_{x\to\infty}\{\log|\phi(\pm ix)|\}/x \leq \pi\{1 - (1/2\rho)\}$ (cf. Carlson [18]), we have

$$\sum_{m_k < r_\mu} 1/m_k \leq \{1 - \frac{1}{2\rho+\epsilon}\}\log r_\mu + O(1) \quad (\mu \to \infty)$$

for some sequence $(r_\mu)_{\mu=1}^{\infty}$ tending to infinity. Hence $(2\rho+\epsilon)D \geq 1$. Since $\epsilon>0$ is arbitrary, we obtain $\rho D \geq 1/2$.

2.3. Pólya-Turán's problem

For $0 \leq \alpha < \beta \leq 2\pi$, $r>0$ and an entire function f(z), we put $M(r,\alpha,\beta,f) = \max\{|f(re^{it})|;\ \alpha \leq t \leq \beta\}$. Pólya [96] shows that if f(z) is of finite order and satisfies Fabry's gap condition, then, for any $0 \leq \alpha < \beta \leq 2\pi$,

$$\limsup_{r\to\infty} \log M(r,\alpha,\beta,f) \,/\, \log M(r,f) = 1. \tag{9}$$

Turán [115] shows that if f(z) satisfies the condition "$n_k \geq k^{1+\epsilon}$ $(k \geq 1)$ for some $\epsilon>0$", then (9) holds. Fuchs [30] improves Pólya's theorem as follows: Under Pólya's condition, $\limsup_{r\to\infty}\log L(r,f)/\log M(r,f) = 1$, where $L(r,f) = \min\{\,|f(re^{it})|\,;\ 0 \leq t < 2\pi\}$ (cf. Sons [109]). Sheremeta [106] shows that under Fejér's gap condition, (9) holds. To compare $\log M(r,\alpha,\beta,f)$ and $\log M(r,f)$ (Pólya-Turán's problem) is very important to study the value-distribution in a sector $S_{\alpha,\beta} = \{z \in \mathbb{C};\ \alpha < \arg z < \beta\}$. For this problem, Wiman-Valiron's method is useful; this is the method to study $\mu(r,f) = \max\{|a_k| r^{n_k};\ k \geq 0\}$ (the maximum term) and $\nu(r,f) = \max\{n_k;\ |a_k|\, r^{n_k} = \mu(r,f)\}$ (the central index). By this method, our problem reduces to the study of the behaviour of polynomials (cf. Hayman [47]).

3. Entire functions (The value-distribution theory)

3.1 The ρD theorems

For $r>0$, $a \in \mathbb{C}$ and an entire function f(z), we put

$$T(r,f) = \frac{1}{2\pi}\int_0^{2\pi} \log^+ |f(re^{it})|\, dt \text{ (the characteristic function),}$$

$$n(r,a,f) = (\text{the number of roots of } f(z) = a \text{ in } 0<|z|<r),$$

$$N(r,a,f) = \int_0^r n(x,a,f)/x\, dx \text{ (the counting function)},$$

$$\delta(a,f) = 1 - \limsup_{r\to\infty} \frac{N(r,a,f)}{T(r,f)} \text{ (the deficiency)},$$

$$\Delta(f) = \sum_{a\in\mathbb{C}} \delta(a,f),$$

where $\log^+ r = \max\{\log r, 0\}$. Fejér [29] (1908) shows the following: If an entire function f(z) satisfies Fejér gap condition, then f(z) takes any complex value at least once. Biernacki [11] improves this theorem as follows: An entire function f(z) satisfying (7) has no finite Picard exceptional value, i.e., $\lim_{r\to\infty} n(r,a,f) = \infty$ for all $a\in\mathbb{C}$. Pfluger-Pólya [93] shows the following: If f(z) has no finite Borel exceptional value and if $\rho^*(f)$ is a positive integer, then $\rho^*(f)D_0(f)$ is a positive integer, where $D_0(f) = \lim_{r\to\infty} \omega(r,f)/r$ (the density). (A complex number $a\in\mathbb{C}$ is called a Borel exceptional value of f(z), if $\limsup_{r\to\infty} N(r,a,f)/\log r < \rho^*(f)$.) Sunyer i Balaguer [112] shows the following: For an entire function f(z), if $\rho^*(f)$ is a positive integer and $\rho^*(f)D^*(f) < 1$, then, for any entire function g(z) of order less than $\rho^*(f)$,

$$\limsup_{r\to\infty} n(r, 0, f+g)/\log M(r, f) \geq C > 0,$$

where $D^*(f) = \limsup_{r\to\infty} \omega(r, f)/r$ (the upper density) and C is a constant depending only on $\rho^*(f)$. Edrei-Fuchs [23] improves this theorem as follows: For an entire function f(z), if $\rho_*(f)$ is a positive integer and $\rho_*(f)D_*(f) < 1$, then $\Delta(f) < 1$, where $D_*(f) = \liminf_{r\to\infty} \omega(r,f)/r$ (the lower density). We write simply $D_{com}(f) = D_{com}(S(f))$ (cf. 1.4). Fuchs [32] shows that, for any $a\in\mathbb{C}$, $\delta(a,f) \leq \text{Const}\ \rho_*(f)\ \{1+\log^+\rho_*(f)\}D_{com}(f)$. Improving his method, we obtain

$$\Delta(f) \leq \text{Const}\ \rho_*(f)D_{com}(f) \text{ ([83])}. \tag{10}$$

<u>[Method 4]</u> To explain Fuchs' method, we here give the proof of $\delta(0,f) \leq \text{Const}\ \rho D$ ($\rho =$

$\rho*(f)$, $D = D_{com}(f))$. Let $m(r,1/f) = (1/2\pi) \int_0^{2\pi} \log^+ 1/| f(re^{it}) | \, dt$. Then $\delta(0,f) = \liminf_{r\to\infty} m(r, 1/f)/T(r,f)$. We may assume that $a_0 = f(0) = 1$. For any ξ larger than πD, we have $<e^{int}>_{n\epsilon S(f)} \neq L^2(-\xi,\xi)$, and hence, there exists $h \epsilon L^2(-\xi,\xi)$ such that

$$\int_{-\xi}^{\xi} h(t)\,dt = 1, \quad \int_{-\xi}^{\xi} h(t)e^{int}\,dt = 0 \ (n \epsilon S(f)-\{0\})$$

(Paley-Wiener's theorem). For any $0 \le \theta < 2\pi$,

$$1 = \left| \int_{-\xi}^{\xi} h(t)\, f(re^{i(\theta+t)})\,dt \right| \le M(r,\theta-\xi,\theta+\xi,f) \int_{-\xi}^{\xi} | h(t) | \,dt,$$

which shows that

$$M(r,\theta-\xi,\theta+\xi,f) \ge \eta \ (0 \le \theta < 2\pi), \tag{11}$$

where η is a constant independent of r and θ. Note that $m(r,1/f) \le m(r,\eta/f) + O(1)$. Let $G_r = \{t \epsilon [0,2\pi); | f(re^{it}) | < \eta\}$. Since G_r is open, we can write $G_r = \cup_{j=1}^m I_j$ with mutually disjoint intervals $\{I_j\}_{j=1}^m$. Then $| f(re^{i\cdot}) | = \eta$ at each endpoint of these intervals. We have, with γ_j = (the midpoint of I_j),

$$m(r,\eta/f) = -\frac{1}{2\pi} \int_{G_r} \log \{| f(re^{it}) | /\eta\}\,dt$$
$$= -\frac{1}{2\pi} \sum_{j=1}^m \int_{I_j} = \frac{1}{2\pi} \sum_{j=1}^m \int_{I_j} (t-\gamma_j) \frac{\partial}{\partial t} \log | f(re^{it}) | \,dt.$$

Poisson-Jensen's formula shows that, for $R>r$,

$$\frac{\partial}{\partial t} \log | f(re^{it}) | = \operatorname{Re} \frac{ire^{it}}{2\pi} \int_0^{2\pi} \frac{2Re^{is}}{(Re^{is} - re^{it})^2} \log | f(Re^{is}) | \,ds$$

$$+ \operatorname{Re} \sum\nolimits_k^* \left\{ \frac{ire^{it}}{re^{it} - b_k} + \frac{ire^{it}}{R^2 - \bar{b}_k re^{it}} \right\} = a(re^{it},R) + b(re^{it}, R),$$

where $\sum_k^*$ is the summation over all zeros $\{b_k\}$ of $f(z)$ in $|z|<R$. Then

$$m(r,\eta/f) = \frac{1}{2\pi} \sum_{j=1}^m \int_{I_j} (t-\gamma_j)\, a(re^{it}, R)\,dt$$
$$+ \frac{1}{2\pi} \sum_{j=1}^m \int_{I_j} (t-\gamma_j)\, b(re^{it}, R)\,dt = A(r, R) + B(r, R).$$

Inequality (11) shows that $|\, I_j \,| \leq 2\xi$ $(1 \leq j \leq m)$. Hence

$$|A(r,R)| \leq \frac{\xi}{2\pi} \sum_{j=1}^{m} \int_{I_j} |\, a(re^{it},R) \,|\, dt$$

$$\leq \frac{\xi}{2\pi} \int_0^{2\pi} |a\ (re^{it},R)\ |\ dt$$

$$\leq \frac{\xi rR}{\pi^2} \int_0^{2\pi} \int_0^{2\pi} |Re^{is} - re^{it}|^{-2} |\log|f(Re^{is})\,|\,|\ ds\ dt$$

$$= \frac{2\xi rR}{\pi(R^2-r^2)} \int_0^{2\pi} |\log|f(Re^{is})\,|\,|\ ds$$

$$\leq \text{Const}\ \xi rR(R^2-r^2)^{-1}T(R,f) + O(1).$$

We now choose $\xi = 2\pi D$ and $R = re^{1/\rho}$. Then $A(r,R) \leq \text{Const}\ \rho D\ T(R,f) + O(1)$. In the analogous manner, $B(r,R) \leq \text{Const}\ \rho D\ T(R,f) + O(1)$ $(R = re^{1/\rho})$ except for a set E of r satisfying $\int_E dr/r < \infty$ (cf. [83]). Since f(z) is of order ρ, we can choose a sequence $(r_\mu)_{\mu=1}^{\infty}$ tending to infinity so that $\{r_\mu\}_{\mu=1}^{\infty} \cap E \neq \emptyset$ and

$$T(R_\mu,f) \leq (R_\mu/r_\mu)^{2\rho} T(r_\mu,f) = e^2\ T(r_\mu,f)\ (R_\mu = r_\mu e^{1/\rho},\ \mu \geq 1).$$

Then

$$m(r_\mu,1/f) \leq A(r_\mu,R_\mu) + B(r_\mu,R_\mu) + O(1)$$

$$\leq \text{Const}\ \rho D\ T(r_\mu,f) + O(1),$$

which shows that $\delta(0,f) \leq \text{Const}\ \rho D$. Our method is analogous to the proof of Hayman's inequality: $\Sigma_{a\in\mathbb{C}}\ \delta(a,f)^{\alpha} < \infty$ $(\alpha > 1/3)$. Lewis-Wu [70] shows that if f(z) is an entire function of finite lower order ρ, then $\Sigma_{a\in\mathbb{C}}\ \delta(a,f)^{(1/3)-\gamma} < \infty$ for some $\gamma > 0$ depending only on ρ. Their method may improve (10). (See W. K. Hayman, Meromorphic Functions, Clarendon, Oxford, 1975).

3.2 The "Δ=0" theorem

It is interesting to improve Fejér-Biernacki's result. If an entire function has a finite Picard exceptional value, then it is an asymptotic value. Hence Macintyre's conjecture is regarded as a test in this direction. Here are some tests to improve Fejér-Biernacki's result from the point of view of the Nevanlinna theory. Kövari [60] shows that if f(z) satisfies $\lim_{k\to\infty}\{\phi(k)\log k\}/n_k = 0$, then f(z) has no finite Borel exceptional value, where $\phi(x)$ is a function in $(0,\infty)$ such that $\int_1^\infty 1/\phi(x)\,dx < \infty$. It is known that

An entire function with (7) satisfies

$$\Delta(f) = 0 \quad ([86]). \tag{12}$$

Here Fejér's gap condition (7) cannot be, in general, replaced by Fabry's gap condition (5). Pólya [96] conjectures that an entire function with (5) has no finite Picard exceptional value (Pólya's conjecture). This problem remains open. It is also interesting to study the value-distribution in sectors. Biernacki [12] shows that if f(z) satisfies "$n_{k+1}-n_k \ge n_k^{(1/2)+\epsilon}$ $(k\ge1)$ for some $\epsilon>0$", then f(z) has no finite Picard exceptional value in any sector $S_{\alpha,\beta}$. (This assertion holds valid with the above gap condition replaced by (7).) Hayman-Rossi [48] studies the value-distribution of entire functions with (5) in sectors.

[Method 5] We show the outline of the proof of (12). For two functions A(r), B(r) in $(0,\infty)$, we write "$A(r) \ge B(r)$ l.f." if $\int_E dr/(1+r) < \infty$, where $E = \{r>0;\ A(r) < B(r)\}$. If f(z) satisfies (7), then $D_{com}(f) = 0$. Hence if f(z) is of finite lower order, then (10) yields $\Delta(f) = 0$. In the case where f(z) is of infinite lower order, we use the approximate degree u_r $(r>0)$. Let

$$\Omega(r,f) = \int_0^r \frac{\omega(x,f)}{x}\, dx \text{ (Hayman's density function)},$$

$$\omega(x,f) = (\text{ the number of elements of } S(f) - \{0\} \text{ less than } x).$$

We fix a small positive number δ to be determined later. For $r>0$, the approximate degree u_r is defined by $\Omega(u_r,f) = \delta \log M(r,f)$. Then Wiman-Valiron's method shows that

$$\log M(R_r,f) \le 2 \log M(r,f) \text{ 1.f.},$$

where $R_r = r\{1 + \Omega(u_r,f)/u_r\}$. We shall show that

$$\log M(r,f) \le 2\, T(r,f) \text{ 1.f.} \tag{13}$$

Once (13) is known, we have

$$T(R_r,f) \le \log\ M(R_r,f) \le 2 \log M(r,f) \le 4\, T(r,f) \text{ 1.f.}$$

Inequality "$T(R_r,f) \le 4\, T(r,f)$ 1.f." plays the same role as Borel's inequality [88, p. 245] in Nevanlinna's second main theorem [88, p. 240]. Using [Method 4] with (r_μ,R_μ) replaced by (r,R_r), we obtain $\Delta(f) = 0$ (cf. [86]). The proof of (13) is as follows. We express f(z) in the form (3). Wiman-Valiron's method shows that

$$\Sigma^*|a_k|\, r^{n_k} = o(1) \text{ 1.f.}, \tag{14}$$

where Σ^* is the summation over all k satisfying $n_k > u_r$. Let $f^*(z) = \{\Sigma_{k=0}^{\infty} - \Sigma^*\}a_k z^{n_k}$. Then the boundedness of the Hilbert transform yields that

$$T(r,f^*) \ge \log^+\max\, \{|a_k|r^{n_k};\ n_k \le u_r\} - \text{Const}\ \Omega(u_r,f).$$

Hence (14) and the definition of u_r give that

$$T(r,f) \ge T(r,f^*) - o(1)$$

$$\ge \log \mu(r,f) - \text{Const}\ \delta \log M(r,f) - o(1) \text{ 1.f.}$$

Since $\log \mu(r,f) \ge (1 - o(1)) \log M(r,f)$ 1.f. (cf. Wiman [121]), we have

$$T(r,f) \ge \{1 - o(1) - \text{Const}\ \delta\}\log M(r,f) \text{ 1.f.}$$

Choosing δ so that Const $\delta \leq 1/3$, we obtain (13) (cf. M. Marden, The geometry of zeros, Math. Survey III, Amer. Math. Soc., New York, 1949).

4. Analytic functions in the unit disk (Approximation theory)

4.1 Paley's theorem

From now, we discuss only gap series of form (3) with radius of convergence 1. Hence gap series which we deal with are analytic function in $\mathbb{D} = \{ |z| < 1\}$. In his letter to Zygmund, Paley noted that if an Hadamard gap series f(z) (i.e., a gap series with (4)) is unbounded and satisfies $\lim_{k\to\infty} a_k = 0$, then, for any $a\epsilon\mathbb{C}$, $\{t \epsilon [0,2\pi); \lim_{r\to 1} f(re^{it}) = a\}$ is dense in $[0,2\pi)$. (Due to an accident that happened to him, Paley was unable to publish the proof.) M. Weiss [117] gives the proof of this theorem. Salem-Zygmund [102] studies absolutely convergent Hadamard gap series. Kahane-G. Weiss-M. Weiss [58] improves their result as follows. For q>1, there exists ϵ>0 depending only on q with the following property: If an Hadamard gap series f(z) satisfies $n_{k+1}/n_k \geq q$, $|a_k| \leq \epsilon \Sigma_{j=k+1}^{\infty} |a_j| < \infty$ $(k\geq 1)$, then "$f(e^{it})$, $0\leq t<2\pi$" is a Peano curve, i.e., the interior of $\{f(e^{it}); 0\leq t<2\pi\}$ is not empty. Gnuschke-Hauschild-Pommerenke [40] studies Hadamard gap series from the point of view of Block functions, and gives a new proof of Paley's theorem.

[Method 6] M. Weiss [117] formulates Paley's idea as follows. For $q > 1$, there exists two constants A, B with the following property: For a polynomial $Q(t) = \Sigma_{j=1}^{k} b_j e^{im_j t}$ satisfying $m_{j+1}/m_j \geq q$ $(1 \leq j \leq k)$, for a straight line $\mathcal{L}$ whose distance from the origin is less than $A \Sigma_{j=1}^{k} | b_j |$ and an interval $I \subset [0, 2\pi)$ of length larger than B/m_1, there exists $\xi \epsilon I$ such that $Q(\xi)$ is contained in the half plane bounded by $\mathcal{L}$ which does not contain the origin. To prove Paley's theorem, we express f(z) as a summation of a suitable sequence $(h_k(z))_{k=1}^{\infty}$ of polynomials. Let $a \in \mathbb{C}$. Using the above property, we inductively choose $(t_k)_{k=1}^{\infty}$ so that $t_{k+1} \in [t_k - (B/n'_{k+1}), t_k + (B/n'_{k+1})]$ $(n'_{k+1} = \min\{n; n \in S(h_{k+1})\}, k \geq 1)$ and $\lim_{k\to\infty} \Sigma_{j=1}^{k} h_j(e^{in_jt_j}) = a$. Then $\lim_{r\to 1} f(re^{it_\infty}) = a$ $(t_\infty = \lim_{k\to\infty} t_k)$ (This method is also applicable to study the interpolation problem, cf. Méla [75]).

4.2 Asymptotic values

There remains to investigate the case where

$$\limsup_{k\to\infty} |\, a_k \,| > 0. \tag{15}$$

Binmore [14] shows that an Hadamard gap series satisfying (15) has no finite asymptotic value. His method is based on Ingham's lemma [50] and has many applications (cf. Gnuschke-Pommerenke [38], Metzger [76], Nicholls-Sons [89]). An analytic function f(z) in $\mathbb{D}$ is called annular if there exists a sequence $(\gamma_n)_{n=1}^{\infty}$ of Jordan curves in $\mathbb{D}$ such that the interior of γ_n contains the origin $(n \geq 1)$, $\lim_{n\to\infty} \min\{|\, z \,|; z \in \gamma_n\} = 1$ and $\lim_{n\to\infty} \min\{|\, f(z) \,|; z \in \gamma_n\} = \infty$. Using Binmore's method, Sons shows that an Hadamard gap series f(z) is annular if and only if $\limsup_{k\to\infty} |\, a_k \,| = \infty$. We say that an analytic function f(z) in $\mathbb{D}$ belongs to MacLane's class $\mathcal{A}$, if $\{t \in [0, 2\pi);$ $\lim_{|z|\to 1,\, z\in\gamma} |\, f(z) \,| = \infty$ for some curve $\gamma \subset \mathbb{D}$ with endpoint $e^{it}\}$ is dense in $[0, 2\pi)$. MacLane [72] asked whether Hadamard gap series with (15) belong to $\mathcal{A}$ (cf. Anderson [1], Binmore-Hornblower [15]). This problem was solved in the affirmative in ([85]). Gnuschke-Pommerenke [39] and Pommerenke [99] study the order of growth of Hadamard gap series along curves. The radial behaviour of gap series is closely related to probability theory (cf. [43], [80]). For an analytic function f(z) in $\mathbb{D}$ and $0 \leq t < 2\pi$, C(t,f) denotes the totality of cluster points of $\{f(re^{it})\}_{0<r<1}$ in $\hat{\mathbb{C}} = \mathbb{C} \cup \{\infty\}$. Anderson [5] shows that if an Hadamard gap series f(z) is unbounded and satisfies $\limsup_{k\to\infty}|a_k| < \infty$, then $C(t,f) = \hat{\mathbb{C}}$ a.e. He [1] also studies the radial behaviour of functions with small gaps .

[Method 7] Here are two fundamental methods to study gap series: the high degree derivative method (Fuchs [31]) and the peak function method (Binmore [14]). Since these two methods are mutually very close, we show only Fuchs' method. To do this, we prove that an Hadamard gap series f(z) with (15) belong to $\mathcal{A}$. If $\limsup_{k\to\infty}|a_k| < \infty$, then Paley's method yields $f \in \mathcal{A}$. Hence we may assume that $\limsup_{k\to\infty}|a_k| = \infty$. For the sake of simplicity, we work only with $a_k = k$, $n_k = 2^k$ $(k\geq 1)$. It is sufficient to construct a curve $\gamma\subset\mathbb{D}$ with endpoint 1 so that $\lim_{|z|\to 1, z\in\gamma} |f(z)| = \infty$. There exist a positive integer L and a constant η with the following property: To any $1/2<r<1$, there corresponds a positive integer $\ell=\ell_r$ less than L so that $|f^{(\ell)}(r)| \geq \eta n_d^{\ell} a_d r^{n'}$, where $d=d_r = [(\log \frac{1}{1-r}) / \log 2]$, $n' = n_d(n_d-1) \cdots (n_d-\ell+1)$. This fact shows that $f^{(\ell)}(z)$ and the ℓ-th derivative of $a_d z^{n_d}$ are comparable at $z=r$. We see that there exists ϵ_r, $0 < \epsilon_r < (1-r)/2$ such that $|f(z)| \geq \eta' a_d |z|^{n_d}$ for any z, $|z-r|=\epsilon_r$, where η' is a constant independent of r. Since $|z|^{n_d} \geq \text{Const} > 0$ $(|z-r|=\epsilon_r)$, we have $|f(z)| \geq \text{Const}\ \eta' a_d$ $(|z-r|=\epsilon_r)$. Now let $\gamma =$ (the boundary of $\cup_{0<r<1}\{z;\ |z-r| = \epsilon_r\}) \cup \{\text{Im } z > 0\}$. Then γ satisfies the required conditions.

5. Analytic functions in the unit disk (The value-distribution theory)

5.1 A proof of Paley's conjecture

Taking account of his theorem in 4.1, Paley conjectured that an unbounded Hadamard gap series has no finite Picard exceptional value. G. Weiss-M. Weiss [120] showed that the conjecture holds under the condition "$n_{k+1}/n_k \geq q_0$ (= about 100)

$(k \geq 1)$". Fuchs [31] showed that the conjecture holds under (15) (cf. Chang [20]). It has been shown in [79] that Paley's conjecture is valid ([79]). The method given in [79] generalizes Kahane-G. Weiss-M. Weiss' theorem in 4.1 also. Later on, a new proof was given by Gnuschke-Pommerenke [37] .

[Method 8] We now give the proof of Paley's conjecture. In the proof, Cartan's lemma [19] plays an essential role: For $\alpha>0$ and a polynomial $Q(\zeta) = \Sigma_{k=1}^{n} c_k \zeta^k$, a set $\{\zeta \in \mathbb{C};\ |Q(\zeta)| > |c_n|\alpha^n\}$ is covered by disks so that the summation of radii is less than or equal to $2e\alpha$. This lemma and the high degree derivative method yield that

(P_1) (Fuchs [31]) Let ℓ be a positive integer and let $F(\zeta)$ be an analytic function in $\mathbb{D}(w,r) = \{\zeta;\ |\zeta-w| < r\}$ such that $|F^{(\ell)}(w)|\ (= y_1) > 0$ and $\sup_{\zeta \in \mathbb{D}(w,r)} |F^{(\ell)}(\zeta)|$ $(= y_2) < \infty$.Then

$$F(\mathbb{D}(w,r)) \supset \mathbb{D}(F(w), \eta_\ell r^\ell y_1^{\ell+1} y_2^{-\ell}),$$

where η_ℓ is a constant depending only on ℓ.

By Fuchs' result, we may assume that $\lim_{k\to\infty} a_k = 0$. Put $F(\zeta) = f(e^\zeta)$, $W = \{\mathrm{Re}\ \zeta < 0\}$ and $W^* = W \cap \{0 \leq \mathrm{Im}\ \zeta < 2\pi\}$. Elementary calculation yields that

(P_2) There exists an increasing sequence $(k_\nu)_{\nu=1}^{\infty}$ of positive integers satisfying $\Sigma_{\nu=1}^{\infty} |a_{k_\nu}| = \infty$ and the following property: To any $w \in \mathbb{C}$ with $\mathrm{Re}\ w = -1/n_{k_\nu}$, there corresponds an integer $L = L_{F,w}$, $\lambda_q \leq L \leq \lambda_q'$ so that

$$|F^{(L)}(w)| \geq C_q n_{k_\nu}^L |a_{k_\nu}|, \qquad \sup_{\zeta \in \mathbb{D}(w,(1-q^{-1})/n_{k_\nu})} |F^{(L)}(\zeta)| \leq C_q' n_{k_\nu}^L |a_{k_\nu}|,$$

where λ_q, λ_q', C_q, C_q' are constants depending only on $q = \inf_{k\geq 1} n_{k+1}/n_k$.

Thus (P_1) shows that:

(P_3) For any $w \in \mathbb{C}$ with $\mathrm{Re}\, w = -1/n_{k_\nu}$, there exists a constant η'_q depending only on q such that

$$F(\mathbb{D}(w,(1-q^{-1})\,/n_{k_\nu})) \supset \mathbb{D}(F(w),\eta'_q|a_{k_\nu}|).$$

We now show that $F(\zeta)$ takes 0 infinitely often in W^*. Put $r_\nu = -1/n_{k_\nu}$, $W_\nu = \{\mathrm{Re}\,\zeta < r_\nu\}$, $W^*_\nu = W_\nu \cap \{0 \le \mathrm{Im}\,\zeta < 2\pi\}$ $(\nu \ge 1)$. Given $\nu' \ge 1$, we assume that $F(\zeta)$ does not take 0 in $W^* - W^*_{\nu'}$. Let $\delta = \min\{|F(\zeta)|;\ \mathrm{Re}\,\zeta = r_{\nu'}\}$, $X_\nu = \overline{W}_\nu - W_{\nu'}$, $\delta_\nu = \min\{|F(\zeta)|;\ \zeta \in X_\nu\}$ $(\nu > \nu')$. Our assumption gives that $\delta > 0$, $\delta_\nu > 0$ $(\nu > \nu')$. Paley's theorem shows that $\lim_{\nu\to\infty} \delta_\nu = 0$. Hence there exists ν'', $\nu'' > \nu'$ such that $\delta_\nu < \delta$ for any $\nu \ge \nu''$. Choose $(w_\nu)_{\nu=\nu''}^{\infty}$ so that $w_\nu \in X_\nu$, $|F(w_\nu)| = \delta_\nu$. Then the minimum modulus principle shows that $\mathrm{Re}\, w_\nu = r_\nu$. Property (P_3) gives that $F(\mathbb{D}(w_\nu,$ $(1-q^{-1})\,/n_{k_\nu})) \supset \mathbb{D}(F(w_\nu),\eta'_q|a_{k_\nu}|)$. Since $r_\nu + (1-q^{-1})\,/n_{k_\nu} \le r_{\nu+1}$, we have $\delta_{\nu+1}$ $\le \min\{|F(\zeta)|;\ \zeta \in \mathbb{D}(w_\nu, (1-q^{-1})/n_{k_\nu})\} \le \delta_\nu - \eta'_q|a_{k_\nu}|$, i.e., $\delta_\nu - \delta_{\nu+1} \ge \eta'_q|a_{k_\nu}|$. Thus $\delta_{\nu''} = \Sigma_{\nu=\nu''}^{\infty}(\delta_\nu - \delta_{\nu+1}) \ge \eta'_q\, \Sigma_{\nu=\nu''}^{\infty}|a_{k_\nu}| = \infty$. This is a contradiction. Since ν' is arbitrary, $F(\zeta)$ takes 0 infinitely often in W^*, i.e., f(z) takes 0 infinitely often in $\mathbb{D}$.

5.2 δ-thin sets, weak δ-thin sets

Sunyer i Balaguer [112] and Sons [108] investigate Borel exceptional values of gap series. We here note deficient values of gap series. A set E of non-negative integers is called δ–thin, if $\Delta(f) = 0$ for any analytic function f(z) of unbounded type (i.e., $\lim_{r\to1} T(r,f) = \infty$) with $S(f) \subset E$. A gap series

$$1 + 2\sum_{k=1}^{\infty} z^{k^2} = \prod_{n=1}^{\infty} \{(1+z^{2n-1})^2(1-z^{2n})\}\ \left(=\theta_3\left(0, \tfrac{1}{i\pi}\log z\right)\right)$$

plays an important role in the value-distribution theory and does not take 0 in $\mathbb{D}$. Hence it is interesting to study analytic functions with (15). A set E is called weak δ–thin, if

$\Delta(f) = 0$ for any analytic function f(z) of unbounded type with (15) and $S(f) \subset E$. It is known that a finite union of Hadamard gap series is weak δ-thin ([78]). If $(n_k)_{k=1}^{\infty}$ satisfies $\inf_{k\geq 1}(\log n_{k+1})/\log n_k > 1$, then $(n_k)_{k=1}^{\infty}$ is δ-thin.

[Method 9] We show that an Hadamard gap series is weak δ-thin. Our method is a variation of the high degree derivative method; we use the Littlewood-Offord derivatives instead of the standard derivative. Let f(z) be an Hadamard gap series satisfying (15). It is sufficient to show that $\liminf_{r\to 1} m(r,1/f) < \infty$ for all $a\in\mathbb{C}$. We only show $\liminf_{r\to 1} m(r,1/(f-a)) < \infty$. Let $\sigma_r = \limsup_{k\to\infty} |a_k| r^{n_k}$ $(1/2<r<1)$. Then $\inf_{1/2<r<1} \sigma_r > 0$. For $1/2<r<1$, we put $\xi'_r = 1 - (2/n_k)$, $\xi''_r = 1 - (1/n_k)$, where $k=k_r$ is the largest integer of j such that $|a_j| r^{n_j} \geq \sigma_r/2$. Since $\lim_{r\to 1} r\, \xi'_r = 1$, it is sufficient to show that

$$\sup_{1/2<r<1} \frac{1}{\xi''_r - \xi'_r} \int_{\xi'_r}^{\xi''_r} m(r\xi,\, 1/f)\, d\xi < \infty \qquad (16)$$

The m-th Littlewood-Offord derivative D_m is defined by $D_m g(z) = \{\frac{d}{dz}(g(z)z^{-m})\} z^{m+1}$.

Fixing $1/2<r<1$, we compute

$$D^*_\nu f(z) = \{\prod_{1\leq|\mu|\leq\nu} D_{n_{k-\mu}}\} f(z) \quad (k=k_r,\ \nu\geq 1).$$

Then $n_k \in S(D^*_\nu f)$, $S(D^*_\nu f) \cap \{n_\mu;\ 1\leq|\mu-k|\leq\nu\}\ \emptyset$ and

$$D^*_\nu f(z) = \{\prod_{1\leq|\mu|\leq\nu} (n_k - n_{k-\mu})\} a_k z^{n_k} + \text{(the other terms)}.$$

We can choose ν_q depending only on $q = \inf_{k\geq 1} n_{k+1}/n_k$ so that, for any z with $|z|=r$,

$$|D^*_{\nu_q} f(z)| \geq \{\prod_{1\leq|\mu|\leq\nu_q} |n_k - n_{k-\mu}|\}(\sigma_r/4).$$

Then we have the following property: For any z with $|z|=r$, there exist $\{\xi_j\}_{j=1}^{2\nu_q+2}$, $\xi'_r = \xi_1 < \cdots < \xi_{2\nu_q+2} = \xi''_r$ such that

$$|f(z\xi)| \geq \kappa_q \sigma_r \Phi(\xi) \quad (\xi'_r \leq \xi \leq \xi''_r),$$

where $\Phi(\xi) = \min\{(\xi_{\mu+1}-\xi)^{2\nu_q}, (\xi-\xi_\mu)^{2\nu_q}\}$ $(\xi_\mu \leq \xi \leq \xi_{\mu+1},\ 1 \leq \mu \leq 2\nu_q+1)$ and

κ_q is a constant depending only on q ([78]). Since

$$\frac{1}{\xi_r''-\xi_r'} \int_{\xi_r'}^{\xi_r''} \log^+ \Phi(\xi)\, d\xi \leq \text{Const}\ \nu_q^2,$$

we obtain (16).

For $0<\rho<1$ and an analytic function f(z) in $\mathbb{D}$, the Borel transform of f(z) is defined by $Bf(w) = \int_1^{\infty} f(\rho e^{-x})\, e^{-wx}\, dx$. The function Bf(w) is meromorphic in $\mathbb{C}$ and T(r,Bf) $= \Omega(r,f) + O(1)$. Hence it seems important to generalize gap series from the point of view of the Borel transform.

6. Problems

In the study of gap series with radius of convergence ∞, it is interesting to improve Müntz-Szász$'$ approximation theorem and Pólya's gap theorems [96]. Hence it seems important to study relations with the completeness problems and quasi-analytic functions. From the point of view of the Nevanlinna theory, it seems interesting;

(A) to study $\Delta(\cdot)$ of gap series without a condition on $\rho_*(\cdot)$,

(B) to improve Pfluger-Pólya's result [93] without a condition on $\rho_*(\cdot)$,

(C) to establish $\cos(\pi\rho D)$ theorems,

(D) to compare $M(r,\cdot)$, $T(r,\cdot)$, $L(r,\cdot)$.

(E) to study relations between gaps and covering surfaces.

In the study of gap series with radius of convergence 1, it seems interesting;

(F) to study gap series from the point of view of probability theory.

(G) to generalize gap series from the point of view of the Borel transform.

Here are some open problems.

(a) Does an entire function with Fabry's gaps have a finite Picard value? (Pólya's conjecture [96])

(b) A set E of non-negative integers is called δ-thin with respect to entire functions, if $\Delta(f) = 0$ for any entire function f(z) with $S(f) \subset E$. Give a good sufficient condition for δ-thinness. Is "$D_{com}(\cdot) = 0$" sufficient?

(c) Does "$\Delta(f) \leq \kappa\{\rho_*(f)D_{com}(f)\}$" hold? Are the analogous inequalities valid for D_0, D_{log}? Here

$$\kappa(x) = \begin{cases} 0 & (0 \leq x < 1/2) \\ 1 - \sin(\pi x) & (1/2 \leq x < 1) \\ 1 & (x \geq 1). \end{cases}$$

(d) For an entire function f(z) such that $\rho_*(f)D_{com}(f)$ is not an integer, estimate $\limsup_{r\to\infty} N(r,0,f)/T(r,f)$ from below.

(e) Give a good sufficient condition for δ-thinness with respect to analytic functions in $\mathbb{D}$. Are Hadamard gap series δ-thin?

(f) Give a good sufficient condition for weak δ-thinness with respect to analytic functions in $\mathbb{D}$.

(g) Let f(z) be an unbounded analytic function in $\mathbb{D}$ such that S(f) is a Sidon set. Does f(z) have a finite Picard exceptional value?

(h) Let f(z) be an analytic function in $\mathbb{D}$ such that $\lim_{r\to 1} \Omega(\frac{1}{1-r},f)/T(r,f) = 0$. Does "$\Delta(f) = 0$" hold?

(i) Is it true that, for an unbounded analytic function f(z) in $\mathbb{D}$ with Hadamard gaps, either "$C(t,f) = \hat{\mathbb{C}}$ a.e." or "$C(t,f) = \{\infty\}$ a.e."?

(j) Is it true that, for an Hadamard gap series $(n_k)_{k=1}^{\infty}$,

$$\{a \in \mathbb{R};\ \liminf_{k\to\infty} |\sum_{j=1}^{k} \cos n_j t - a| = 0\} = \mathbb{R} \text{ a.e.?}$$

(k) Is it true that, for an Hadamard gap series $(n_k)_{k=1}^{\infty}$,

$$\{a \in \mathbb{C};\ \liminf_{k\to\infty} |\sum_{j=1}^{k} e^{in_j t} - a| = 0\} = \mathbb{C} \text{ a.e.?}$$

References

This is not a complete list but a sampling.

[1] Anderson, J. M., Boundary properties of analytic functions with gap power series, Quart. J. Math. 21 (1970), pp. 247 - 256.

[2] Anderson, J. M., The growth of entire functions of finite lower order, with density conditions, J. London Math. Soc. 4 (1971), pp. 118 - 122.

[3] Anderson, J. M., Bounded analytic functions with Hadamard gaps, Mathematika 23 (1976), pp. 142 - 146.

[4] Anderson, J. M., Müntz-Szász theorems and lacunary entire functions, Linear spaces and approximation (Proc. Conf. Math. Res. Inst., Oberwolfach, 1977), pp. 491 - 501, Birkhäuser, Basel, 1978.

[5] Anderson, J. M., Le comportent radial des séries lacunaires, C. R. Acad. Sci. Paris 291 (1980), pp. 83 - 85.

[6] Anderson, J. M., K. F. Barth, and D. A. Brannan, Research problems in complex analysis, Bull. London Math. Soc. 9 (1977), pp.129 - 162.

[7] Anderson, J. M., and K. G. Binmore, Coefficient estimates for lacunary power series and Dirichlet series I, Proc. London Math. Soc. 18 (1968), pp.36 - 48.

[8] Bari, N. K., A treatise on trigonometric series, Pergamon, Oxford, 1964.

[9] Barth, K. F., D. A. Brannan, and W. K. Hayman, Research problems in complex analysis, Bull. London Math. Soc. 16 (1984), pp.490 - 517.

[10] Beurling, A., and P. Malliavin, On the closure of characters and the zeros of entire functions, Acta Math. 118 (1967), pp. 79 - 93.

[11] Biernacki, M., Sur les équations algébriques contenant des paramètres arbitraires, Bull. Int. Acad. Polon. Sci. Lett. 3 (1927), pp. 542 - 685.

[12] Biernacki, M., Sur les fonctions entières à séries lacunaires, C. R. Acad. Sci. Paris 187 (1928), pp. 477 - 479.

[13] Bingham, N. H., Variants of the law of the iterated logarithm, Bull. London Math. Soc. 18 (1986), pp. 433 - 467.

[14] Binmore, K. G., Analytic functions with Hadamard gaps, Bull. London Math. Soc. 1 (1969), pp. 211 - 217.

[15] Binmore, K. G., and R. Hornblower, Boundary behaviour of functions with

Hadamard gaps, Nagoya Math. J. 48 (1972), pp. 173 - 181.

[16] Brannan, D. A., and J. G. Clunie, Aspects of contemporary complex analysis (Proc. NATO Adv. Study Inst., Univ. Durham, Durham, 1979), Academic Press, London, 1980.

[17] Carleman, T., Les fonctions quasi analytiques, Gauthier-Villars, Paris, 1926.

[18] Carlson, F., Sur une classe des séries de Taylor, Thesis, Uppsala Univ., 1914.

[19] Cartan, H., Sur les systèmes de fonctions holomorphes à variétés linéars lacunaires et leurs applications, Ann. Sci. École Norm. Sup. 45 (1928), pp. 255 - 346.

[20] Chang, I. L., On the zeros of power series with Hadamard gaps-distribution in sectors, Trans. Amer. Math. Soc. 178 (1973), pp. 393 - 400.

[21] Drury, S. W., Sur les ensembles de Sidon, C. R. Acad. Sci. Paris 271 (1970), pp. 162 - 163.

[22] Edrei, A., Gap and density theorems for entire functions, Scripta Math. 23 (1957), pp. 1 - 25.

[23] Edrei, A., and W. H. J. Fuchs, Valeurs déficientes et valeurs asymptotiques des fonctions méromorphes, Comment. Math. Helv. 33 (1959), pp. 258 - 295.

[24] Erdös, P., On trigonometric sums with gaps, Magyar Tud. Akad. Mat. Kutato Int. Közl. 7 (1962), pp. 37 - 42.

[25] Erdös, P., and I. S. Gál, On the law of the iterated logarithm, Nederl. Akad. Wetensch. Proc. 58 (1955), pp. 65 - 84.

[26] Erdös, P., P. Macintyre, and A. J. Macintyre, Integral functions with gap power series,Proc. Edinburgh Math. Soc. 2 (1954), pp. 62 - 70.

[27] Essén, M. R., The cos $\pi\lambda$ theorem, Lecture Notes in Math. 467, Springer-Verlag, Berlin, 1975.

[28] Fabry, E., Sur les séries de Taylor qui ont une infinité de points singuliers, Acta Math. 22 (1898-1899), pp. 65 - 87.

[29] Fejér, L., Über die Wurzel vom kleinsten absoluten Betrage einer algebraischen Gleichung, Math. Ann. 65 (1908), pp. 413 - 423.

[30] Fuchs, W. H. J., Proof of a conjecture of G. Pólya concerning gap series, Illinois J. Math. 7 (1963), pp. 661 - 667.

[31] Fuchs, W. H. J., On the zeros of power series with Hadamard gaps, Nagoya Math. J. 29 (1967),pp. 167-174.

[32] Fuchs, W. H. J., Nevanlinna theory and gap series, Symposia on Theoretical Physics and Mathematics 9 (Inst. Math. Sci., Madras, 1968), pp. 177 - 181, Plenum Press, New York, 1969.

[33] Fuchs, W. H. J., The development of the theory of deficient values since Nevanlinna, Ann. Acad. Sci. Fenn. 7 (1982), pp. 33 - 48.

[34] Gapoškin, V. F., Lacunary series and independent functions, (Russian) Uspehi Mat. Nauk 21 (1966), no. 6 (132), pp. 3 - 82.

[35] Gerver, J., The differentiability of the Riemann function at certain rational multiples of π, Proc. Nat. Acad. Sci. USA 62 (1969), pp. 668 - 670.

[36] Gethner, R. M., Zeros of the successive derivatives of Hadamard gap series in the unit disk, preprint.

[37] Gnuschke, D., and Ch. Pommerenke, On the absolute convergence of power series with Hadamard gaps, Bull. London Math. Soc. 15 (1983), pp. 507 - 512.

[38] Gnuschke, D., and Ch. Pommerenke, On annular functions with Hadamard gaps, Complex Variables 3 (1984), pp. 125 - 134.

[39] Gnuschke, D., and Ch. Pommerenke, On the growth of functions with Hadamard gaps, J. London Math. Soc. 30 (1984), pp. 441 - 450.

[40] Gnuschke-Hauschild, D., and Ch. Pommerenke, On Block functions and gap series, J. Riene Angew. Math. 367 (1986), pp. 172 - 186.

[41] Hadamard, J., Essai sur l'étude des fonctions données par leur développement de Taylor, J. Math. 8 (1892), pp. 101 - 186.

[42] Hardy, H., Weierstrass's non-differentiable function, Trans. Amer. Math. Soc. 17 (1916), pp. 322 - 323.

[43] Hawkes, J., Probabilistic behaviour of some lacunary series, Z. Wahrscheinlich keitstheorie verw. Geb. 53 (1980), pp. 21 - 33.

[44] Hayman, W. K., Research problems in function theory, Athlone Press, London, 1967.

[45] Hayman, W. K., Angular value distribution of power series with gaps, Proc. London Math. Soc. 24 (1972), pp. 590 - 624.

[46] Hayman, W. K., Research problems in function theory: New problems, Proceedings of the Symposium on Complex Analysis (Univ. Kent, Canterbury, 1973), pp. 155 -180, Cambridge Univ. Press, London, 1974.

[47] Hayman, W. K., The local growth of power series: A survey of the Wiman-Valiron method, Canad. Math. Bull. 17 (1974), pp. 317 - 358.

[48] Hayman, W. K., and J. F. Rossi, Characteristic, maximum modulus and value distribution, Trans. Amer. Math. Soc. 284 (1984), pp. 651 - 664.

[49] Hewitt, E., and K. A. Ross, Abstract harmonic analysis II, Springer-Verlag, Berlin, 1970.

[50] Ingham, A. E., On the high indices theorem of Hardy and Littlewood, Quart. J. Math. 8 (1937), pp. 1- 7.

[51] Kahane, J. P., Sur les fonctions moyenne-périodiques bornées, Ann. Inst. Fourier 7 (1957), pp. 293 - 314.

[52] Kahane, J. P., Travaux de Beurling et Malliavin, Séminaire Bourbaki, Exposé 225, Vol. 1961/1962, Benjamin, New York, 1966.

[53] Kahane, J. P., Pseudo-périodicité et séries de Fourier lacunaires, Ann. Sci. École Norm. Sup. 79 (1962), pp. 93 - 150.

[54] Kahane, J. P., Lacunary Taylor and Fourier series, Bull. Amer. Math. Soc. 70 (1964), pp. 199 - 213.

[55] Kahane, J. P., Some random series of functions Second edition, Cambridge Univ. Press, Cambridge, 1985.

[56] Kahane, J. P., Géza Freud and lacunary Fourier series, J. Approx. Theory 46 (1986), pp. 51 - 57.

[57] Kahane, J. P., and R. Salem, Ensembles parfaits et séries trigonométriques, Hermann, Paris, 1963.

[58] Kahane, J. P., G. Weiss, and M. Weiss, On lacunary power series, Ark. Mat. 5 (1963), pp. 1 - 26.

[59] Kawata, T., Gap theorems in Fourier analysis, (Japanese) Seminar on Mathematical Sciences 1, Keio Univ., Dept. of Math., Yokohama, 1980.

[60] Kövari, T., On the Borel exceptional values of lacunary integral functions, J. Analyse Math. 9 (1961/62), pp. 71 - 109.

[61] Kövari, T., A gap-theorem for entire functions of infinite order, Michigan Math. J. 12 (1965), pp. 133 - 140.

[62] Kövari, T., On the asymptotic path of entire functions with gap power series, J. Analyse Math. 15 (1965), pp. 281 - 286.

[63] Kövari, T., Asymptotic values of entire functions of finite lower order with density conditions, Acta Sci. Math. Szeged 26 (1965), pp. 233 - 237.

[64] Kövari, T., On the growth of entire functions of finite order with density conditions, Quart. J. Math. 17 (1966), pp. 22 - 30.

[65] Korevaar, J., Approximation on curves by linear combinations of exponentials, Approximation theory (Proc. Internat. Sympos., Univ. Texas, Austin, 1973), pp.387 - 393, Academic Press, New York, 1973.

[66] Korevaar, J., Zero distribution of entire functions and spanning radius for a set of complex exponentials, Aspects of contemporary complex analysis (Proc. NATO Adv. Study Inst., Univ. Durham, Durham, 1979), pp. 293 - 312, Academic Press, London, 1980.

[67] Korevaar, J. and M. Dixon, Lacunary polynomial approximation, Linear spaces and approximation (Proc. Conf., Math. Res. Inst., Oberwolfach, 1977), pp. 479 - 489, Birkhäuser, Basel, 1978.

[68] Levinson, N., Gap and density theorems, Amer. Math. Soc. Colloq. Publ. 26, 1940.

[69] López, J. M., and K. A. Ross, Sidon sets, Marcel Dekker, New York, 1975.

[70] Lewis, J. L., and J. M. Wu, On conjectures of Arakelyan and Littlewood: To Walter Hayman, Preprint.

[71] Macintyre, A. J., Asymptotic paths of integral functions with gap power series, Proc. London Math. Soc. 2 (1952), pp. 268 - 296.

[72] Maclane, G. R., Asymptotic values of holomorphic functions, Rice Univ. Studies 49, 1963.

[73] Makarov, N., On the distortion of boundary sets under conformal mappings, Proc. London Math. Soc. 51 (1985), pp. 369 - 384.

[74] Mandelbrojt, S., Séries lacunaires, Hermann, Paris, 1936.

[75] Méla, J. F., Approximation diophantienne et ensembles lacunaires, Bull. Soc. Math. France 19 (1969), pp. 26 - 54.

[76] Metzger, T. A., Gaps in the Fourier series of automorphic forms II, Analytic

number theory (Proc. Con., Temple Univ., Philadelphia, 1980), pp. 382 - 395, Lecture Notes in Math. 899, Springer-Verlag, Berlin, 1981.

[77] Müntz, Ch., Über den Approximationssatz von Weierstraß, Math. Abhandlungen H. A. Schwarz gewidmet, pp. 303 - 312, Berlin, 1914.

[78] Murai, T., Sur la distribution des valeurs des séries lacunaires, J. London Math. Soc. 21 (1980), pp. 93 -110.

[79] Murai, T., The value-distribution of lacunary series and a conjecture of Paley, Ann. Inst. Fourier 31 (1981), pp. 135 - 156.

[80] Murai, T., On lacunary series, Nagoya Math. J. 85 (1982), pp. 87 - 154.

[81] Murai, T., The central limit theorem for trigonometric series, Nagoya Math. J. 87 (1982), pp. 79 - 94.

[82] Murai, T., Lacunary series and ranges of non-synthesis, Tôhoku Math. J. 34 (1982), pp. 289 - 294.

[83] Murai, T., The deficiency of gap series, Analysis 2 (1982), pp. 347 - 361.

[84] Murai, T., The deficiency of analytic functions in the unit disk, Bull. Inst. Math. Acad. Sinica 10 (1982), pp. 133 - 148.

[85] T. Murai, The boundary behaviour of Hadamard lacunary series, Nagoya Math. J. 89 (1983), pp. 65 - 76.

[86] Murai, T., The deficiency of entire functions with Fejèr gaps, Ann. Inst. Fourier 33 (1983), pp. 39 - 58.

[87] Murai, T., Gap series, (Japanese) Sūgaku 35 (1983), pp. 35 - 49.

[88] Nevanlinna, R., Analytic functions, Springer-Verlag, Berlin, 1970.

[89] Nicholls, P. J., and L. R. Sons, Automorphic functions with gap power series, Illinois J. Math. 25 (1981), pp. 383 - 389.

[90] Paley, R. E. A. C., On lacunary power series, Proc. Nat. Acad. Sci. USA 19 (1933), pp. 271 - 272.

[91] Paley, R. E. A. C., and R. Wiener, Fourier transforms in the complex domain, Amer. Math. Soc. Colloq. Publ. 19, 1934.

[92] Pavlov, A. I., Growth along curves of entire functions that are defined by gap power series, (Russian) Sibirsk. Mat. Zh. 13 (1971), pp. 1169 - 1181.

[93] Pfluger, A., and G. Pólya, On the power series of an integral function having an

exceptional value, Proc. Cambridge Philos. Soc. 31 (1935), pp. 153 - 155.

[94] Philipp, W., and W. Stout, Almost sure invariance principles for partial sums of weakly dependent random variables, Mem. Amer. Math. Soc. 161, 1975.

[95] Pisier, G., Ensembles de Sidon et processus gaussiens, C. R. Acad. Sci. Paris 286 (1978), pp. 671 - 674.

[96] Pólya, G., Untersuchungen über Lücken und Singularitäten von Potenzreihen, Math. Z. 29 (1929), pp. 549 - 640.

[97] Pólya, G., On converse gap theorems, Trans. Amer. Math. Soc. 52 (1942), pp. 65 - 71.

[98] Pommerenke, Ch., Lacunary power series and univalent functions, Michigan Math. J. 11 (1964), pp. 219 - 223.

[99] Pommerenke, Ch., On gap series and the Lehto-Virtanen maximum principle, Ann. Acad. Sci. Fenn. 10 (1985), pp. 455 - 460.

[100] Rudin, R., Trigonometric series with gaps, J. Math. Mech. 9 (1960), pp. 203 - 228.

[101] Salem, R., Oeuvres Mathématiques, Hermann, Paris, 1967.

[102] Salem, R., and A. Zygmund, Lacunary series and Peano curves, Duke Math. J. 12 (1945), pp. 569 - 578.

[103] Salem, R., and A. Zygmund, On lacunary trigonometric series I, Proc. Nat. Acad. Sci. USA 33 (1947), pp. 333 - 338.

[104] Salem, R., and A. Zygmund, La loi du logarithm itéré pour les séries trigonométriques lacunaires, Bull. Sci. Math. 74 (1950), pp. 1 - 16.

[105] Schmeisser, G., Zur Theorie der Potenzreihen mit Lücken, Math. Z. 126 (1972), pp. 40 - 46.

[106] Sheremeta, M. N., Growth in a sector of entire functions presented by lacunary series, (Russian) Sib. Math. Zh. 21 (1978), pp. 197 - 208.

[107] Sidon, S., Verallgemeinerung eines Satzes über die absolute Konvergenz von Fourierreihen mit Lücken, Math. Ann. 97 (1927), pp. 675 - 676.

[108] Sons, L. R., Value distribution and power series with moderate gaps, Michigan Math. J. 13 (1966), pp. 425 - 433.

[109] Sons, L. R., An analogue of a theorem of W. H. J. Fuchs on gap power series, Proc. London Math. Soc. 21 (1970), pp. 525 - 539.

[110] Sons, L. R., Zeros of sums of series with Hadamard gaps, Pacific J. Math. 43 (1972), pp. 515 - 524.

[111] Stečkin, S., On absolute convergence of Fourier series III, (Russian) Izv. Akad. Nauk SSSR 20 (1956), pp. 385 - 412.

[112] Sunyer i Balaguer, F., Sur la substitution d'une valeur exceptionnelle par une propriété lacunaires, Acta Math. 87 (1952), pp. 17 - 31.

[113] Szász, O., Über die Approximation stetiger Funcktionen durch lineare Aggregate von Potenzen, Math. Ann. 77 (1916), pp. 482 - 496.

[114] Takahashi, S., Probability limit theorems for trigonometric series, Colloq. Math. Soc. Janos Bolyai 11. (1974), pp. 381 - 397.

[115] Turán, P., Über lakunäre Potenzreihen, Rev. Math. Pures Appl. 1 (1956), pp. 27 - 32.

[116] Weierstraß, K., Über continuirliche Functionen eines reellen Arguments, die für keinen Werth des letzteren einen bestimmten Differentialquotienten besitzen, Küigl. Akad. Wiss. (1872): Math. Werke II, pp. 71 - 74, Johnson, New York, 1967.

[117] Weiss, M., Concerning a theorem of Paley on lacunary trigonometric series, Acta Math. 102 (1959), pp. 225 - 238.

[118] Weiss, M., On the law of the iterated logarithm for lacunary trigonometric series, Trans. Amer. Math. Soc. 91 (1959), pp. 444 - 469.

[119] Weiss, M., A theorem on lacunary trigonometric series, Orthogonal Expansions and their Continuous Analogues (Proc. Conf., Edwardsville, Illinois, 1967), pp. 227 - 230, Southern Illinois Univ. Press, Carbondale, 1968.

[120] Weiss, M. and G. Weiss, On the Picard property of lacunary power series, Studia Math. 22 (1963), pp. 221 - 245.

[121] Wiman, A., Über den Zusammenhang zwischen dem Maximalbetrage einer analytischen Funktion und dem grössten Gliede der zugehörigen Taylorschen Reihe, Acta Math. 37 (1914), pp. 305 - 326.

[122] Zygmund, A., Trigonometric series I, II, Cambridge Univ. Press, Cambridge, 1959.

Department of Mathematics
Faculty of Science,
Nagoya University

MEROMORPHIC SOLUTIONS OF SOME POLYNOMIAL DIFFERENCE EQUATIONS

Yoshikuni Nakamura and Niro Yanagihara

1. Introduction

In this note, we will study the difference equation

$$y(x+1)^m = P(y(x)), \tag{1.1}$$

where m is an integer, $m \geq 2$, and P(y) is a polynomial of degree p.

When $m = 1$, the equation (1.1) has been investigated to some extent, see e.g., [2], [6], [7]. Hence we will consider here the case $m \geq 2$.

In the case of the differential equations,

$$y' = P(y)$$

may admit only elementary functions as solutions, while the equation

$$y'^2 = 4y^3 - g_2 y - g_3$$

gives elliptic functions. Therefore, it would be natural to ask what situations would occur in the case of difference equation (1.1).

We know that the equation (1.1) may admit meromorphic solutions of finite order (in the sense of Nevanlinna) only if $m = p$ [3].

When $m = p$, the equation (1.1) possesses entire solutions (of finite order) if and only if either

$$y(x+1)^m = (ay(x)+b)^m, \tag{1.2}$$

or

$$m = 2 \text{ and } y(x+1)^2 = A^2 - y(x)^2, \tag{1.2'}$$

where a, b, and A are constants, $A \neq 0$ [4].

If we allow solutions to have poles, then situations become much more complicated. We will show in another paper [8] that, even if we allow solutions with poles, we do not obtain any other types of equations, supposed it possesses a solution of finite order. Of course in this case we have m = p. The equations (1.2) and (1.2′) admit such solutions, e.g., (1.2′) possesses solutions

$$y(x) = \frac{A}{2} \{ \kappa (x) e^{(\pi i/2)x} + [\kappa(x) e^{(\pi i/2)x}]^{-1} \},$$

in which $\kappa(x)$ is a periodic meromorphic function with period 1 and of finite order.

We will consider in this note the case $m \neq p$. Then, we know at first that there are no solutions which have poles, i.e., we obtain the following theorem:

Theorem 1. Suppose $m \neq p$ in (1.1). Then any meromorphic solutions are entire.

Solutions must be of infinite order [3].

By Theorem 1, we can restrict ourselves to the case in which the equation (1.1) admits entire solutions.

Theorem 2. Suppose $m > p$ and $p \geq 2$. The equation (1.1) admits entire solutions if and only if

$$y(x+1)^m = Ay(x)^p, \tag{1.3}$$

where A is a constant. The solutions of (1.3) are written as

$$y(x) = A^{\frac{1}{m-p}} \exp[\, \kappa(x)\, (p/m)^x \,], \tag{1.3'}$$

in which $\kappa(x)$ is an entire periodic function with period 1.

Now we consider the case when $2 \leq m < p$. We suppose that P(y) in (1.1) is not of the form $P_1(y)^m$ for any polynomial $P_1(y)$,

Theorem 3. Suppose $m < p$ and $m \geq 3$. Then the equation (1.1) admits entire solutions if and only if

$$y(x+1)^m = Ay(x)^p, \tag{1.4}$$

where $p = mq + a$, $1 \leq a < m$, $q \geq 1$. The solutions of (1.4) are written as

$$y(x) = A^{\frac{1}{m-p}} \exp[\ \kappa(x)\ (p/m)^x\ ], \tag{1.4'}$$

in which $\kappa(x)$ is an entire periodic function with period 1.

We note that (1.3), (1.3′) and (1.4), (1.4′) are apparently of the same form.

Phenomena of completely differents kind appear in the case $m = 2 < p$.

Theorem 4. Suppose $m = 2 < p$.

(a) When p is even and $p = 2q + 2$, $q \geq 1$, then the equation (1.1) admits entire solutions if and only if

$$y(x+1)^2 = Q(y(x))^2\ (A^2 - y(x)^2), \tag{1.5}$$

where A is a constant $\neq 0$, and $Q(y)$ is a polynomial of degree q such that

$$Q(y) = \sum_{k=0}^{q/2} \binom{q+1}{2k} (iy/A)^{2k}\ \{1 - (y/A)^2\}^{\frac{q}{2}-k} \tag{1.5$_e$}$$

if q is even, and

$$Q(y) = (1/i) \sum_{k=1}^{(q+1)/2} \binom{q-1}{2k-1} (iy/A)^{2k-1} \left\{1 - (y/A)^2\right\}^{\frac{q+1}{2}-k} \tag{1.5$_o$}$$

if q is odd.

(b) When p is odd and $p = 2q + 1$, $q \geq 1$, then the equation (1.1) admits entire solutions if and only if either

$$y(x+1)^2 = Ay(x)^p \tag{1.6}$$

or

$$y(x+1)^2 = Q(y(x))^2 (y(x) - c),\ c \neq 0, \tag{1.6'}$$

in which $Q(y) = b_q y^q + \cdots + b_0$ is a polynomial of degree q such that

$$Q(y) = b_q c^q\ Q^*(y/c), \qquad b_q{}^2\ c^{2q-1} = (-2)^{2q-1}, \tag{1.7}$$

where

$$Q^*(x) = 2^{-q}[U_q(x) + U_{q-1}(x)] \qquad (1.7')$$

with Tchebicheff polynomials U_q and U_{q-1} of the second kind [1, p. 183], i.e.,,

$$U_q(\cos\theta) = \sin(q + 1)\, \theta/\sin\theta,$$

i.e.,

$$U_q(x) = \sum_{n=0}^{[q/2]} \frac{(-1)^n (q-n)\,!}{n!\,(q-2n)\,!} (2x)^{q-2n}. \qquad (1.7'')$$

Therefore, if

$$Q^*(x) = \sum_{n=0}^{q} b_k^* \, x^k , \qquad (1.7^*)$$

then

$$b^*_{q-2n} = (-1)^n \frac{(q-n)...(q-2n+1)}{2^{2n}\; n!} , \qquad (1.7^\#)$$

$$b^*_{q-2n-1} = (-1)^n \frac{(q-n-1)\; ...\; (q-2n)}{2^{2n\,+\,1}\; n!}$$

We remark that, taking y(x) / c as y(x), the equation $(1.6')$ can be written as

$$y(x + 1)^2 = (-2)^{2q-1}\, Q^*(y(x)\,)^2(y(x) - 1) =$$
$$= (1/2)\, [U_q(y(x)\,) + U_{q-1}(y(x)\,)\,]^2(1 - y(x)\,). \qquad (1.6^*)$$

Existence parts of Theorem 4 are contained in the following Theorems 5 and 6.

Theorem 5. When $m = 2$ and $p = 2q + 2$, $q \geq 1$, then solutions of (1.5) are written as

$$y(x) = (A/2i)\, [\alpha(x) - 1/\alpha(x)], \qquad (1.8)$$

where $\alpha(x)$ is given by

$$\alpha(x) = (\pm i)^{1/q} \exp[\, \kappa(x)\, (q + 1)^x \,] \qquad (1.8_e)$$

or

$$\alpha(x) = (\pm i)^{1/q} \exp[\, \kappa(x)\, (-q-1)^x \,] \qquad (1.8_e')$$

if q is even, and

$$\alpha(x) = (\pm 1)^{1/q} \exp[\ \kappa(x)\ (q + 1)^x\ ] \tag{1.8$_o$}$$

or

$$\alpha(x) = (\pm 1)^{1/q} \exp[\ \kappa(x)\ (-\ q - 1)^x\ ] \tag{1.8$'_o$}$$

if q is odd. In (1.8_e) - ($1.8'_o$), $\kappa(x)$ is an entire periodic function with period 1.

Finally we will show the following theorem.

Theorem 6. The equation (1.6′) possesses an entire solution.

We remark that solutions of (1.6) are given by (1.3′) or (1.4′) with $m = 2$.

2. Preliminary lemmas

Theorem 1 is implied by the following lemma.

Lemma 2.1. Suppose that a meromorphic function y(x) satisfies the equation (1.1) with $\deg[P(y)] = p \neq m$. Then y(x) does not have any poles, i.e., y(x) must be entire.

Proof. Suppose y(x) would have a pole x_0 of order s.

First consider the case when $p > m$. By (1.1), $x_0 - 1$ is also a pole for y(x), with order $ms/p < s$. Repeating this procedure, we obtain a contradiction that y (x) would have a pole of order s', $0 < s' < 1$. The case $p < m$ can be treated analogously.

Q.E.D.

Remark. When $p = m$, some solution of (1.1) may have poles, as shown in Introduction.

Corollary 2.2. If $p \neq m$ in (1.1), then any solution is transcendental.

Lemma 2.3. Suppose an entire function y (x) satisfies the equation

$$y(x + 1)^m = Ay(x)^p, \quad A \neq 0, \quad p \neq m, \tag{2.1}$$

then y(x) does not have any zero points.

Proof. When $p > m$. Suppose y(x) would have a zero x_0 of order s. Then $x_0 - 1$

would be also a zero of order $ms/p < s$, as seen by (2.1). Repeating this procedure, we will obtain a contradiction, as in the proof of lemma 2.1. The case when $p < m$ is treated analogously. Q.E.D.

Lemma 2.4. Suppose, in the equation (1.1), that $p \gtreqless m$ and $P(y)$ is not of the form $P_1(y)^m$ with any polynomial $P_1(y)$. If (1.1) is satisfied by an entire function $y(x)$, then (1.1) must be of the form either

$$y(x+1)^m = Q(y(x))^m (y(x) - c)^a, \ 1 \leq a < m. \tag{2.2}$$

where $Q(y)$ is a polynomial of degree q, $q \geq 0$, or

$$m = 2 \ \text{and} \ y(x+1)^2 = Q(y(x))^2 (y(x) - c_1)(y(x) - c_2), \tag{2.3}$$

where $c_1 \neq c_2$.

Proof. Put

$$P(y) = A(y - c_1)^{p_1} \cdots (y - c_h)^{p_h}, \ c_i \neq c_j \ \text{if} \ 1 \neq j.$$

Suppose $m \nmid p_k$. Then obviously $y(x)$ must be totally ramified at c_k, if $y(x)$ takes the value c_k. By a theorem of Nevanlinna [4, p. 277], there are at most two totally ramified values for entire functions and no such values for an entire function with one (finite) Picard exceptional value. Therefore, $m \mid p_k$ except at most two k.

If there is only one such k, i.e., $m \nmid p_k$, then (1.1) is of the form (2.2).

If $m \nmid p_1$ and $m \nmid p_2$, then we can write

$$P(y) = Q(y)^m (y - c_1)^{p_1} (y - c_2)^{p_2}, \ 1 \leq p_1, p_2 < m.$$

Put $k_i = \text{G.C.D.}(m, p_i)$ and $p_i' = p_i / k_i$, $m_i = m / k_i$, $i = 1, 2$. Obviously $m_i \geq 2$, and

$$(1 - 1/m_1) + (1 - m_2) \leq 1,$$

hence $m_1 = m_2 = 2$. Therefore $p_1' = p_2' = 1$ and $k_1 = k_2$. Thus (1.1) must be of the form (2.3).

3. Proof of Theorem 2

Since m > p, we cannot have the case (2.3) in Lemma 2.4. Therefore we have the case (2.2), with q = 0 and a = p. That is

$$y(x+1)^m \;=\; A(y(x)-c)^p. \tag{3.1}$$

Suppose $c \neq 0$. If y(x) would not take c, then y(x) would not take 0 also, which contradicts the theorem of Picard. Therefore y(x) takes 0 as well as c. Since $p \geq 2$, y(x) is totally ramified at 0 as well as at c. As in the proof of Lemma 2.4 we have, by a theorem of Nevanlinna on ramified values [4, p. 277], that m = p = 2, which contradicts the assumption m > p. Hence c = 0, and we obtain (1.3).

By Lemma 2.3, y(x) has no zeros, hence u(x) = log[y(x)] is entire and satisfies

$$u(x+1) \;=\; (p/m)\, u(x) \;+\; (1/m) \log A,$$

hence we get the solution (1.3′).

4. Proof of Theorem 3

Since $m \geq 3$, we obtain (2.2) in Lemma 2.4.

$$y(x+1)^m \;=\; Q(y(x)\,)^m (y(x)-c)^a, \quad 1 \leq a < m. \tag{2.2}$$

We can suppose that G.C.D. (a, m) = 1.

Obviously, either y(x) does not take c, or y(x) is totally ramified at c with order m. In either case, we can put

$$y(x) = c + z(x)^m \tag{4.1}$$

with an entire function z(x). Then by (2.2)

$$z(x+1)^m \;=\; Q(z(x)^m + c)\,(z(x)\,)^a - c. \tag{4.2}$$

The right hand side is a polynomial of z(x) of degree mq+a, hence it is not of the form $P_1(z(x)\,)^m$. Thus (4.2) must be of the form (2.2), i.e.,

$$Q(z^m + c)z^a - c = Q_1(z)^m(z - c')^a, \tag{4.3}$$

with a polynomial $Q_1(z)$ of degree q and a constant c'.

We can repeat this procedure indefinitely. Thus

$$Q_{h-1}(z^m + c^{(h-1)})\, z^a - c^{(h-1)} = Q_h(z)^m(z - c^{(h)})^a, \tag{4.3$_h$}$$

$h = 2, 3, \ldots$ Put

$$Q(z) = \sum_{k=0}^{q} b_k z^k \quad , \quad Q_1(z) = \sum_{k=0}^{q} b_k' z^k. \tag{4.4}$$

<u>Lemma 4.1</u> <u>Suppose, in (4.3), that</u> $cc' = 0$. <u>Then we get</u> $c = c' = 0$, <u>and</u> $Q(z) = b_q z^q$, i.e., $b_{q-1} = \ldots = b_0 = 0$.

<u>Proof</u>. If $c' = 0$, then we have obviously that $c = 0$.

Suppose $c = 0$, then (4.3) is written as

$$Q(z^m)z^a = Q_1(z)^m(z - c')^a. \tag{4.3'}$$

Put $Q(z) = z^{\ell} Q^{(1)}(z)$, $\ell \geq 0$, $Q^{(1)}(0) \neq 0$. If $c' \neq 0$, then we must have $Q_1(z) = z^{\ell_1} Q_1^{(1)}(z)$, $\ell_1 > 0$, $Q_1^{(1)}(0) \neq 0$. Thus $m\ell_1 - m\ell = a$, which is a contradiction since G.C.D. $(m , a) = 1$. Hence we get $c' = 0$.

Thus, if $cc' = 0$, then $c = c' = 0$ in (4.3), and

$$Q(z^m)z^a = Q_1(z)^m\, z^a \text{ , i.e., } Q(z^m) = Q_1(z)^m.$$

Thus we get $Q_1(z) = z^{\ell_1} q_1(z^m)$, $q_1(0) \neq 0$. Suppose $q_0 = \deg[q_1(z)] \geq 1$. Consider (4.3_2). We get $c'' = 0$ since $c' = 0$. Thus

$$Q_1(z^m) = Q_2(z)^m, \text{ hence } z^{m\ell_1} q_1(z^{m^2}) = Q_2(z)^m.$$

Therefore we have $Q_2(z) = z^{\ell_2} q_2(z^{m^2})$, $q_2(0) \neq 0$, and $\deg[q_2(z)] = q_0/m$. Repeating this procedure h-times, we get

$$Q_h(z) = z^{\ell_h} q_h(z^{m^h}), \qquad q_h(0) \neq 0, \qquad \deg[q_h(z)] = q_0/m^{h-1} < 1$$

if h is sufficiently large, which is a contradiction. Hence $q_0 = 0$ and $Q\,(z) = b_q z^q$.

Q.E.D.

Now we will show that $cc' = 0$ in (4.3). Put

$$L(z) = Q(z^m + c)\, z^a, \text{ i.e., } L(z) - c = Q_1(z)^m (z - c')^a. \tag{4.5}$$

L(z) is of the form

$$L(z) = \sum_{k=0}^{q} A_k z^{mk+a}. \tag{4.5'}$$

Let

$$Q_1(z) = b_q'(z - \alpha_1)^{q_1} \cdots (z - \alpha_h)^{q_h}, \ \alpha_i \neq \alpha_j \quad \text{if } i \neq j,$$

$$q_1 + \cdots + q_h = q. \tag{4.5''}$$

Suppose $cc' \neq 0$. Then $\alpha_j \neq 0$, j = 1, ..., h. α_j is the mq_j-ple or $(mq_j + a)$ -ple root of $L(z) - c = 0$, according to $\alpha_j \neq c'$ or $\alpha_j = c'$. Therefore α_j is at least $(mq_j - 1)$ -ple root of

$$L'(z)/z^{a-1} = \sum_{k=0}^{q} (mk + a)\, A_k z^{mk} = 0.$$

Since $L'(z)/z^{a-1}$ is a polynomial of z^m,

$$\alpha_j^{(\nu)} = \alpha_j \exp[2\pi\nu i/m] \quad , \qquad \nu = 0, 1, ..., m-1,$$

are also at least $(mq_j - 1)$ -ple roots of $L'(z)/z^{a-1} = 0$. On the other hand, we have

$$L(\alpha_j^{(\nu)}) - c = L(\alpha_j)\exp[2\pi\nu a i/m] - c \neq 0$$

for $\nu = 1, ..., m - 1$, since G.C.D. (m, a) = 1. Hence

$$\alpha_j^{(\nu)} \neq \alpha_j^{(\nu')}, \ \nu, \nu' = 0, \ \ldots, \ m - 1, \quad \text{if } i \neq j.$$

Therefore the equation $L'(z)/z^{a-1} = 0$ admits t roots, where

$$t \geq \sum_{j=1}^{h} (mq_j - 1)\, m \ = \ m(mq - h) > mq$$

since $m \geq 3$, which is a contradiction because mq is the degree of $L'(z)/z^{a-1}$. Thus we get $cc' = 0$ which was to be shown.

By Lemma 4.1, we have that $Q(z) = b_q z^q$, and the equation must be of the form (1.4). The solution (1.4′) is obtained as in §3.

5. Proofs of Theorem 4(a) and Theorem 5

Since p is even, we must have the case (2.3), not (2.2), i.e.,

$$y\,(x+1)^2 = Q(y(x)\,)^2(y(x) - c_1)\,(c_2 - y(x)\,), \quad c_1 \neq c_2. \tag{5.1}$$

Put

$$y\,(x) = u(x) + c, \quad \text{where } c = \tfrac{1}{2}\,(c_1 + c_2). \tag{5.2}$$

Then

$$(u(x+1) + c)^2 \; = \; Q^{\sim}(u(x)\,)^2[A^2 - u(x)^2], \quad A = \tfrac{1}{2}(c_2 - c_1), \tag{5.2'}$$

where $Q^{\sim}(u) = Q(u + c)$. We can write

$$A^2 - u(x)^2 = f(x)^2$$

with an entire function f(x), as seen by (5.2′). Put

$$u(x) = (A/2i)\;[\alpha(x) - \alpha(x)^{-1}]. \tag{5.3}$$

Then

$$\alpha(x) = (iu(x)\,/\,A) \pm (1 - (u(x)\,/\,A)^2)^{1/2}. \tag{5.3'}$$

Since

$$(1 - (u(x)\,/\,A)^2)^{1/2} = \pm\;(f(x)\,/\,A)$$

and

$$(iu/A) \pm (1 - (u/A)^2)^{1/2} \neq 0,$$

we see that $\alpha(x)$ in (5.3) is entire and does not take 0. We get

$$f(x) = \pm(A/2)[\alpha(x) + \alpha(x)^{-1}\,], \tag{5.3''}$$

and by (5.2′)

$$(A/2i)\ [\alpha(x+1) - \alpha(x+1)^{-1}] =$$

$$= Q^{\sim}\left(\frac{A}{2i}[\alpha(x) - \alpha(x)^{-1}]\right)\left(\frac{A}{2}[\alpha(x) + \alpha(x)^{-1}]\right) - c. \tag{5.4}$$

We write the right hand side of (5.4) as $F(\alpha(x))$. If

$$Q^{\sim}(u) = b_q u^q + \cdots + b_0,$$

then

$$F(\alpha) = (A/2i)^q (A/2)\ b_q \alpha^{q+1} + (-A/2i)^q (A/2) b_q \alpha^{-q-1} + \cdots$$

By (5.4), we get

$$\alpha(x + 1) = i\ F(\alpha(x)\ )\ /\ A \pm \left\{1 - [F(\alpha(x)\ )\ /\ A]^2\right\}^{1/2}. \tag{5.4'}$$

Put

$$1 - [F(\alpha)\ /\ A]^2 = \alpha^{-2q-2}\ G(\alpha),$$

with a polynomial $G(\alpha)$, $G(0) = -\ (A/2i)^{2q}(1/4)\ b_q{}^2 \neq 0$. Let

$$G(\alpha) = B(\alpha - d_1)^{e_1} \cdots (\alpha - d_h)^{e_h}\ ,\ d_i \neq d_j \text{ if } i \neq j.$$

Obviously $d_j \neq 0$, for $G(0) \neq 0$. Since $\alpha(x+1)$ is entire, we must have, by (5.4′), that $G(\alpha)^{1/2}$ is entire. Since $\alpha(x)$ does not take 0, $\alpha(x)$ has not any totally ramified values. Hence any e_k, $1 \leq k \leq h$, is divided by 2. Therefore by (5.4′)

$$\alpha(x + 1) = \alpha(x)^{-q-1}\ T(\alpha(x)),$$

where $T(\alpha)$ is a polynomial of α of degree t, with $0 \leq t \leq 2q+2$. Thus

$$(A/2i)[\alpha^{-q-1}\ T(\alpha) - \alpha^{q+1}/T(\alpha)\ ] = \alpha^{-q-1}\ F_1(\alpha), \tag{5.5}$$

where $F_1(\alpha)$ is a polynomial of degree 2q+2. By (5.5) we see that $T(\alpha) \mid \alpha^{2q+2}$, hence $T(\alpha) = T_0\alpha^t$ with a constant T_0. Since $F_1(\alpha)$ is of degree 2q+2, we must have by (5.5) that either $t = 2q+2$ or $t = 0$, i.e., either

$$\alpha(x + 1) = T_0\ \alpha(x)^{q+1} \tag{5.6}$$

or

$$\alpha(x + 1) = T_0 \alpha(x)^{-q-1} \tag{5.6'}$$

If we have (5.6), then (5.4) we get

$$\alpha(x+1) = (A/2i)^{q-1}(A/2)\ b_q\alpha(x)^{q+1} \tag{5.7$_1$}$$

and

$$\alpha(x+1) = (-2i/A)^{q-1}(2/A)\ (1/b_q)\ \alpha(x)^{q+1}. \tag{5.7$_2$}$$

On the other hand, if we have (5.6′), then also by (5.4) we get

$$\alpha(x+1) = -(2i/A)^{q-1}(2/A)\ (1/b_q)\ \alpha(x)^{-q-1} \tag{5.7$'_1$}$$

and

$$\alpha(x+1) = (-1)^q(A/2i)^{q-1}\ (A/2)\ b_q\alpha(x)^{-q-1}. \tag{5.7$'_2$}$$

Further, we get easily that

$$c = 0. \tag{5.7$''$}$$

We obtain by (5.7$_1$) and (5.7$_2$) as well as (5.7$'_1$) and (5.7$'_2$),

$$b_q = [(-1)^{q-1}]^{1/2}(2i/A)^{q-1}(2/A) \tag{5.8}$$

and

$$T_0 = \pm\, i \qquad \text{if q is even,}$$

$$T_0 = \pm\, 1 \qquad \text{if q is odd.} \tag{5.8$'$}$$

We also see by (5.7$''$) that u(x) = y(x) and Q˜(x) = Q(x) in (5.2′). Thus the solution y(x) is written as in (1.8) with $\alpha(x)$ such that

$$\alpha(x+1) = \pm\, i\ \alpha(x)^{q+1} \qquad \text{if q is even,} \tag{5.9$_1$}$$

$$\alpha(x+1) = \pm\ \alpha(x)^{q+1} \qquad \text{if q is odd,} \tag{5.9$_2$}$$

or

$$\alpha(x+1) = \pm\, i\ \alpha(x)^{-q-1} \qquad \text{if q is even,} \tag{5.9$'_1$}$$

$$\alpha(x+1) = \pm\ \alpha(x)^{-q-1} \qquad \text{if q is odd.} \tag{5.9$'_2$}$$

Hence by (5.9$_1$) and (5.9$_2$)

$$\alpha(x) = (\pm\, i)^{1/q} \exp[\,\kappa(x)\,(q+1)^x\,] \qquad \text{if q is even,} \qquad (5.10_1)$$

$$\alpha(x) = (\pm\, 1)^{1/q} \exp[\,\kappa(x)\,(q+1)^x\,] \qquad \text{if q is odd.} \qquad (5.10_2)$$

Similarly, by $(5.9'_1)$ and $(5.9'_2)$

$$\alpha(x) = (\pm\, i)^{1/q} \exp[\,\kappa(x)\,(-q-1)^x\,] \qquad \text{if q is even,} \qquad (5.10'_1)$$

$$\alpha(x) = (\pm\, 1)^{1/q} \exp[\,\kappa(x)\,(-q-1)^x\,] \qquad \text{if q is odd.} \qquad (5.10'_2)$$

In these formulas, $\kappa(x)$ denotes a periodic entire function with period 1. Thus we know that solutions of the equation (1.5) are given by (1.8), (1.8_e) – $(1.8'_o)$.

Polynomial Q(y) in (1.5) is also determined by (5.4) with (5.9_1) or (5.9_2), as well as by (5.4) with $(5.9'_1)$ of $(5.9'_2)$. For example, if q is even, then

$$(A/2i)\; i\,(\alpha^{q+1}) + \alpha^{-q-1}) = (A/2)\,(\alpha + \alpha^{-1})\; Q\Big(\tfrac{A}{2i}(\alpha - \alpha^{-1})\Big). \qquad (5.11)$$

By (5.3′) with y instead of u, we get

$$\alpha + \alpha^{-1} = \pm\, 2(1 - (y/A)^2)^{1/2}, \qquad (5.11')$$

and

$$\alpha^{q+1} + \alpha^{-q-1} =$$

$$= \pm\, 2 \sum_{k=0}^{q/2} \binom{q+1}{2k} (iy/A)^{2k}\, [(1 - (y/A)^2)^{1/2}]^{q+1-2k}. \qquad (5.11'')$$

Since

$$y = (A/2i)\,(\alpha - \alpha^{-1}),$$

we obtain (1.5_e) for Q (y) by (5.11) , (5.11′), and 5.11″).

6. Proof of Theorem 4(b)

Since p is odd, we have (2.2) for m = 2, i.e.,

$$y(x+1)^2 = Q(y(x)\,)^2\,(y(x) - c). \qquad (2.2)$$

As in §4, we have

$$y(x) = c + z(x)^2 \qquad (6.1)$$

with an entire function z (x), which satisfies the equation

$$z(x+1)^2 = Q_1(z(x))^2 (z(x) - c') \tag{6.2}$$

with a polynomial $Q_1(y)$ and a constant c', such that

$$Q(z^2 + c)\, z - c = Q_1(z)^2(z - c'). \tag{6.3}$$

Also, as in (4.3_h), we can repeat this procedure indefinitely. In particular

$$Q_1(z^2 + c')\, z - c' = Q_2(z)^2(z - c''). \tag{6.3'}$$

Lemma 4.1 holds for m = 2 also, as easily seen.

Hence, if c = 0, we have (1.6).

Suppose $c \neq 0$. Then $c'c'' \neq 0$.

Put

$$L(z) = Q(z^2 + c) = \sum_{k=0}^{q} A_k z^{2k+1},$$
$$L(z) - c = Q_1(z)^2(z - c'), \tag{6.4}$$
$$L_1(z) = Q_1(z^2 + c')\, z = \sum_{k=0}^{q} A'_k z^{2k+1},$$
$$L_1(z) - c' = Q_2(z)^2(z - c''). \tag{6.4'}$$

Write $Q_1(z)$ as in (4.5''):

$$Q_1(z) = b'_q\, (z - \alpha_1)^{q_1} \cdots (z - \alpha_h)^{q_h},\ \alpha_i \neq \alpha_j \quad \text{if } i \neq j, \tag{4.5''}$$

and $Q_2(z)$ as

$$Q_2(z) = b'_q\, (z - \beta_1)^{p_1} \cdots (z - \beta_\ell)^{p_\ell},\ \beta_i \neq \beta_j \quad \text{if } i \neq j,$$
$$p_1 + \cdots + p_\ell = q, \tag{6.5}$$

Since $cc'c'' \neq 0$, we have $\alpha_j\beta_k \neq 0$. For each j, α_j is $2q_j$-ple or $(2q_j + 1)$-ple root of L (z) − c = 0, according to $\alpha_j \neq c'$ or $\alpha_j = c'$ respectively. Hence $-\alpha_j$ is also a $2q_j$-ple or $(2q_j + 1)$-ple root of L (z) + c = 0. Therefore, α_j and $-\alpha_j$ are at least $(2q_j - 1)$-ple roots of

$$L'(z) = 0,$$

and $\alpha_i \neq \alpha_j$, i, j = 1, ..., h.

Similarly, β_k is a $2p_k$-ple or $(2p_k + 1)$-ple root of $L_1(z) - c' = 0$ and $-\beta_k$ is a $2p_k$-ple or $(2p_k + 1)$-ple root of $L_1(z) + c' = 0$. Hence β_k and $-\beta_k$ are at least $2p_k$-ple roots of

$$L_1(z)^2 - c'^2 = Q_1(z^2 + c')^2 z^2 - c'^2 = 0.$$

Therefore, $\beta_k^2 + c'$ is an at least $2p_k$-ple root of

$$Q_1(z)^2(z - c') - c'^2 = 0,$$

hence of

$$L(z) - c - c'^2 = 0.$$

Hence $-(\beta_k^2 + c')$ is an at least $2p_k$-ple root of

$$L(z) + (c + c'^2) = 0.$$

Thus $(\beta_k^2 + c')$ and $-(\beta_k^2 + c')$ are $(2p_k - 1)$-ple roots of

$$L'(z) = 0.$$

If $c + c'^2 \neq -c$ and $c + c'^2 \neq 0$, then α_j, $-\alpha_j$, $(\beta_k^2 + c')$, $-(\beta_k^2 + c')$, j = 1, ..., h, k = 1, ..., ℓ, are all distinct. Therefore the polynomial $L'(z)$ of degree 2q would possess t zeros, where

$$t \geq 2\sum_{j=1}^{h} (2q_j - 1) + 2\sum_{k=1}^{\ell} (2p_k - 1) = 8q - 2(h + \ell) \geq 4q,$$

which is a contradiction.

If $c + c'^2 = 0$, then $(\beta_k^2 + c')$ may be equal to some $-(\beta_s^2 + c')$. But, since $c + c'^2 \neq \pm c$, we have that $(\beta_k^2 + c')$, α_j, $-\alpha_j$ are all distinct. Hence $L'(z)$ would possess t' zeros, where

$$t' \geq 2\sum_{j=1}^{h} (2q_j - 1) + \sum_{k=1}^{\ell} (2p_k - 1) = 6q - (2h + \ell) \geq 3q,$$

which is also a contradiction.

Therefore, we must have that

$$c + c'^2 = -c, \text{ i.e., } 2c + c'^2 = 0, \text{ if } cc' \neq 0. \tag{6.6}$$

Put, as in (4.4),

$$Q(z) = \sum_{k=0}^{q} b_k z^k, \; Q_h(z) = \sum_{k=0}^{q} b_k{}^{(h)} z^k, \; h = 1, 2, \ldots$$

and

$$B_k{}^{(h)} = \sum_{\substack{\nu_1+\nu_2=2 \\ k_1\nu_1+k_2\nu_2=k}} \frac{2!}{\nu_1!\,\nu_2!} (b_{k_1}{}^{(h)})^{\nu_1} (b_{k_2}{}^{(h)})^{\nu_2}, \; h = 1, 2, \ldots ,$$

$$C_k{}^{(h)} = \sum_{\nu=k}^{q} b_\nu{}^{(h)} \binom{\nu}{k} (C^{(h)})^{\nu-k} \quad , \quad h = 0, 1, \ldots$$

with

$$C_k{}^{(0)} = C_k \,,\; b_k{}^{(0)} = b_k \,,\; b_k{}^{(1)} = b'_k \,,\; b_k{}^{(2)} = b''_k \text{ , etc.}$$

Then by (6.3) and (6.3_h) :

$$Q_{h-1}(z^2 + c^{(h-1)})\, z - c^{(h-1)} = Q_h(z)^2 (z - c^{(h)}), \tag{6.3_h}$$

we get

$$-c'\, B'_{2k} + B'_{2k-1} = 0, \quad k = 1, \ldots, q, \tag{6.7-k}$$

$$-c'\, B'_{2k+1} + B'_{2k} = C_k = \sum_{\nu=k}^{q} b_\nu \binom{\nu}{k} c^{\nu-k}, \; k = 0, \ldots, q-1, \tag{6.8-k}$$

$$b'^{\,2}_q = b_q, \tag{6.8-q}$$

and

$$-c^{(h)} B^{(h)}_{2k} + B^{(h)}_{2k-1} = 0, \qquad k = 1, \ldots, q, \tag{6.7_h-k}$$

$$-c^{(h)} B^{(h)}_{2k+1} + B^{(h)}_{2k} = C_k^{(2k-1)} = \sum_{\nu=k}^{q} b_v{}^{(h-1)} \binom{\nu}{k} (c^{(h-1)})^{\nu-k},$$

$$k = 0, \ldots, q-1 \tag{6.8_h-k}$$

$$({b_q}^{(h)})^2 = {b_q}^{(h-1)}, \qquad (6.8_h\text{-}q)$$

with

$$2c^{(h-1)} + (c^{(h)})^2 = 0. \qquad (6.6_h)$$

Write

$$b_k = b_q c^{q-k}\, \lambda_k\,, \qquad k = 0, 1, \ldots, Q-1,$$
$$b^{(h)} = {b_q}^{(h)}\, (c^{(h)})^{q-k}\, {\lambda_k}^{(h)}\,, \qquad k = 0, 1, \ldots, q-1,$$

$${\Lambda_k}^{(h)} = \sum_{\substack{\nu_1+\nu_2=2 \\ k_1\nu_1+k_2\nu_2=k}} \frac{2!}{\nu_1!\,\nu_2!} ({\lambda_{k_1}}^{(h)})^{\nu_1} ({\lambda_{k_2}}^{(h)})^{\nu_2},$$

$${\Gamma_k}^{(h)} = \sum_{\nu=k}^{q} {\lambda_\nu}^{(h)} \binom{\nu}{k},$$

then

$${B_k}^{(h)} = ({b_q}^{(h)})^2 (c^{(h)})^{2q-k}\, {\Lambda_k}^{(h)},$$
$${C_k}^{(h)} = {b_q}^{(h)}\, (c^{(h)})^{q-k}\, {\Gamma_k}^{(h)}.$$

Hence by (6.6_h)

$$-\Lambda'_{2k} + \Lambda'_{2k-1} = 0, \qquad k = 1, \ldots, q, \qquad (6.7'\text{-}k)$$

$$-\Lambda'_{2k+1} + \Lambda'_{2} = (-1/2)^{q-k}\, \Gamma_k, \qquad k = 0, 1, \ldots, q-1, \qquad (6.8'\text{-}k)$$

and

$$-\Lambda^{(h)}_{2k} + \Lambda^{(h)}_{2k-1} = 0, \qquad k = 1, \ldots, q \qquad (6.7'_h\text{-}k)$$

$$-\Lambda^{(h)}_{2k+1} + \Lambda^{(h)}_{2k} = (-1/2)^{q-k}\, {\Gamma_h}^{(h-1)}, k = 1, \ldots, q-1. \qquad (6.8'_h\text{-}k)$$

By $(6.7'_h\text{-}q)$, we set

$$\lambda_{q-1}^{(h)} = 1/2, \qquad h = 1, 2, \ldots$$

By $(6.8'_h\text{-}\ (q-1)\ )$, $h \geq 2$, we get

$$\lambda_{q-2}^{(h)} = -\ (q-1)\ /\ 4, \qquad h = 2, 3, \ldots$$

By $(6.7'_h\text{-}\ (q-1)\ )$, $h \geq 2$, we get

$$\lambda_{q-3}^{(h)} = -\ (q-2)\ /\ 8, \qquad h = 2, 3, \ldots$$

By $(6.8'_h\text{-}\ (q-2)\ )$, $h \geq 3$, we get

$$\lambda_{q-4}^{(h)} = (q-2)\ (q-3)\ /\ 32, \qquad h = 3, 4, \ldots$$

By $(6.7'_h\text{-}\ (q-2)\ )$, $h \geq 3$, we get

$$\lambda_{q-5}^{(h)} = (q-3)\ (q-4)\ /\ 64, \qquad h = 3, 4, \ldots$$

By $(6.8'_h\text{-}\ (q-3)\ )$, $h \geq 4$, we get

$$\lambda_{q-6}^{(h)} = -\ (q-3)\ (q-4)\ (q-5)\ /\ (128 \times 3), \qquad h = 4, 5, \ldots$$

By $(6.7'_h\text{-}\ (q-3)\ )$, $h \geq 4$, we get

$$\lambda_{q-7}^{(h)} = -\ (q-4)\ (q-5)\ (q-6)\ /\ (256 \times 3), \qquad h = 4, 5, \ldots$$

Repeating this procedure

$$\lambda_k^{(h)}, \ k = 0, 1, \ldots, q-1, \ \ h = q, q+1, \ldots$$

are determined, which are independent of h, $h \geq q$. Since $\lambda_k^{(h)}$, $0 \leq h \leq q$, satisfy the equations of the same form, we know that $\lambda_k^{(h)}$ are independent of h, $h \geq 0$. Hence we write

$$\lambda_k^{(h)} = b_k^*$$

and put

$$Q^*(z) = \sum_{k=0}^{q} b_k^* z^k.$$

Since

$$Q_h(z) = b_q^{(h)} (c^{(h)})^q \ Q^*(z/c^{(h)}),$$

we get by (6.3_h)

$$[b_q^{(h)}(c^{(h)})^q]^2\ Q^*(z/c^{(h)})^2\ (z - c^{(h)})$$
$$= b_q^{(h-1)}(c^{(h-1)})^q\ Q^*((z^2 + c^{(h-1)})\ /\ c^{(h-1)})\ z - c^{(h-1)}. \qquad (6.9)$$

If we write

$$\zeta = z/c^{(h)},$$

then (6.9) is written, noting (6.6_h) and (6.8_h-q), as

$$(b_q^{(h)})^2(c^{(h)})^{2q-1}\ Q^*(\zeta)^2(\zeta - 1)$$
$$= (-2)^{-q}(b_q^{(h)})^2(c^{(h)})^{2q-1}\ Q^*\ (-2\zeta^2 + 1)\ \zeta + 1/2. \qquad (6.9')$$

Therefore

$$A\ =\ (b_q^{(h)})^2(c^{(h)})^{2q-1} = (-1/2)Q^*(0)^{-2}$$

is a constant independent of h, and hence

$$(b_q^{(h)})^2(c^{(h)})^{2q-1}\ =\ (b_q^{(h-1)})^2(c^{(h-1)})^{2q-1}$$

$$=\ (b_q^{(h)})^4[-(c^{(h)})^2/2]^{2q-1},$$

from which we obtain

$$(b_q^{(h)})^2(c^{(h)})^{2q-1} = (-2)^{2q-1} \qquad (6.10)$$

and, by (6.9′),

$$(-2)^{q-1}Q^*(-2\zeta^2 + 1)\ \zeta + \tfrac{1}{2} = (-2)^{2q-1}Q^*(\zeta)^2(\zeta - 1). \qquad (6.10')$$

Since the equation (6.10′) is satisfied by

$$Q^*(\zeta) = 2^{-q}[\ U_q(\zeta) + U_{q-1}(\zeta)\ ],$$

in which U_q and U_{q-1} are Tchebicheff polynomials of the second kind [1, p. 184], and such $Q^*(\zeta)$ is unique, we obtain (1.6), (1.7), and (1.7′).

7. Proof of Theorem 6

By (1.7) and (1.7′), we can take the equation (1.6′) in the form

$$y(x + 1)^2 = P(y(x)) \tag{7.1}$$

with

$$P(y) = (-2)^{2q-1}\, Q^*(y)^2\, (y - 1). \tag{7.1'}$$

We also note that $Q^*(y)$ satisfies the equation (6.10′):

$$(-2)^{q-1} Q^*(-2y^2 + 1)\, y + \tfrac{1}{2} = (-2)^{2q-1} Q^*(y)^2 (y - 1). \tag{6.10'}$$

Using (1.7′) and (6.10′), we obtain easily

$$P(-1) = (-1)^2, \tag{7.2}$$

$$P'(-1) = -(2q+1)^2/2. \tag{7.3}$$

Put

$$y(x) = -1 + u(x),$$

then by (7.2) we get

$$u(x + 1) = bu(x) + b_2 u(x)^2 + \ldots \tag{7.4}$$

with

$$b = P'(-1) / (-2) = (2q+1)^2/4 > 1. \tag{7.4'}$$

It is well known that the equation (7.4) possesses a solution $u_1(x)$ which is holomorphic in

$$D = \left\{ x \,;\, |\, b^x \,| \leq \rho \right\} \text{ for sufficiently small } \rho > 0 \tag{7.5}$$

and is expanded there as

$$u(x) = \sum_{k=1}^{\infty} \gamma_k\, b^{kx}, \tag{7.5'}$$

in which γ_1 may be arbitrarily fixed. $u_1(x)$ is determined uniquely if γ_1 is prescribed.

Thus (7.1) possesses a solution

$$y_1(x) = -1 + u_1(x), \tag{7.6}$$

which is holomorphic in D of (7.5). By (6.10′), we see easily that

$$y_2(x) = -1 + \sum_{k=1}^{\infty} \gamma'_k \, b^{kx},$$

in which we see by (7.7)

$$\gamma'_1 = 4\gamma_1.$$

Therefore we have that

$$y_2(x) = y_1(x + \delta) \qquad \text{with } \delta = (\log 4) \, / \, (\log b) > 0.$$

Thus we get

$$y_1(x + \delta) = 1 - 2y_1(x)^2 \quad , \quad \delta > 0.$$

Hence $y_1(x)$ can be continued analytically to $\{| \, x \, | < \infty\}$. and we obtain an entire solution for (7.1).

References

[1] Erdelyi, A. et al. : Higher Transcendental Functions, vol. II. McGraw-Hill Co., Inc., New York 1953.

[2] Kimura, T., On the iteration of analytic functions. Funkcial. Ekvac., 14 (1971), pp. 197 - 238.

[3] Nakamura, Y., On the order of méromorphic solutions of some difference equations. J. Coll. Arts & Sci., Chiba Univ., B-18 (1985), pp. 5 - 9.

[4] Nakamura, Y., Polynomial difference equations which have entire solutions of finite order. Proc. Japan Acad., Ser. A, 61 (1985), pp. 144 - 146.

[5] Nevanlinna, R., Analytic Functions. Springer-Verlag, Berlin 1970.

[6] Shimomura, S., Entire solutions of some polynomial difference equations. J. Fac. Sci., Univ. Tokyo, Sec. IA Math., 28 (1981), pp. 253 - 266.

[7] Yanagihara, N., Meromorphic solutions of some difference equations. Funkcial. Ekvac., 25 (1980), pp. 309 - 326.

[8] Yanagihara, N., Polynomial difference equations which have meromophic solutions of finite order. To appear.

Department of Mathematics
College of General Education
Chiba University

Department of Mathematics
Faculty of Science
Chiba University

ZEROS OF CERTAIN DIFFERENTIAL POLYNOMIALS

Mitsuru Ozawa

1. Introduction.

It is very interesting to investigate whether a differential polynomial P(g) does have zeros or not and to decide g for which P(g) does not have any zero. This kind of problems for various P(g) was studied by Hayman [2]. Mues [3] treated the problem in the case $P(g) = gg'' - ag'^2$ with a constant a. He proved the following.

Theorem A. *Let* $a \neq 1$ *be a constant and* g *an entire function. Assume that* $gg'' - ag'^2$ *does not have any zero. Then* $g = Be^{\alpha z}$, $\alpha \neq 0$, $B = \text{const.} \neq 0$ *in general or* $g(z) = Az + B$, $A \neq 0$ if $a \neq 0$ or $g(z) = \alpha z^2 + \beta z + \gamma$, $4\alpha\gamma - \beta^2 \neq 0$ *if* $a = 1/2$.

He also mentioned that the case $a = 1$ is really exceptional.

In this paper we shall consider the same problem for several differential polynomials. Firstly let us take $P(g) = gg'' - (B + Ae^{-z})\,g'^2$, where $A(\neq 0)$, B are constants. We have the following

Theorem 1. *Assume that* g *is an entire function of finite order. Then* $gg'' - (B + Ae^{-z})\,g'^2$, $A(\neq 0)$, *does not have any zero when and only when* $g = c(e^z + A)$, $c \neq 0$ *if* $B \neq 0, 1$ *or* $g = ce^z + M$, $M \neq Ac$ *and* $g = ce^{\alpha z}$ *if* $B = 1$ *or* $g = c(e^z + A)$ *and* $g = cz+d$ *if* $B = 0$.

If we omit the order-finiteness of g then $gg'' - (B + Ae^{-z})\,g'^2 \neq 0$ by $\exp(e^z/A)$ if $B \neq 1$ and by $\exp(e^z/A + ce^{2z}/2)$ $c \neq 0$ if $B = 1$. If further we allow meromorphic g, then $g = (e^z/n - A)^{-n}$ satisfies $gg'' - (B + Ae^{-z})\,g'^2 \neq 0$ if $B \neq 1$, where n is a positive integer.

Secondly let us take $P(g) = gg'' - a(z)g'^2$ with a transcendental entire function a(z)

of order less than one. Then we have

Theorem 2. *Let* g *be an entire function of finite order and* a(z) *a transcendental entire function of order less than one. Then* $gg'' - a(z)g'^2$ *has always zeros, if* g *is not a constant.*

Thirdly we take $P(g) = zgg'' - zg'^2 + gg'/2$.

Theorem 3. *Let* g *be an entire function of finite order. If* $zgg'' - zg'^2 + gg'/2$ *has no zero, then either* $g = B \cos \sqrt{kz}$ *with non-zero constants* k, B *or* $g = Be^{\alpha z}$ *with non-zero constants* α, B *and vice versa.*

We need another idea to attack the problem for a wider class of functions such as of any order or meromorphic.

In the final section we take $P(g) = g' + (B - Ae^{-z})\, g^2$.

2. Proof of Theorem 1. Let us consider $gg'' - (B + Ae^{-z})\, g'^2 = e^P$ with a polynomial P. Suppose that $g = e^L$ with a polynomial L. Then $L'' + (1-B)\, L'^2 - Ae^{-z}L'^2 = \exp\,(P-2L)$. Hence $P-2L = \alpha z+\beta$. Thus $L'' + (1-B)\, L'^2 - Ae^{-z}\, L'^2 = e^{\beta}e^{\alpha z}$. If $\alpha = 0$, then $L' = 0$, that is, $L = c$ and $L'' + (1 - B)\, L'^2 - e^{\beta} = -e^{\beta} = 0$, which is a contradiction. If $\alpha = -1$, then

$$L'' + L'^2\,(1 - B) - (AL'^2 + e^{\beta})\, e^{-z} = 0.$$

Hence

$$L'^2\,(1 - B) + L'' = 0, \qquad AL'^2 + e^{\beta} = 0.$$

In this case $B = 1$ and $L'' = 0$, that is, $L' = c$. Hence $Ac^2 + e^{\beta} = 0$ and $L = cz+d$. Thus $g = De^{cz}$, $B = 1$. If $\alpha \neq 0, -1$, then by the Borel unicity theorem [4], [5] $AL'^2 = 0$ and $e^{\beta} = 0$, which is untenable.

Suppose that g is a polynomial. If the degree of g' is greater than 0, then $gg'' - Bg'^2 - Ag'^2e^{-z}$ has at least one zero. Hence $g' = c =$ a constant. In this case

$$gg'' - Bg'^2 - Ag'^2e^{-z} = -c^2(B + Ae^{-z})$$

has infinitely many zeros if $B \neq 0$. Therefore B should be equal to zero. Hence

$g = cz + d$, $B = 0$.

From now on we may assume that g is transcendental and g has zero. Consider

$$gg'' - (B + Ae^{-z})\, g'^2 = e^P \tag{2.1}$$

where P is a polynomial. Let z_0 be a zero of g. Then

$$-(B + Ae^{-z_0})\, g'(z_0)^2 = e^{P(z_0)} \neq 0.$$

Hence z_0 is a simple zero of g(z). By differentiation of (2.1) we have

$$gg''' + g'g''(1 - 2B - 2Ae^{-z}) + Ae^{-z}g'^2 = P'e^P$$

$$= P'gg'' - (B + Ae^{-z})\, P'g'^2. \tag{2.2}$$

Therefore

$$g''(z_0)(1 - 2B - 2Ae^{-z_0}) + (BP'(z_0) + A(1 + P'(z_0))\, e^{-z_0})\, g'(z_0) = 0.$$

This shows that

$$g''(1 - 2B - 2Ae^{-z}) + (BP' + Ae^{-z} + AP'e^{-z})\, g' = kg \tag{2.3}$$

with an entire function k. This together with (2.2) gives

$$gg''' + kgg' = P'gg''.$$

Hence

$$g''' - P'g'' + kg' = 0, \tag{2.4}$$

which implies that k is a polynomial. (2.3) and (2.4) are called the linearization of (2.1).

By differentiation of (2.3)

$$g'''(1 - 2B - 2Ae^{-z}) + g''(BP' + A(3 + P')\, e^{-z})$$

$$+ g'(BP'' - k - A(1 + P' - P'')\, e^{-z}) - k'g = 0.$$

Hence by (2.4)

$$g'' \left\{(1 - B)\, P' + A\, (3 - P')\, e^{-z}\right\}$$

$$+\, g' \left\{2\, (B - 1)\, k + BP'' + A\, (2k - 1 - P' + P'')\, e^{-z}\right\} - k'\, g = 0. \qquad (2.4)'$$

Eliminating g'' we have

$$g'\, (A_1 + A_2 e^{-z} + A_3 e^{-2z}) + g\, (B_1 + B_2 e^{-z}) = 0. \qquad (2.5)$$

with

$$A_1 = B\, (1 - B)\, P'^2 - B\, (1 - 2B)\, P'' + 2\, (1 - B)\, (1 - 2B)\, k,$$

$$A_2 = A \left\{1 - 2B + 2P' + (1 - 2B)\, P'^2 - (1 - 4B)\, P'' - 4\, (1 - B)\, k\right\},$$

$$A_3 = A^2 (-P'^2 + 2P'' + 1 + 4k),$$

$$B_1 = (1 - 2B)\, k' - (1 - B)\, P'k$$

and

$$B_2 = A\, \{(P' - 3)\, k - 2k'\}.$$

Let V be $A_1 + A_2 e^{-z} + A_3 e^{-2z}$. Then

$$N\, (r, 0, g) \le N\, (r, 0, V) \le m\, (r, V)$$

$$\le 2m\, (r, e^{-z}) + O\, (\log r).$$

Hence the order of N (r, 0, g) satisfies

$$\operatorname{ord} N\, (r, 0, g) \le 1.$$

Let us put $g = He^L$. Then ord $H \le 1$. Then by (2.1)

$$H\, (H'' + 2H'\, L' + HL'' + HL'^2) - (B + Ae^{-z})\, (H' + HL')^2 = e^{P-2L}.$$

Since L is a polynomial, P − 2L should be $\alpha z + \beta$. Thus $P' - \alpha = 2L'$, $P'' = 2L''$. Let us denote g' by De^L with $D = H' + HL'$. Then

$$D'' + 2D'L' + DL'' + DL'^2 - P'\, (D' + DL') + kD = 0,$$

that is,

$$D'' - \alpha D' + \left(k - \frac{P'^2}{4} + \frac{P''}{2} + \frac{\alpha^2}{4}\right) D = 0.$$

If D is transcendental, then by Wiman-Valiron's theory [6] and by ord D ≤ 1 we have that

$$P'^2 - 2P'' - 4k = \gamma$$

is a constant. If D is a polynomial, then

$$4k - P'^2 + 2P'' + \alpha^2 = 0.$$

In both cases

$$4k = P'^2 - 2P'' - \gamma \tag{2.6}$$

with a constant γ. By differentiation of (2.5)

$$g''(A_1 + A_2e^{-z} + A_3e^{-2z}) + g'\left\{A_1' + B_1 + (A_2' - A_2 + B_2)\, e^{-z} + (A_3' - 2A_3)\, e^{-2z}\right\}$$
$$+ g\left\{B_1' + \left(B_2' - B_2\right) e^{-z}\right\} = 0.$$

Rewriting (2.3) as

$$g''(C_1 + C_2e^{-z}) + g'\,(D_1 + D_2e^{-z}) - kg = 0$$

and eliminating g'' we have

$$g'\,[E_1 + E_2e^{-z} + E_3e^{-2z} + E_4e^{-3z}] + g\,[F_1 + F_2e^{-z} + F_3e^{-2z}] = 0, \tag{2.7}$$

where

$$E_1 = (A_1' + B_1)\, C_1 - A_1D_1,$$

$$E_2 = (A_1' + B_1)\, C_2 + (A_2' - A_2 + B_2)\, C_1 - A_2D_1 - A_1\, D_2,$$

$$E_3 = (A_2' - A_2 + B_2)\, C_2 + (A_3' - 2A_3)\, C_1 - A_2D_2 - A_3\, D_1,$$

$$E_4 = (A_3' - 2A_3)\, C_2 - A_3D_2,$$

$$F_1 = B_1'\, C_1 + kA_1,$$

$$F_2 = B_1'\, C_2 + (B_2' - B_2)\, C_1 + kA_2,$$

$$F_3 = (B_2' - B_2)\, C_2 + kA_3.$$

By (2.5) and (2.7) we have

$$(A_1 + A_2e^{-z} + A_3e^{-2z})\,(F_1 + F_2e^{-z} + F_3e^{-2z})$$

$$- (B_1 + B_2e^{-z})\,(E_1 + E_2e^{-z} + E_3e^{-2z} + E_4e^{-3z}) = 0.$$

This is a polynomial $Q(e^{-z})$ of e^{-z} with polynomial coefficients.

Now we have

$$\begin{aligned} Q(e^{-z}) = A_1F_1 - B_1E_1 \;&+\; e^{-z}(A_1F_2 + A_2F_1 - B_1E_2 - B_2E_1) \\ &+\; e^{-2z}(A_1F_3 + A_2F_2 + A_3F_1 - B_1E_3 - B_2E_2) \\ &+\; e^{-3z}(A_2F_3 + A_3F_2 - B_1E_4 - B_2E_3) \\ &+\; e^{-4z}(A_3F_3 - B_2E_4) = 0. \end{aligned}$$

By the Borel unicity theorem all the coefficients must reduce to zero. By a simple computation

$$E_1B_1 - F_1A_1 = \frac{\gamma P'^4}{16}\,(1 - B)^2\,(1 - 2B)^2 + S,$$

where S is a polynomial of degree less than the one of P'^4 when $\deg P' \geq 1$ and $B \neq 1$, $B \neq 1/2$. Suppose that $B \neq 1$, $B \neq 1/2$. Then by $E_1B_1 - F_1A_1 = 0$, P' should be equal to zero. This is a contradiction. Thus $\deg P' = 0$, that is, P' is a constant c. Hence $4k = c^2 - \gamma$ is again a constant. In this case (2.4) has the following form:

$$g'' - cg'' + \frac{c^2 - \gamma}{4}\, g' = 0.$$

This has solutions of the following form:

$$g' = c_1e^{\lambda_1 z} + c_2e^{\lambda_2 z},\ \lambda_1 = (c + \sqrt{\gamma})\,/2,\ \lambda_2 = (c - \sqrt{\gamma})\,/2 \text{ if } \gamma \neq 0$$

and

$$g' = c_1e^{cz/2} + c_2ze^{cz/2}, \quad \text{if } \gamma = 0.$$

Here c_1 and c_2 are constants. We may assume that at least one of c_1, c_2 is not equal to zero. Firstly we assume that $c_1c_2 \neq 0$. Suppose that $\gamma \neq 0$. Since $k' = 0$, we have by

(2.4)′

$$g''\left\{(1-B)c + A(3-c)e^{-z}\right\}$$

$$+ g'\left\{2(B-1)\frac{c^2-\gamma}{4} + A\left(\frac{c^2-\gamma}{2} - 1 - c\right)e^{-z}\right\} = 0.$$

This is equivalent to

$$c_1e^{\lambda_1 z}\left\{\lambda_1 c(1-B) - (1-B)\frac{c^2-\gamma}{2}\right\} + c_2e^{\lambda_2 z}\left\{\lambda_2 c - \frac{c^2-\gamma}{2}\right\}(1-B)$$

$$+ c_1Ae^{(\lambda_1-1)z}\left\{\lambda_1(3-c) + \frac{c^2-\gamma}{2} - 1 - c\right\}$$

$$+ c_2Ae^{(\lambda_2-1)z}\left\{\lambda_2(3-c) + \frac{c^2-\gamma}{2} - 1 - c\right\} = 0.$$

If $\lambda_1 = 0$, then $\lambda_2 = c$, $c^2 = \gamma \neq 0$. Thus we have

$$c_2c^2(B-1)e^{cz} + c_1A(c+1)e^{-z} + (c^2+4c+1)Ae^{(c-1)z} = 0.$$

This gives easily a contradiction by Borel's unicity theorem. Hence we may assume that $\lambda_1\lambda_2 \neq 0$, $(\lambda_1 - c)(\lambda_2 - c) \neq 0$ and $c^2 \neq \gamma$. If $\lambda_2 = 1$, then

$$c_1e^{\lambda_1 z}\left(\lambda_1 c - \frac{c^2-\gamma}{2}\right)(1-B) + c_2e^{z}\left(c - \frac{c^2-\gamma}{2}\right)(1-B)$$

$$+ c_1Ae^{(\lambda_1-1)z}\left\{\lambda_1(3-c) + \frac{c^2-\gamma}{2} - 1 - c\right\} + c_2A\left(2 - 2c + \frac{c^2-\gamma}{2}\right) = 0.$$

In this case by $\lambda_2 = 1$ we have $c^2 - \gamma = 4c - 4$. Hence we have

$$c_1e^{\lambda_1 z}(\lambda_1 c - 2c + 2)(1-B) + c_2e^{z}(2-c)(1-B)$$

$$+ c_1Ae^{(\lambda_1-1)z}(3-c)(\lambda_1 - 1) = 0.$$

By $\lambda_1 \neq \lambda_2$, $\lambda_1 \neq 1$. If $\lambda_1 = 2$, then B = 1, which is a contradiction. If $\lambda_1 \neq 2$, then c = 2 and c = 3, which is a contradiction. If $\lambda_2 \neq 1$, then we consider the subcase $\lambda_1 = \lambda_2 - 1$, that is, $\gamma \neq 1$. In this case we have

$$c_1 e^{\lambda_1 z}\left(\lambda_1 c - \frac{c^2 - 1}{2}\right)(1 - B) + c_2 e^{(\lambda_1 + 1)z}\left(\lambda_1 c + \frac{2c - c^2 + 1}{2}\right)(1 - B)$$

$$+ c_1 A e^{(\lambda_1 - 1)z}(c - 3)\left(\lambda_1 - \frac{c+1}{2}\right) + c_2 A e^{\lambda_1 z}(3 - c)\left(\lambda_1 + 1 - \frac{c+1}{2}\right) = 0.$$

Thus c = 3 or $\lambda_1 = (c + 1)/2$. Further $\lambda_1 c = (c^2 - 2c - 1)/2$. If c = 3 and $\lambda_1 c = (c^2 - 2c - 1)/2$, then $\lambda_1 = 1/3$. Hence $\sqrt{\gamma} = -7/3$, which contradicts $\gamma = 1$. If $\lambda_1 c = (c^2 - 2c - 1)$ and $\lambda_1 = (c + 1)/2$, then $c = -1/3$ and $\lambda_1 = 1/3$. Now returning back to the original functional and the original problem,

$$gg'' - (B + Ae^{-z})\, g'^2$$

$$= e^{(2/3)z}\left\{c_1^2(1 - B) - 2Ac_1c_2\right\} + e^{5z/3}\left(c_1c_2\frac{17}{4} - Ac_2^2 - 2Bc_1c_2\right)$$

$$+ e^{8z/3} c_2^2(1 - B) - c_1^2 A e^{-z/3} + \frac{Dc_1}{3} e^{z/3} + \frac{4}{3} Dc_2 e^{4z/3}$$

should be e^P. However $c_2^2(1 - B) \neq 0$ and $c_1^2 A \neq 0$ imply that the above functional contains at least two exponentials $e^{8z/3}$ and $e^{-z/3}$. This is a contradiction.

If $\lambda_2 \neq 1$ and $\lambda_2 = \lambda_1 - 1$, then $\sqrt{\gamma} = 1$ and hence $\lambda_1 = (c + 1)/2$ and $\lambda_2 = (c - 1)/2$. In this case we have

$$c_1 e^{\lambda_1 z}\left(\lambda_1 c - \frac{c^2 - 1}{2}\right)(1 - B) + c_2 e^{(\lambda_1 - 1)z}\left\{(\lambda_1 - 1)c - \frac{c^2 - 1}{2}\right\}(1 - B)$$

$$+ \; c_2 e^{(\lambda_1 - 2)z} \left\{ (\lambda_1 - 1)(3 - c) + \frac{c^2 - 2c - 3}{2} \right\} = 0.$$

Hence

$$\lambda_1 c = \frac{c^2 - 1}{2}, \quad (\lambda_1 - 1)(3 - c) = - \frac{(c - 3)(c + 1)}{2}$$

and

$$(\lambda_1 - 1) \, c = \frac{c^2 - 1}{2}.$$

Thus $c = 0$, which implies that $\lambda_1 = 1/2$, $\lambda_2 = -1/2$. However $3\lambda_1 - 3 = 3/2$, that is, $\lambda_1 = 3/2$. This is a contradiction. In a general case we have

$$\lambda_1 c = (c^2 - \gamma)/2, \quad \lambda_1(3 - c) = c + 1 - (c^2 - \gamma)/2$$

and

$$\lambda_2 c = (c^2 - \gamma)/2, \quad \lambda_2(3 - c) = c + 1 - (c^2 - \gamma)/2.$$

Thus $\lambda_1 c = \lambda_2 c$ and $\lambda_1(3 - c) = \lambda_2(3 - c)$. Hence $\lambda_1 = \lambda_2$, which may be excluded. Next we assume that $c_1 \neq 0$, $c_2 = 0$. The case $c_1 = 0$, $c_2 \neq 0$ is similar. In this case

$$g' = c_1 e^{\lambda_1 z}, \quad g = \delta + \frac{c_1}{\lambda_1} e^{\lambda_1 z}, \quad \lambda_1 \neq 0.$$

Hence

$$gg'' - (B + Ae^{-z}) g'^2 = \delta c_1 \lambda_1 e^{\lambda_1 z} + c_1^2 (1 - B) e^{2\lambda_1 z} - Ac_1^2 e^{(2\lambda_1 - 1) z}.$$

If $2\lambda_1 - 1 \neq \lambda$, then there appear three exponentials. Therefore the right hand side does not reduce to the form e^P. If $2\lambda_1 - 1 = \lambda_1$, then $\lambda_1 = 1$ and

$$gg'' - (B + Ae^{-z}) g'^2$$
$$= \delta c_1 e^z + c_1^2 (1 - B) e^{2z} - Ac_1^2 e^z.$$

Since $c_1^2 (1 - B) \neq 0$, $\delta c_1 = Ac_1^2$. Thus $g = c_1 (A + e^z)$.

Suppose that $\gamma = 0$. Then

$$g' = c_1 e^{cz/2} + c_2 z e^{cz/2},$$

$$g = \delta + \frac{2c_1}{c} e^{cz/2} + \frac{2c_2}{c}\left(z - \frac{2}{c}\right) e^{cz/2},$$

$$g = \frac{cc_1}{2} e^{cz/2} + c_2\left(1 + \frac{c}{2}z\right) e^{cz/2}.$$

Since $P' = c$,

$$e^{cz + D} = gg'' - (B + Ae^{-z})\, g'^2$$

$$= \delta(c_1 + c_2 z)\, e^{cz/2} - A(c_1 + c_2 z)^2 e^{(c-1)z}$$

$$+ (c_1 + c_2 z)\left\{\frac{2c_1}{c} + \frac{2c_2}{c} z - \frac{4c_2}{c^2} - B(c_1 + c_2 z)\right\} e^{cz}.$$

If $c = 1$, then $c_1 + c_2 z = 0$, that is, $c_1 = c_2 = 0$, which may be omitted off. If $c - 1 = c/2$, then $c = 2$ and

$$\delta(c_1 + c_2 z) - A(c_1 + c_2 z)^2 = 0.$$

Hence $c_2 = 0$ and $\delta = Ac_1$. This gives

$$g = c_1\, (A + e^z).$$

Suppose that $B = 1$. Then we have

$$A_1 = P'', \; A_2 = A\left(-\frac{P'^2}{2} + 2P' - 2P'' - 1\right),$$

$$A_3 = A^2\, (1 - P'^2 + 4k) = A^2(1 - \gamma),$$

$$B_1 = -\frac{P'P''}{2} - \frac{P''}{2}, \; B_2 = A\{(P' - 3)\, k - 2k'\}$$

and

$$C_1 = -1, C_2 = -2A, \; D_1 = P', \; D_2 = A\,(1 + P').$$

Therefore

$$E_1 = -\frac{P'P''}{2} - \frac{3}{2}P'',$$

$$F_1 = \frac{P'^2P''}{4} + \tfrac{1}{2}P'P'' - \tfrac{\gamma}{4}P'' - \frac{P^{(4)}}{4}.$$

Hence

$$E_1B_1 - A_1F_1 = -\tfrac{3}{4}P''^2 + \tfrac{\gamma}{4}P''^2 + \tfrac{1}{2}P''P^{(4)}$$

and this should be vanish identically. Therefore $P'' = 0$ if $\gamma \neq 0$. Firstly we consider the case $\gamma = 0$. Now let us consider $A_3F_3 - B_2E_4$, which is the coefficient of e^{-4z} in $Q(e^{-z})$ and hence is equal to zero. However it has the following form:

$$\tfrac{1}{4}\, A^4(1 - \gamma)P'^4 + R,$$

where R is a polynomial of lower degree than the one of P'^4 if $\deg P' \geq 1$. This is a contradiction. Hence $\deg P' = 0$, that is, P' is a constant c. Thus we have the same conclusion $P' = c$ in both cases $\gamma \neq 0$ and $\gamma = 0$. As in the former case ($B \neq 1$, $B \neq 1/2$) we have

$$g = \delta + \frac{c_1}{\lambda_1} e^{\lambda_1 z}, \quad \lambda_1 \neq 0.$$

Then with $B = 1$

$$gg'' - (B + Ae^{-z})\, g'^2 = \delta c_1\lambda_1 e^{\lambda_1 z} - Ac_1^2 e^{(2\lambda_1 - 1)\, z}.$$

Hence $2\lambda_1 - 1 = \lambda_1$, $\delta c_1\lambda_1 \neq Ac_1^2$, that is, $\lambda_1 = 1$, $\delta \neq Ac_1$. Thus $g = \delta + c_1e^z$, $\delta \neq Ac_1$. Suppose that $B = 1/2$. Let us consider two coefficients $A_3F_3 - B_2E_4$ and $A_2F_3 + A_3F_2 - B_1E_4 - B_2E_4 - B_2E_3$ of $Q(e^{-z})$, which are equal to zero. By $A_3F_3 - B_2E_4 = 0$ we have either $\gamma = 1$ or $P' = c$ (const.). When $\gamma = 1$, we have $A_3 = 0$, $E_4 = 0$ and $B_2E_3 - A_2F_3 = A^3P'^4/2 + T$, where T is a polynomial of lower degree than that of P'^4 if $\deg P' \geq 1$. This is a contradiction, since $B_2E_3 - A_2F_3 = 0$ in the present case. Thus P' reduces to a constant c. Now we have similarly the final result: $g = c_1(A + e^z)$.

Here a remark should be mentioned. Let g be a meromorphic function of the following form $g = 1/f^n$ with entire f, which has only simple zeros and satisfies

$$gg'' - Ae^{-z}g'^2 \neq 0.$$

Then by

$$gg'' - Ae^{-z}g'^2 = \frac{-n}{f^{2n+2}}[ff'' - (n+1-nAe^{-z})\, f'^2]$$

we have

$$ff'' - (n+1-nAe^{-z})\, f'^2 \neq 0.$$

If f is of finite order, then we can obtain

$$f = c\,(e^z/n - A).$$

This gives

$$g = D\left(\frac{e^z}{n} - A\right)^{-n}.$$

3. Proof of Theorem 2. Assume that there is a point z_0 for which $g(z_0) = 0$. Then as in §2 we can linearize the given differential equation $gg'' - a(z)\, g'^2 = e^P$ with a polynomial P, which may be a constant. Then we have

$$g''(1-2a) + (P'a - a')\, g' = kg \tag{3.1}$$

and

$$g''' - P'g'' + kg' = 0 \tag{3.2}$$

with a polynomial k. By differentiation of (3.1) and by (3.2)

$$g''(P' - P'a - 3a') + g'(-2k + 2ak + P'a + Pa' - a'') - k'g = 0.$$

This equation together with (3.1) gives, $Vg' + Ug = 0$ with

$$V = (P' - P'a - 3a')\,(P'a - a') + (2a-1)\,(2ka - 2k + P'a + Pa' - a'')$$

and

$$U = k'\,(1-2a) - k(P' - P'a - 3a').$$

Since g and g' do not have any common zero, we can prove

$$N\,(r, 0\; g) \le N\,(r, 0, V) \le m\,(r, V) \le O\,(m\,(r, a)\,) + O\,(\log r).$$

In fact, k and P are polynomials. Hence ord N (r, 0, g) $\leq$ ord a < 1. Let us put g = He^L, ord H < 1, with a polynomial L. Then by $gg'' - a(z)\, g'^2 = e^P$

$$H\,(H'' + 2H'L' + HL'' + HL'^2) - a\,(H' + HL')^2 = e^{P-2L}.$$

Thus P – 2L reduces to a constant α. Then $P' = 2L'$. Let us put $g' = De^L$, $D = H' + HL'$. Then by (3.2)

$$D'' + 2D'L' + DL'' + DL'^2 - P'\,(D' + DL') + kD = 0,$$

that is,

$$D'' + D\Big(k + \frac{P''}{2} - \frac{P'^2}{4}\Big) = 0.$$

If D is transcendental, then we make use of Wiman-Valiron's theory. Then by ord D $\leq$ ord H < 1

$$4k + 2P'' - P'^2 = 0.$$

Hence $D'' = 0$. This contradicts the transcendency of D. If D is a polynomial, then again

$$4k + 2P'' - P'^2 = 0.$$

and $D'' = 0$. Thus $D = H' + HL' = cz + d$ with constants c, d. If H is transcendental, then again by Wiman-Valiron's theory

$$n\,(r) = -\frac{\zeta L'(\zeta)}{1 + \epsilon_1} + \frac{c\zeta^2 + d\zeta}{(1 + \epsilon_1)\, H\,(\zeta)},$$

where n (r) is the central index of H, $\epsilon_1 \to 0$ as $r \to \infty$ and $H(\zeta) = \max_{|z| = r} |H(z)|$, $|\zeta| = r$. Here $H(\zeta) \to \infty$ transcendentally as $r \to \infty$. The above equality holds for all $r = |\zeta|$ excepting a set E of r of finite logarithmic measure. Thus by ord H < 1, $L'(\zeta) \to 0$ as $\zeta \to \infty$ $|\zeta| \notin E$ and then for any ζ. Thus $L' = 0$. This shows that $H' = cz + d$, which is a contradiction. If H is a polynomial, then deg $HL' = 1$ unless deg $H' = 1$, $L' = 0$. If $L' = 0$, deg $H' = 1$ then $H = cz^2/2 + dz + b$, that is,

$g=(cz^2/2 + dz + b)e^{\beta}$, $L = \beta$. However

$$gg'' - a(z)\, g'^2$$

has infinitely many zeros. if deg $HL' = 1$, then either deg $H = 1$, deg $L' = 0$ or deg $H = 0$, deg $L' = 1$. If deg $H = 0$, then H is a constant. Thus g does not have any zero, which may be omitted off. If deg $H = 1$ and deg $L' = 0$, then L' is a constant β and $H\beta + \gamma = cz + d$, $\gamma = H'$. Thus $g = \delta z + \epsilon$. Hence

$$gg'' - a(z)\, g'^2 = -\delta\, a(z),$$

which has infinitely many zeros.

Suppose that $g = e^L$. Then

$$L'' + L'^2(1 - a(z)) = e^{P-2L}.$$

Thus $P - 2L$ should be a constant β. Then

$$-L'^2 a(z) + L'^2 + L''$$

has infinitely many zeros. Thus the case $g = e^L$ does not occur. Thus we have the desired result.

If a (z) is a non-constant polynomial, then it is possible to get a similar result as in Theorem 2 quite similarly.

4. Proof of Theorem 3. Firstly we consider

$$zgg'' - zg'^2 + \tfrac{1}{2} g'g = C$$

with a non-zero constant C. By differentiation

$$zgg''' - zg'g'' + \tfrac{3}{2} gg'' - \tfrac{1}{2} g'^2 = 0.$$

Let z_0 be a point at which $g(z_0) = 0$. Then $-z_0 g'(z_0)^2 = C \neq 0$ and $-z_0 g'(z_0)\, g''(z_0) - g'(z_0)^2/2 = 0$, that is, $-z_0 g''(z_0) - g'(z_0)/2 = 0$.

Hence

$$zg'' + \frac{1}{2} g' + kg = 0$$

with a polynomial k. Further

$$zg''' + \frac{3}{2} g'' + kg' = 0.$$

These two equations give $k'g = 0$ and hence $k' = 0$. Therefore k is a constant. Let $z = \zeta^2$ and $G(\zeta) = g(\zeta^2)$. Then

$$\zeta^2 g''(\zeta^2) + \frac{1}{2} g'(\zeta^2) + kg(\zeta^2) = 0$$

shows that

$$G''(\zeta) + 4kG(\zeta) = 0.$$

Hence

$$G(\zeta) = c_1 \exp(\sqrt{4k}\ i\zeta) + c_2 \exp(-\sqrt{4k}\ i\zeta).$$

This is equivalent to

$$g(z) = c_1 \exp(\sqrt{4kz}\ i) + c_2 \exp(-\sqrt{4kz}\ i).$$

By a rotation of z around the origin g (z) is transformed into g (z) itself. Hence $c_1 = c_2$.

Therefore

$$g\ (z) = B \cos \sqrt{4kz}.$$

Suppose that g (z) does not have any zero. Then $g = e^L$. This gives

$$zL'' + \frac{1}{2} L' = Ce^{-2L}$$

Hence L should be a constant. Hence we have $C = 0$, which is a contradiction.

Next suppose that with a non-constant polynomial P

$$zgg'' - zg'^2 + \frac{1}{2} gg' = e^P$$

By differentiation

$$\begin{aligned} zgg''' - zg'g'' - \frac{1}{2} g'^2 + \frac{3}{2} gg'' &= P' e^P \\ &= zP'gg'' - zP'g'^2 + \frac{1}{2} P'gg'. \end{aligned} \tag{4.1}$$

Suppose that there is a point z_0 for which $g(z_0) = 0$. Then

$$-z_0 g'(z_0)^2 = e^{P(z_0)} \neq 0$$

and

$$-z_0 g''(z_0) + \left(z_0 P'(z_0) - \tfrac{1}{2}\right) g'(z_0) = 0.$$

Hence

$$zg'' + \left(\tfrac{1}{2} - zP'\right) g' + kg = 0 \tag{4.2}$$

with a polynomial k. Further by (4.1) and (4.2)

$$zg''' + \left(\tfrac{3}{2} - zP'\right) g'' + \left(k - \tfrac{1}{2}P'\right) g' = 0. \tag{4.3}$$

(4.2) and (4.3) give

$$\left(\tfrac{1}{2} P' + zP''\right) g' = k'g. \tag{4.4}$$

If $k' \neq 0$, then $zP'' + P'/2 \neq 0$. Since $N(r, 0, g) = O(\log r)$ in this case, g has only finitely many zeros. Let us put $g = He^L$. $\deg H \geq 1$, since g has at least one zero. Further L is not a constant, since P is not a constant. By

$$zH(H'' + H'L' + HL'' + HL'^2) - z(H' + HL')^2 + \tfrac{1}{2}H(H' + HL') = e^{P-2L},$$

$P - 2L$ should be a constant α. Hence $P' = 2L'$, $P'' = 2L''$. By (4.4) we have

$$Hk' = (H' + HL')\left(\tfrac{1}{2}P' + zP''\right). \tag{4.5}$$

By (4.2)

$$z(H'' + H'L' + HL'' + HL'^2) + \left(\tfrac{1}{2} - zP'\right)(H' + HL') + kH = 0. \tag{4.6}$$

Let us put $t = \deg k$, $s = \deg H$, $p = \operatorname{dep} P'$. Suppose that $p \geq 1$. Put $k = a_t z^t + \cdots$, $H = b_s z^s + \cdots$ and $P' = c_p z^p + \cdots$. Then by (4.6)

$$\begin{aligned} 0 &= kH - \tfrac{1}{4}zHP'^2 + \cdots \\ &= \left(a_t z^t + \cdots - \tfrac{1}{4}c_p^2 z^{2p+1} + \cdots\right)\left(b_s z^s + \cdots\right) + \cdots. \end{aligned}$$

Hence $t = 2p + 1$ and $a_t = c_p^2/4$. Further by (4.5)

$$0 = Hk' - \frac{1}{2}HL'P' - HzL'P'' + \cdots$$
$$= z^{s+2p}c_p^2 b_s(t - 1 - p) + \cdots.$$

Thus $t = p + 1$ but $t = 2p + 1$. Thus $p = 0$, which contradicts $p \geq 1$. Therefore $p = 0$, that is, P' is a constant c. $c \neq 0$ since $zP'' + P'/2 \neq 0$. Then by (4.5)

$$H\left(k' - \frac{c^2}{4}\right) = \frac{1}{2}cH'$$

we have

$$k' = \frac{c^2}{4},$$

since $\deg H \geq 1$. Hence $H' = 0$. Then we have

$$g' = \frac{c}{2}g, \qquad g = \exp\left\{\frac{c}{2}z + d\right\}.$$

This g does not have any zero. Hence we may omit this case. If $k' = 0$ and $zP'' + P'/2 = 0$ imply $P = \text{const.}$, which may be omitted off.

Suppose that $g = e^L$. Then

$$zL'' + \frac{1}{2}L' = e^{P - 2L}.$$

Hence $P - 2L$ should be a constant c. Then

$$zL'' + \frac{1}{2}L' = e^c$$

This implies that L' is a constant α. Hence $L = \alpha z + \beta$. Therefore

$$g = Be^{\alpha z}, \quad B \neq 0.$$

Thus we have the desired result.

5. An application of linearization.

Let us consider the following problem: When does not $g' + (B - Ae^{-z})g^2$ have any zero? Here $A\ (\neq 0)$, B are constants and g is an entire function.

Firstly, suppose that g has no zero, that is, $g = e^L$. Then

$$g' + (B - Ae^{-z})\, g^2 = L'\, e^L + Be^{2L} - Ae^{2L - z}$$

has the form e^P. Assume that L is transcendental. Then

$$L' + (B - Ae^{-z})\, e^L = e^{P-L}.$$

Let us put $P - 2L = p$. If p is a polynomial, then

$$L' + (B - Ae^{-z} - e^p)\, e^L = 0.$$

This gives $L' = 0$, that is , L is a constant. This is untenable. If p is transcendental, then, by making use of the Borel unicity theorem to $L' + (B - Ae^{-z})\, e^L - e^{p + L} = 0$, we have a contradiction. Hence L is a polynomial. Let us put $s = \deg L$. If p is linear, then $L' + (B - Ae^{-z} - e^P)\, e^L = 0$ implies $L' = 0$, which is untenable. If p is not linear, then again by the Borel unicity theorem we have a contradiction. If $s = 1$, that is, $L = \alpha z + \beta$, $\alpha \neq 0$, then $P - L = \gamma z + \delta$. Hence

$$\alpha + (B - Ae^{-z})\, e^\beta e^{\alpha z} = e^\delta e^{\gamma z}.$$

Suppose $B \neq 0$. Then only one possible case is $\alpha = \gamma = 1$. This gives $g = e^z/A$.

Suppose $B = 0$. Then

$$\alpha - Ae^\beta e^{(\alpha - 1)\, z} = e^\delta e^{\gamma z}.$$

In this case $\alpha = 1$, $\gamma = 0$ and $1 - Ae^\beta = e^\delta$. Thus

$$g = (1 - e^\delta)\, e^z/A, \quad e^\delta \neq 1.$$

Next assume that $s = 0$, that is, g is a constant. This gives

$$C^2 B - C^2 Ae^{-z}, \quad g = c, \quad c \neq 0.$$

Thus $B = 0$.

Next suppose that g has at least one zero z_0. Then by

$$g' + (B - Ae^{-z})\, g^2 = e^P$$

$g'(z_0) = e^{P(z_0)} \neq 0$. Hence g has only simple zeros. By differentiation

$$g'' + 2(B - Ae^{-z})\, gg' + Ae^{-z}g^2 = P'e^P$$
$$= P'g' + P'(B - Ae^{-z})\, g^2. \tag{5.1}$$

Hence $g''(z_0) - P'(z_0)\, g'(z_0) = 0$. Thus

$$g'' - P'g' = kg \tag{5.2}$$

with a suitable entire function k. By (5.1)

$$g'(2B - 2Ae^{-z}) + g\,(-BP' + (1 + P')\, Ae^{-z}) = -k. \tag{5.3}$$

(5.2) and (5.3) are the linearization of the given Riccati differential equation $g' + (B - Ae^{-z})\, g^2 = e^P$. By differentiation of (5.3)

$$g''(2B - 2Ae^{-z}) + g'(-BP' + (3 + P')\, Ae^{-z})$$
$$+ g(-BP'' + (P'' - P' - 1)\, Ae^{-z}) + k' = 0.$$

Therefore by (5.2)

$$g'(BP' + (3 - P')\, Ae^{-z}) + g(2Bk - BP'' + (P'' - P' - 1 - 2k)\, Ae^{-z}) + k' = 0. \tag{5.4}$$

Hence by elimination of g'

$$g(A_1 + Ae^{-z}A_2 + A^2e^{-2z}A_3) = B_1 + Ae^{-z}B_2), \tag{5.5}$$

where

$$A_1 = -B^2\,(P'^2 + 4k - 2P''),$$
$$A_2 = B\,(2 + 2P'^2 - 4P'' + 8k),$$
$$A_3 = 1 - P'^2 + 2P'' - 4k,$$
$$B_1 = B\,(-kP' + 2k'),$$
$$B_2 = -(3 - P')\,k - 2k'.$$

Suppose $A^2e^{-2z}A_3 + Ae^{-z}A_2 + A_1 \not\equiv 0$ and hence $B_1 + Ae^{-z}B_2 \not\equiv 0$. By the original equation

$$m(r, e^P) \le 3m(r, g)\,(1 + o(1)\,) + \tfrac{r}{\pi}(1 + o(1)\,) + O(1)$$

for $r \notin E$, where E is a set of finite linear measure. Further

$$\begin{aligned} m(r, P') &= m(r, P'e^P/e^P) \\ &= O(\log m\,(r, e^P) + \log r) \\ &= O(\log m\,(r, g) + \log r) \end{aligned}$$

for $r \notin E$. Therefore by (5.2)

$$\begin{aligned} m(r, k) &\leq m\left(r, \frac{g''}{g}\right) + m\left(r, \frac{g'}{g}\right) + m(r, P') + O(1) \\ &= O(\log m(r, g) + \log r) \end{aligned}$$

for $r \notin E$. Further

$$m(r, A_j) = O(\log m(r, g) + \log r),$$

$$m(r, B_j) = O(\log m(r, g) + \log r).$$

Thus by (5.5)

$$m(r, g) \leq 4m(r, e^{-z}) + O(\log m(r, g) + \log r).$$

Therefore

$$m(r, g)\,(1 + o(1)\,) \leq 4m(r, e^{-z}),\ r \notin E.$$

Hence this holds without exception set. Thus g is an entire function of order at most one. Suppose that $A_1 + Ae^{-z}A_2 + A^2e^{-2z}A_3 \equiv 0$, $B_1 + Ae^{-z}B_2 \equiv 0$. In this case (5.3) and (5.4) are the same equation. Hence coefficients are proportional. By $B_1 + Ae^{-z}B_2 = 0$

$$B(2k' - P'k) + Ae^{-z}(\,(P' - 3)\,k - 2k') = 0.$$

Assume that $2k' - P'k = 0$ at ζ. Then $(3 - P')\,k + 2k' = 0$ at ζ. Thus $k(\zeta) = 0$ and $k'(\zeta) = 0$. Hence $k/k' = 0$ at ζ. On the other hand by the proportionality of coefficients of (5.3) and (5.4)

$$\frac{2B - 2Ae^{-z}}{BP' + (3 - P')\,Ae^{-z}} = \frac{k}{k'},$$

we have $Ae^{-\zeta} = B$. Thus

$$\zeta = \log \frac{A}{B} + 2p\pi i$$

with an integer p. Thus at ζ

$$g(-BP' + Ae^{-z}(1 + P')\,) = -k = 0,$$

that is,

$$Bg(\zeta) = 0.$$

Now assume that $B \neq 0$. Then $g(\zeta) = 0$. Hence by (5.4)

$$g'(\zeta)\, 3B + g(\zeta)\, K(\zeta) = -k'(\zeta) = 0,$$

$$K(\zeta) = 2Bk(\zeta) - BP''(\zeta) + Ae^{-\zeta}(-2k(\zeta) - 1 - P'(\zeta) + P''(\zeta)\,).$$

This implies that $g'(\zeta) = 0$. This is a contradiction, since g and g' do not have any common zero. Assume that $B = 0$. Then $A_1 = A_2 = B_1 = 0$. This case gives $Ae^{-z}A_3g = B_2$ as (5.5). Hence $A_3 = B_2 = 0$ in the present case. Therefore we have by $B_2 = 0$

$$k(3 - P') + 2k' = 0,$$

which implies that

$$k = e^D e^{\frac{1}{2}(P - 3z)}.$$

By $A_3 = 0$ we have

$$4k = 4e^D e^{\frac{1}{2}(P - 3z)} = 1 + 2P'' - P'^2.$$

By differentiation

$$\frac{P' - 3}{2}(1 + 2P'' - P'^2) = 2P''' - 2P'P''.$$

By Clunie's lemma [1]

$$m(r, P') < O(\log r) + o(T\,(r, P')\,),\ r \notin E.$$

This shows that P' is a polynomial and so is k. Hence $P - 3z$ is a constant C. Thus $k = -2$, $P' = 3$. Then by (5.2) $g'' - 3g' + 2g = 0$, which shows that g is of order 1. Suppose that $2k' - kP'$ has no zero. Then $2k' + 3k - P'k$ has no zero either. Let us put

$$2k' - kP' = e^{s},\ 3k + 2k' - P'k = e^{T}.$$

Hence $Be^{s} = Ae^{-z+T}$ by $B_1 + Ae^{-z}B_2 = 0$. Thus T-s-z reduces to a constant c and $e^{c} = B/A$. Further

$$3k = e^{s}\,(Be^{z} - A)\,/\,A,\quad 3k' = e^{s}(B\,(s' + 1)\,e^{z} - As')\,/\,A.$$

Hence

$$P' = \frac{3A + 2s'A - 2B\,(s' + 1)\,e^{z}}{A - Be^{z}}.$$

Suppose that ζ is a zero of $A - Be^{z}$. Then

$$0 = 3A + 2s'(\zeta)\,A - 2B\,(s'(\zeta) + 1)\,e^{\zeta} = A.$$

This is a contradiction.

From now on we may consider the case that g is of finite order. Then P and k are polynomials. By (5.5) $A_1 + Ae^{-z}A_2 + A^2e^{-2z}A_3$ is a divisor of $B_1 + Ae^{-z}B_2$. If $A_1 \neq 0$, $A_3 \neq 0$, then

$$N(r,\,0,\,A_1 + Ae^{-z}A_2 + A^2e^{-2z}) \sim 2m(r,\,e^{-z}).$$

But

$$N(r,\,0,\,B_1 + Ae^{-z}B_2) \leq m(r,\,e^{-z})\,(1 + o(1)\,).$$

This is impossible. Hence either A_1 or A_3 is equal to zero. Suppose $A_3 = 0$. Then $A_1 = -B^2$, $A_2 = 4B$ and

$$B_1 = B(P'^3 - 6P'P'' - P' + 4P''')\,/\,4,$$

$$B_2 = (-P'^3 + 6P'P'' + P' - 4P''' + 3P'^2 - 6P''' - 3)\,/\,4.$$

Thus

$$A_1 + Ae^{-z}A_2 + A^2e^{-2z}A_3 = -B^2 + 4BAe^{-z} = 0$$

implies that $Ae^{-z}B_2 + B_1 = 0$, then $B \neq 0$. $4Ae^{-z} - B$ vanishes at a countably infinite number of points $-\log(B/4A) + 2p\pi i$ with an integer p. At these points $Ae^{-z}B_2 + B_1 = 0$. Hence

$$-P'^3 + 6P'P'' + P' - 4P''' + 3P'^2 - 6P'' - 3 = 4\,(P'^3 - 6P'P'' - P' + 4P''')$$

holds at $-\log(B/4A) + 2p\pi i$. Hence this holds identically. Therefore P' should be a constant c. Hence $5(c^3 - c) = 3c^2 - 3$, that is, $c = 1$ or $c = -1$ or $c = 3/5$.

If $c^2 = 1$, then $k = 0$. Hence by $g'' - cg' = 0$

$$g' = Me^{cz}, \qquad g = \frac{M}{c}\,e^{cz} + D.$$

Returning back to the original problem

$$g' + (B - Ae^{-z})\,g^2 = Me^{cz} + (B - Ae^{-z})\left(\frac{M^2}{c^2}e^{2cz} + 2\frac{MD}{c}e^{cz} + D^2\right).$$

If $c = 1$, then the above expression has the following form:

$$BM^2e^{2z} + (M + 2MDB - AM^2)\,e^z + BD^2 - 2AMD - AD^2e^{-z}.$$

In this expression four terms should vanish and only one should remain. There appear only three cases:

$$D = 0,\ AM = 1 \quad \text{or} \quad D = 0,\ B = 0 \quad \text{or} \quad D \neq 0,\ B = 0,\ M = 0.$$

If $D = 0$, then g does not have any zero, which is a contradiction. If $M = 0$, $g' = 0$ so $g = D$, which is again a contradiction. If $c = -1$, then we have a contradiction easily. If $c = 3/5$, then $k = 4/25$. By $g'' - 3g'/5 - 4g/25 = 0$ we have

$$g = c_1e^{4z/5} + c_2e^{-z/5}.$$

Then by (5.3)

$$Bc_1e^{4z/5} - Bc_2e^{-z/5} + 2Ac_2e^{-6z/5} = -\frac{4}{25}.$$

This is clearly a contradiction. Suppose $A_1 = 0$. Then we have either $B = 0$ or $4k = 2P'' - P'^2$. If $4k = 2P'' - P'^2$, then $A_2 = 2B$, $A_3 = 1$ and

$$B_1 = \frac{B}{4}\left(P'^3 - 4P'P'' + 2P'''\right),$$

$$B_2 = \frac{1}{4}\left(-P'^3 + 6P'P'' - 10P''' + 3P'^2\right).$$

By (5.5) $gAe^{-z}(2B + Ae^{-z}) = B_1 + Ae^{-z}B_2$. $2B + Ae^{-z}$ is a divisor of $B_1 + Ae^{-z}B_2$. Hence at any ζ for which $2B + Ae^{-\zeta} = 0$

$$B_1 + Ae^{-\zeta}B_2 = 0.$$

Then either $B = 0$ or $3P'^3 - 6P'^2 - 16P'P'' + 22P''' = 0$. Suppose $B \neq 0$. Then P' reduces to a constant c and $c^3 - 2c^2 = 0$, that is, $c = 0$ or $c = 2$. If $c = 0$, then $k = 0$ and hence $g'' = 0$, that is, $g = \alpha z + \beta$. However this does not give $g' + (B - Ae^{-z})\, g^2 = e^P$. If $c = 2$, then $k = -1$. Thus by (5.2)

$$g'' - 2g' + g = 0.$$

Hence

$$g = c_1e^z + c_2ze^z.$$

Then

$$\begin{aligned} & g' + \left(B - Ae^{-z}\right)g^2 \\ &= \left\{c_1 + c_2 - Ac_1^2 + (c_2 - 2Ac_1c_2)\, z - Ac_2^2z^2\right\} e^z \\ &+ B(c_1^2 + 2c_1c_2z + c_2^2z^2)\, e^{2z}. \end{aligned}$$

Only one term should remain. Hence $c_2 = 0$ and $Ac_1 = 1$. Then $g = c_1e^z$, which has no zero. Hence we may omit.

Still there remains the case $B = 0$ to be investigated under the condition that $g(z)$ has at least one zero. In this case $A_1 = A_2 = B_1 = 0$ and hence

$$AgA_3 = B_2e^z.$$

A_3 is a divisor of B_2. So we put $B_2 = A_3t$. It is sufficient to consider the case deg t $\geq$ 1, since g has at least one zero. By (5.3) $-g'2Ae^{-z} + gA(1 + P')\, e^{-z} = -k$. Thus

$$-2Ag' + Ag(1 + P') = -ke^{z}.$$

Therefore by $Ag = te^z$ we have

$$-2\,(t' + t) + t\,(1 + P') = -k.$$

Further we have

$$-(3 - P')\,k - 2k' = B_2 = tA_3 = (-4k - P'^2 + 2P'' + 1)\,t.$$

By elimination of k and k'

$$4tP' + 4t'P' - 4t - 8t' - 4t'' = 4t^2P' - 4t^2 - 8tt'.$$

Let us put $s = \deg t \geq 1$, $\deg P' = p$. Assume that $p \geq 1$. Now we shall compare the degrees of both sides. Then $s = 0$, which is a contradiction. Hence $p = 0$. Put $P' = c$. Then

$$4t\,(c - 1) + 4t'\,(c - 2) - 4t'' = 4t^2\,(c - 1) - 8tt'.$$

Again we shall compare the degrees of both sides. If $c \neq 1$, then $s = 0$, which is a contradiction. If $c = 1$, then again $s = 0$, which is untenable.

If $A_3 \equiv 0$, then $B_2 \equiv 0$. Hence

$$4k = 2P'' - P'^2 + 1, \quad (3 - P')\,k + 2k' = 0.$$

Hence

$$P'^3 - 3P'^2 - 6P'P'' - P' + 6P'' + 4P''' + 3 = 0$$

Thus $P' = c$ and $c = 1, -1$ or 3. Then $k = 0, 0$ or -2, correspondingly. If $c = 1$, then $g'' - g' = 0$ shows that $g' = e^d e^z$ and $g = e^d e^z + \beta$. $\beta \neq 0$, since g has zeros. In this case

$$g' - Ae^{-z}g^2 = e^d\,(1 - Ae^d)\,e^z - \beta^2Ae^{-z} - 2\beta Ae^d$$

has infinitely many zeros. Hence $g' - Ae^{-z}g^2 \neq e^P$. If $c = -1$, then $g'' + g' = 0$ shows that $g' = e^d e^{-z}$, $g = -e^d e^{-z} + \beta$, $\beta \neq 0$. In this case

$$g' - Ae^{-z}g^2 = (e^d - A\beta^2)\, e^{-z} + 2\beta Ae^d e^{-2z} - Ae^{2d} e^{-3z}$$

does not have the form e^P. If $c = 3$, then $k = -2$ and by (5.2)

$$g'' - 3g' + 2g = 0.$$

Therefore

$$g = c_1 e^z + c_2 e^{2z}.$$

In this case

$$g' - Ae^{-z}g^2 = c_1(1 - Ac_1)\, e^z + 2c_2(1 - Ac_1)\, e^{2z} - Ac_2^2\, e^{3z}.$$

If $c_2 = 0$, the $g = c_1 e^z$, which has no zero. This case may be omitted off. Hence $c_2 \neq 0$. Then $c_1 = 1/A$ gives

$$g' - Ae^{-z}g^2 = -Ac_2^2\, e^{3z}.$$

This case is only one possible case. Hence we have

$$g = \frac{1}{A}e^z + c_2 e^{2z}, \qquad c_2 \neq 0.$$

Summing up the above results, we have the following

Theorem 4. *Assume that* g *is entire. Then* $g' + (B - Ae^{-z})\, g^2$ $(A \neq 0)$ *does not have any zeros when and only when*

1) $g = e^z/A$ *if* $B \neq 0$,

2) $g = c \neq$ *or* $g = (1 - c)\, e^z/A$, $c \neq 0, 1$ *or* $g = e^z/A + ce^{2z}$, $c \neq 0$ *if* $B = 0$.

Corollary. *The following Riccati differential equation*

$$g' + (B - Ae^{-z})\, g^2 = e^P,\ A \neq 0$$

has entire solutions

1) $g = e^z/A$ *if* $B \neq 0$ *and* $P = 2z + D + 2p\pi i$, $e^D = B/A^2$,

2) $g = c \neq 0$ *if* $B = 0$ *and* $P = -z + D + 2p\pi i$, $e^D = -Ac^2$,

3) $g = (1 - c)\, e^z/A$ *if* $B = 0$ *and* $P = z + D + 2p\pi i$, $e^D = c(1 - c)/A$,

4) $g = e^z/A + ce^{2z}$ *if* $B = 0$ *and* $P = 3z + D + 2p\pi i$, $e^D = -Ac^2$,

where p *is an arbitrary integer. There is no other entire solution.*

References

[1] Clunie, J., On integral and meromorphic functions. J. London Math. Soc. 37 (1962), pp. 17 - 27.

[2] Hayman, W. K., Picard values of meromorphic functions and their derivatives. Ann. of Math. 70 (1959), pp. 9 - 42.

[3] Mues, E., Über die Nullstellen homogener Differentialpolynome. Manuscripta Math. 23 (1978), pp. 325 - 341.

[4] Nevanlinna, R., Einige Eindeutigkeitssätze in der Theorie der meromorphen Funcktionen. Acta Math. 48 (1926), pp. 367 - 391.

[5] Nevanlinna, R., Le théorème de Picard-Borel et la théorie des fonctions méromorphes. Paris, Gauthier-Villars (1929).

[6] Valiron, G., Lectures on the general theory of integral functions. New York, Chelsea (1949).

Department of Mathematics,
Science University of Tokyo

SURVEYS OF ANALYTIC MAPPINGS BETWEEN TWO RIEMANN SURFACES

Mitsuru Ozawa and Kiyoshi Niino

§ 1. Introduction.

In 1925 R. Nevanlinna [1] established the value distribution theory of meromorphic functions f: $\mathbb{C} \to P_1\mathbb{C}$. He proved so-called the first fundamental theorem, the second fundamental theorem and the defect relation. His theory is a far-reaching generalization of Picard-Borel theorem.

After Nevanlinna there appeared many extensions of his theory. Main extensions are ones to algebroid functions by H. L. Selberg [1] (cf. [2]), E. Ullrich [1] and G. Valiron [1], to the system of entire functions f: $\mathbb{C} \to f = (f_0(z), f_1(z), \cdots, f_n(z))$ by H. Cartan [4], to holomorphic curves x: $\mathbb{C} \to P_n\mathbb{C}$ by H. and J. Weyl [1] and L. V. Ahlfors [1] (cf. Weyl [1], Wu [1]) and to analytic mappings between two Riemann surfaces by L. Sario (cf. Sario-Noshiro [1]).

We now pay attention to analytic mappings φ: R $\to$ S between two Riemann surfaces. In general, Sario established the second fundamental theorem like Nevanlinna's. When S is a closed Riemann surface, he obtained a defect relation like Nevanlinna's from the second fundamental theorem. However if S is open and of infinite genus, then a defect relation could not be obtained from the second fundamental theorem. This was shown by Rodin-Sario [1]. It seems to us from the result of Rodin-Sario and our Theorem 4.2 that when S is of finite genus, we shall not be able to have a defect relation like Nevanlinna's, that is, we shall not be able to establish Nevanlinna theory when S is of infinite genus. So we have to try another treatment.

In this paper we shall state the study on analytic mappings created by the first author.

First of all we have a problem: when do there exist any non-constant analytic mappings of a given R into a given S?

If $R \notin O_{AB}$, then there are many non-trivial analytic mappings. So we may assume $R \in O_{AB}$, that is, there is no bounded analytic function on R.

We have a criterion of non-existence of analytic mappings between two Riemann surfaces. In order to state the criterion, we now introduce Picard constant P(R) of R. Let R be an open Riemann surface. Let $\mathcal{M}(R)$ be the family of non-constant meromorphic functions on R. Let P(f) be the number of values which are not taken by $f \in \mathcal{M}(R)$. We define

$$P(R) = \sup \{P(f) \ ; f \in \mathcal{M}(R)\}.$$

Then we have the following criterion:

Theorem 1.1 (Ozawa [1]). *If* $P(R) < P(S)$, *then there is no non-constant analytic mapping of* R *into* S.

Proof. Contrarily suppose that there is an analytic mapping φ of R into S. For any $f \in \mathcal{M}(S)$, $f(\varphi) \in \mathcal{M}(R)$ and we have $P(f) \leq P(f(\varphi)) \leq P(R)$. It follows that $P(S) \leq P(R)$, which is a contradiction. q.e.d.

In general $P(R) \geq 2$. In fact, there is a non-constant analytic function f on every open Riemann surface by the existence theorem due to Behnke-Stein [1] and then the composition of the exponential function and f does not take at least two values. It is very difficult in general to calculate P(R) of a given open Riemann surface R, because it depends on the theory of value distributions on R.

From the value distribution theory of algebroid functions every n-valued algebroid function has at most 2n Picard exceptional values. Hence we have $P(R_n) \leq 2n$ for an n-sheeted algebroid surface R_n, which is the proper existence domain of an n-valued algebroid function.

It is assumed that the reader is familiar with the Nevanlinna-Selberg theory of meromorphic functions and algebroid functions and the notations, T(r, f), m(r, f), N(r, 0, f), N(r, $\mathfrak{X}$), etc. (cf. Nevanlinna [3], [4], Hayman [1]; Selberg [2], Hiong [1].)

§ 2. Preparation.

In this section we state our important tools used in order to obtain our Theorems.

2.1 Borel's formulation of Picard's theorem.

It is well known that Picard's theorem is equivalent to the impossibility of the identity $A(z)e^{H(z)} + B(z)e^{L(z)} = C(z)$, $(C \not\equiv 0)$, where A, B and C are polynomials and H and L are non-constant entire functions. In fact, if f(z) is a meromorphic functions in $|z| < +\infty$ having three Picard's exceptional values, 0, 1, ∞, then $f(z) = P(z)e^{H(z)}$ and $f(z) - 1 = Q(z)e^{L(z)}$, where P and Q are rational functions. Inversely we have $P(z)e^{H(z)} - Q(z)e^{L(z)} = 1$, where $P = A/C$ and $Q = -B/C$. Then $f(z) = P(z)e^{H(z)}$ and $f(z) = 1 + Q(z)e^{L(z)}$. Hence f(z) has three Picard's exceptional values 0, 1, ∞.

We have the following, which is an extension of Nevanlinna [2] and Wittich [1]:

Theorem 2.1 (Hiromi-Ozawa [1]). *Let* $a_0, a_1, \ldots, a_n$ *be meromorphic functions and* $g_1, \ldots, g_n$ *entire functions. Further suppose*

$$T(r, a_j) = o\Big(\sum_{\nu=1}^{n} m(r, e^{g_\nu})\Big), \qquad r \to \infty,$$

holds for every j, *outside a set of* r *of finite measure. If an identity*

$$\sum_{\nu=1}^{n} a_\nu(z)e^{g_\nu(z)} = a_0(z)$$

holds, then we have an identity

$$\sum_{\nu=1}^{n} c_\nu a_\nu(z)e^{g_\nu(z)} = 0,$$

where c_ν *are constants which are not all zeros.*

The following are useful to study analytic mappings of algebroid surfaces with maximal Picard constants.

Lemma 2.2 (Niino [5]). *Let* $a_{\mu,\ \nu}$ $(\mu,\ \nu = 0,\ 1,\ \cdots,\ m)$ *be meromorphic functions and* H *and* M *two non-constant entire functions such that*

$$m(r,\ e^H) \sim m(r,\ e^M) \quad \textit{and} \quad T(r,\ a_{\mu,\ \nu}) = o\big(m(r,\ e^H)\big) \quad (r \to \infty)$$

hold for $\mu,\ \nu = 0,\ 1,\ \cdots,\ m$, *outside a set of* r *of finite measure. Further suppose that the following identity holds:*

$$\sum_{\mu,\ \nu=0}^{m} a_{\mu,\ \nu}(z)e^{\mu H(z) + \nu M(z)} = 0.$$

(I) *If* $a_{m,\ m}(z) \not\equiv 0$, *then* $a_{m,\ 0}(z) \equiv a_{0,\ m}(z) \equiv 0$ *and* $m(r,\ e^{H+M}) = o(m(r,\ e^H))$ $(r \to \infty)$ *outside a set of* r *of finite measure.*

(II) *If* $a_{\nu,\ \nu}(z) \equiv 0$ $(\nu = 0,\ 1,\ \cdots,\ m-1)$ *and* $a_{m,\ 0}(z) \not\equiv 0$ *or* $a_{0,\ m}(z) \not\equiv 0$, *then* $a_{m,\ m}(z) \equiv 0$ *and* $m(r,\ e^{H-M}) = o(m(r,\ e^H))$ $(r \to \infty)$ *outside a set of* r *of finite measure.*

Theorem 2.3 (Niino [5]). *Let* m *and* k *be two integers satisfying* $1 \le k \le m/2$. *Suppose that two non-constant entire functions* M *and* H *satisfying the following functional equation*

$$(e^{M(z)} - \gamma)^k (e^{M(z)} - \delta)^{m-k} = F(z)\, (e^{H(z)} - \sigma)^k (e^{H(z)} - \tau)^{m-k},$$

$$M(0) = H(0) = 0,\ \gamma\delta\sigma\tau(\gamma - \delta)(\sigma - \tau) \ne 0$$

with four constants $\gamma,\ \delta,\ \sigma$ *and* τ *and a meromorphic function* $F(z) = f_1(z)^m f_2(z)$ *where* f_1 *and* f_2 *are meromorphic in* $|z| < +\infty$ *and* f_2 *has no zero and no pole in* $|z| > r_0$ *for a positive number* r_0. *Then we have one of the following four cases:*

(i) $H(z) = M(z),\ F(z) = 1, \quad \gamma = \sigma,\ \delta = \tau,$

(ii) $H(z) = M(z),\ F(z) = 1, \quad \gamma = \tau,\ \delta = \sigma,$

(iii) $H(z) = -M(z),\ F(z) = (-1)^m \gamma^k \delta^{m-k} e^{mM(z)}, \ \gamma\sigma = \delta\tau = 1,$

(iv) $H(z) = -M(z),\ F(z) = (-1)^m \gamma^{m-k} \delta^k e^{mM(z)}, \ \gamma\tau = \delta\sigma = 1,$

(ii) *and* (iv) *may occur in the case* $k = m/2$ *only.*

2.2. Zeros of an entire function

We can deduce from Nevanlinna's second fundamental theorem

Lemma 2.4 (Ozawa [2]). *Let* $H(z)$ *be a non-constant entire function and* γ *a non-zero constant. Then the function* $e^{H(z)} - \gamma$ *has an infinite number of simple zeros in such a manner that*

$$N_2(r, 0, e^H - \gamma) \sim T(r, e^H), \qquad r \to \infty,$$

outside a set of finite measure, where $N_2(r, 0, f)$ *is the counting functions of simple zeros of* f.

2.3. Algebroid functions.

Let f(z) be a k-valued transcendental algebroid function of z in $|z| < +\infty$ whose defining equation is

$$F(z, f) := A_0(z)f^k + \cdots + A_{k-1}(z)\, f + A_k(z) = 0$$

where $A_0, \cdots, A_k$ are entire functions with no common zeros.

In the first place we assume that f is an entire algebroid function, that is, $A_0(z) \equiv 1$.

Definition. If $F(z, \alpha)$ reduces to a polynomial, then α is called an exceptional value of the first kind. If $F(z, \alpha)$ reduces to the form $P(z)e^{H(z)}$, where P is a polynomial and H is a non-constant entire function, then α is called an exceptional value of the second kind.

We have

Theorem 2.5 (Rémoundos [1]). *Let* f *be a* k-*valued transcendental entire algebroid function. Then there are at most* k−1 *exceptional values of the first kind and at most* k *exceptional values of the second kind.*

Theorem 2.6 (cf. Selberg [2]). *Let* f *be a* k-*valued transcendental algebroid function and* $w_1, \cdots, w_q$ *denote arbitrary* q *finite or infinite values. Then*

$$(q - 2)\, T(r, f) < \sum_{\nu=1}^{q} N(r, w_\nu, f) + N(r, \mathfrak{X}) - N(r, \mathfrak{Z}_f) + S(r, f),$$
$$N(r, \mathfrak{X}) < (2k - 2)\, T(r, f) + O(1).$$

Corollary 2.7. f *has at most* 2k *Picard's exceptional values and at most* 4k *completely ramified values.*

2.4. Functions on an ultrahyperelliptic surface.

Let R be the ultrahyperelliptic surface defined by $y^2 = G(z)$, where G is an entire function having an infinite number of simple zeros and no other zeros. Put $p = (z, y)$, $q = (z, -y) \in R$. Let f be an arbitrary function on R. We put

$$f_1(z) = \frac{1}{2}(f(p) - f(q)) \quad \text{and} \quad f_2(z) = \frac{1}{2y}(f(p) - f(q)).$$

Then f_1 and f_2 are single-valued functions of z. Hence we have the following representation of the function f on R:

$$f(p) = f_1(z) + f_2(z)\, y, \qquad p = (z, y) \in R, \tag{2.1}$$

and consequently, the defining equation of 2-valued function f of z is

$$f^2 - 2f_1(z)\, f + f_1(z)^2 - f_2(z)^2 G(z) = 0. \tag{2.2}$$

§ 3. Picard constant.

It is difficult to calculate the Picard constant of even a given algebroid surface.

However, we can characterize a regularly branched algebroid surface by maximal Picard constant, where an n-sheeted covering surface is called regularly branched if all its branch points are of order n−1. We have

Theorem 3.1 (Aogai [1], Ozawa [2] in the case n=2, Hiromi-Niino [1] in the case n=3). *Let* R *be an* n-*sheeted regularly branched algebroid surface. If* P(R) > 3n/2, *then* P(R) = 2n *and* R *can be defined by an algebroid function* y *such that*

$$y^n = (e^{H(z)} - \alpha)(e^{H(z)} - \beta)^{n-1}, \quad H(0) = 0, \ \alpha\beta(\alpha-\beta) \neq 0,$$

where H(z) *is a non-constant entire function and* α *and* β *are constants.*

Proof of Theorem 3.1 in the case when n=2. Let R be an ultra-hyperelliptic surface defined by $y^2 = G(z)$, where G is an entire function having no zero other than an infinite number of simple zeros. Assume that P(R) = 4. Then we may assume that there is an entire function f in $\mathcal{M}(R)$ such that $f \neq 0, 1, a, \infty$ ($a \neq 0, 1, \infty$). By (2.2) the defining equation of f is

$$F(z, f) := f^2 - 2f_1(z)f + f_1(z)^2 - f_2(z)^2 G(z) = 0,$$

where f_1 and f_2 are entire functions, and by Rémoundos' Theorem 2.5 either

$$\begin{matrix} F(z,0) \\ F(z,1) \\ F(z,a) \end{matrix} = \left\{\begin{matrix} c \\ \beta_1 e^{H_1(z)} \\ \beta_2 e^{H_2(z)} \end{matrix}\right. \quad \text{or} \quad \left\{\begin{matrix} \beta_1 e^{H_1(z)} \\ c \\ \beta_2 e^{H_2(z)} \end{matrix}\right. \quad \text{or} \quad \left\{\begin{matrix} \beta_1 e^{H_1(z)} \\ \beta_2 e^{H_2(z)} \\ c \end{matrix}\right.$$

holds, where β_1, β_2 and c are non-zero constants and H_1, H_2 are non-constant entire functions with $H_1(0) = H_2(0) = 0$.

In the first case we have

$$\begin{cases} f_1^2 - f_2^2 G = c, \\ 1 - 2f_1 = \beta_1 e^{H_1} - c, \\ 1 - 2a_1 = \beta_2 e^{H_2} - c, \end{cases}$$

Then

$$a\beta_1 e^{H_1(z)} - \beta_2 e^{H_2(z)} = (a - c)(1 - a).$$

By the impossibility of the Borel identity we have

$$H_1(z) \equiv H_2(z), \quad a = c \text{ and } a\beta_1 = \beta_2.$$

Therefore, putting $H_1 \equiv H_2 \equiv H$,

$$4f_2^2 G = \beta_1^2 e^{2H} - 2(1 + a)\beta_1 e^H + (1 - a)^2,$$

and consequently, with an entire function F and non-zero constants α and β we have

$$F(z)^2 G(z) = (e^{H(z)} - \alpha)(e^{H(z)} - \beta). \tag{3.1}$$

Since G has a simple zero, $\alpha \neq \beta$.

In the other cases we have quite similarly a representation (3.1) with $\alpha\beta(\alpha - \beta) \neq 0$. Here we remark that $e^{H(z)} - \alpha$ and $e^{H(z)} - \beta$ have an infinite number of zeros by Lemma 2.4.

Conversely, we assume that R is an ultrahyperelliptic surface defined by $y^2 = G(z)$, where G is the entire function having an infinite number of simple zeros only which satisfies (3.1). Then

$$f(z) := \sqrt{\frac{e^{H(z)} - \alpha}{e^{H(z)} - \beta}} = \frac{F(z)}{e^{H(z)} - \beta}\sqrt{G(z)}$$

omits four values 1, −1, $\sqrt{\alpha/\beta}$, $-\sqrt{\alpha/\beta}$ and f belongs to $\mathcal{M}(R)$. Hence $P(R) = 4$. q.e.d.

We now still state some results on the Picard constant of an ultrahyperelliptic surface R defined by

$$y^2 = G(z), \tag{3.2}$$

where G is an entire function having an infinite number of simple zeros and no other zeros.

Theorem 3.2 (Hiromi-Ozawa [1]). *If* R *is an ultrahyperelliptic surface and* P(R) = 3, *then with non-constant entire functions* H, L *satisfying* H(0) = L(0) = 0 *and non-zero constants* β_1, β_2,

$$F(z)^2G(z) = 1-2\beta_1 e^{H(z)}-2\beta_2 e^{L(z)}+\beta_1^2 e^{2H(z)}-2\beta_1\beta_2 e^{H(z)+L(z)}+\beta_2^2 e^{2L(z)}, \tag{3.3}$$

where F *is an entire function.*

Conversely,

Theorem 3.3 (Hiromi-Ozawa [1]). *Let* R *be an ultrahyperelliptic surface defined by* (3.2) *with the entire function* G *satisfying* (3.3). *Then we have* P(R) ≥ 3.

Proof. $f(z) := \frac{1}{2}(1 + \beta_1 e^{H(z)} - \beta_2 e^{L(z)}) + \frac{F(z)}{2}\sqrt{G(z)}$ is an entire function on R and its defining equation as two-valued algebroid function is

$$F(z, f) := f^2 - (1 + \beta_1 e^{H(z)} - \beta_2 e^{L(z)})\, f + \beta_1 e^{H(z)} = 0.$$

Then

$$F(z, 0) = \beta_1 e^{H(z)} \quad \text{and} \quad F(z, 1) = \beta_2 e^{L(z)}.$$

This shows that $f \neq 0, 1, \infty$ on R. Hence P(f) ≥ 3, that is, P(R) ≥ 3. q.e.d.

As corollary of Theorems 3.1 and 3.2, we have

Corollary 3.4. *If* G(z) *is of finite non-integral order, then the surface* R *defined by* $y^2 = G(z)$ *has* P(R) = 2.

It is quite difficult to determine all ultrahyperelliptic surfaces with P(R) = 3 and we did not yet succeed it perfectly. However we have the following:

Theorem 3.5 (Hiromi-Ozawa [1]). *The surface* R *defined by* (3.2) *with* G *satisfying* (3.3) *satisfies* P(R) = 3, *if*

$$m(r, e^H) = o\left(m(r, e^L)\right) \qquad r \to \infty,$$

outside a set of finite measure.

Theorem 3.6 (Ozawa [5]). *Let* R *be a surface defined by* (3.2) *with* G *satisfying* (3.3). *If* H *and* L *are polynomials, then* P(R) = 3 *with the following four exceptional cases:* (i) H = L; (ii) H = 2L, $16\beta_1 = \beta_2^2$; (iii) 2H = L, $\beta_1^2 = 16\beta_2$; (iv) H = –L, $16\beta_1\beta_2 = 1$. *In these exceptional cases* P(R) = 4.

Now let R and $\tilde{R}$ be two ultrahyperelliptic surfaces defined by two equations

$$y^2 = G(z) \tag{3.2}$$

and

$$y^2 = \tilde{G}(z), \tag{3.4}$$

where G and $\tilde{G}$ are two entire functions having no zero other than an infinite number of simple zeros, respectively. If G and $\tilde{G}$ have the same zeros in $\{ \mid z \mid \geq r_0\}$ for a suitable r_0, then we call $\tilde{R}$ a finite modification of R.

Then we have the following:

Theorem 3.7 (Niino [2]). *Let* R *be an unltrahyperelliptic surface with* P(R) = 4 *defined by* (3.2) *with* G *satisfying* (3.1). *Then we have* $P(\tilde{R})$ = 2 or 3 *for its finite modification* $\tilde{R}$. *Further if* H *in* (3.1) *is a polynomial, then* $P(\tilde{R}) = 2$.

The other results on Picard constants of algebroid surfaces are found in Niino [1].

§ 4. Analytic mappings between two ultrahyperelliptic surfaces.

An ultrahyperelliptic surface is the simplest example of Riemann surface of infinite genus. Firstly we discuss on analytic mappings between two ultrahyperelliptic surfaces.

Let R and S be ultrahyperelliptic surfaces defined by

$$y^2 = G(z) \tag{4.1}$$

and

$$u^2 = g(w), \tag{4.2}$$

respectively, where G and g are entire functions having an infinite number of simple zeros and no other zeros.

Let φ be an analytic mapping of R into S and $\mathfrak{P}_R$ (resp. $\mathfrak{P}_S$) be the projection map of R (resp. S) into z-plane (resp. w-plane) such

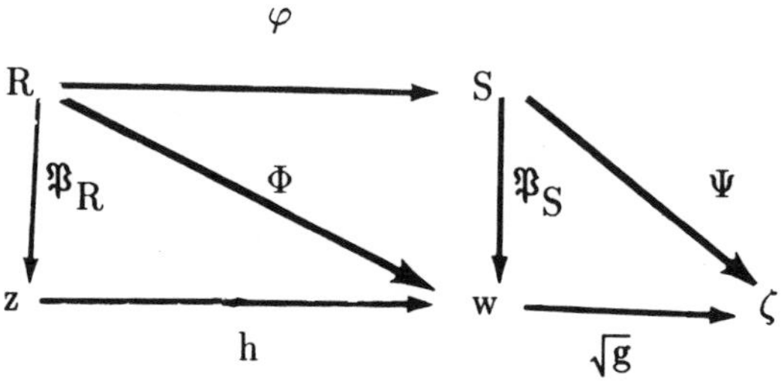

that (z, y)→ z (resp. (w, u)→ w). Put $\Phi = \mathfrak{P}_S \circ \varphi = h \circ \mathfrak{P}_R$. Then Φ is a holomorphic function. Hence from (2.1) we have $h(z) = h_1(z) + h_2(z)\sqrt{G(z)}$ where h_1 and h_2 are single-valued entire functions of z. Let Ψ be an analytic mapping of S into the complex plane satisfying $\Psi = \sqrt{g} \circ \mathfrak{P}_S$. Then $\Psi \circ \varphi = \sqrt{g} \circ \mathfrak{P}_S \circ \varphi = \sqrt{g} \circ h \circ \mathfrak{P}_R$ is a holomorphic function on R. Hence we have

$$\sqrt{g} \circ h(z) = f_1(z) + f_2(z)\sqrt{G(z)},$$

where f_1 and f_2 are single-valued entire functions of z. Therefore we have

$$\sqrt{g} \circ (h_1(z) + h_2(z)\sqrt{G(z)}) = f_1(z) + f_2(z)\sqrt{G(z)},$$

that is,

$$g \circ (h_1(z) + h_2(z)\sqrt{G(z)}) = (f_1(z) + f_2(z)\sqrt{G(z)})^2.$$

Assume that $h_2(z) \not\equiv 0$. Let w_μ be a zero of g. Then w_μ is a simple zero of g by the assumption. Therefore w_μ is either a Picard's exceptional value or a multiple value of two-valued algebroid function $h_1 + h_2\sqrt{G}$. From Corollary 2.7 a two-valued algebroid function has at most four Picard exceptional values and has at most eight totally ramified values. On the other hand there are an infinite number of such points w_μ. This is a contradiction. Hence $h_2 \equiv 0$, that is, $h(z) \equiv h_1(z)$ is a single-valued entire function of z. Then we have

$$g(h(z)) = f_1(z)^2 + f_2(z)^2 G(z) + 2f_1(z)\, f_2(z) \sqrt{G(z)}.$$

Since g(h(z)) is single-valued, we have $f_1 f_2 \equiv 0$. If $f_2 \equiv 0$, then $g(h(z)) = f_1(z)^2$. This again contradicts the Nevanlinna's ramification relation for h. Therefore $f_1 \equiv 0$, that is, $g(h(z)) = f_2(z)^2 G(z)$. Thus we have a desired functional equation

$$f(z)^2 G(z) = g(h(z)) \tag{4.3}$$

with a suitable entire function f.

Conversely if G and g satisfy the equation (4.3) with suitable entire functions h and f. Let $\varphi = \mathfrak{P}_S^{-1} \circ h \circ \mathfrak{P}_R$. We take z_0 and w_0 such that $g(w_0) \neq 0$, $h(z_0) = w_0$. Take one branch of $\mathfrak{P}_S^{-1}(w_0)$ and we do analytic continuation of φ from $p_0 = (z_0, y_0) \in R$. Then φ is well-defined mapping of R into S and evidently φ is analytic at every ordinary point. If p_1 is a branch point of R, then $z_1 = \mathfrak{P}_R p_1$ is a zero point of G(z). Thus $g \circ h(z_1) = 0$. This shows that $h(z_1)$ is a zero point of g with odd multiplicity. Thus $\mathfrak{P}_S^{-1} h(z_1)$ is a branch point of S. Let t and T be the local parameters around p_1 and $\mathfrak{P}_S^{-1} h(z_1)$, respectively. Then we may put $z - z_1 = t^2$, $h(z) - h(z_1) = T^2 = a_k (z-z_1)^{2k+1} (1 + \cdots)$. Hence we have $T^2 = a_k t^{4k+2} (1 + t^2 + \cdots)$ and so $T = a_k^{1/2} t^{2k+1} (1 + t^2 + \cdots)$. This implies the analyticity of T(t). Hence φ is an analytic mapping of R into S.

Thus we have

Theorem 4.1 (Ozawa [4]). *There is a non-trivial analytic mapping* φ *of* R *into* S, *if and only if there is a non-constant entire function* h(z) *satisfying*

$$f(z)^2 G(z) = g(h(z)), \tag{4.3}$$

where f *is a suitable entire function.*

We call h(z) the projection of the analytic mapping φ. From this when we investigate the property of an analytic mapping of R into S, it is sufficient to investigate the property of its projection.

Let N(r, R) be the quantity N(r, $\mathfrak{G}$) defined in Selberg's paper [20]. In this case 2N(r, R) = N(r, 0, G). This is essentially the same as the integrated Euler characteristic in Sario (cf. Sario-Noshiro [1]).

We have

Theorem 4.2 (Ozawa [4]). *If there is an analytic mapping* φ *of* R *into* S, *then*

$$\limsup_{r\to\infty} \frac{N(r, R)}{T(r, h)} = \infty.$$

If N(r, 0, G) is of finite order ρ_G, then we may assume that G is the canonical product of ρ_G over these zeros; a similar remark applies to g. Then we have

Theorem 4.3 (Hiromi-Mutō [1]). *Assume that* $\rho_G < \infty$ *and* $0 < \rho_g < \infty$. *If there is an analytic mapping* φ *of* R *into* S, *then* $\rho_G = p\rho_g$, *where* p *is an integer and the projection* h *of* φ *is a polynomial of degree* p.

The example in Ozawa [3, p. 164 - 165] shows that the condition $\rho_g > 0$ in this theorem cannot be omitted.

Theorem 4.4 (Hiromi-Mutō [1]). *Every non-trivial analytic mapping of* R *into itself is an automorphism whose projection has the form* $e^{2\pi ip/q}z + b$ *with a suitable rational number* p/q *and a constant* b.

Now we consider the family of analytic mapping of R into S. Let $\mathfrak{H}$(R, S) be the family of projections of analytic mappings of R into S and $\mathfrak{H}_P$(R, S) (resp. $\mathfrak{H}_T$(R, S)) the subfamily of $\mathfrak{H}$(R, S) consisting of polynomials (resp. transcendental entire functions). With respect to the structure of $\mathfrak{H}$(R, S), the following theorems are obtained:

Theorem 4.5 (Mutō [3]). $\mathfrak{H}$(R, S) *is at most a countable set.*

Theorem 4.6 (Niino [3], [4]). *If* $\mathfrak{H}_P(R, S)$ *is not empty, then it consists of polynomials of the same degree.*

Theorem 4.7 (Mutō [4]). *If* $\mathfrak{H}(R, S) \neq \phi$, *then* $\mathfrak{H}(R, S) = \mathfrak{H}_P(R, S)$ *or* $\mathfrak{H}(R, S) = \mathfrak{H}_T(R, S)$.

Theorem 4.8 (Niino [3], [4]). *Assume that* $h(z) = a_p z^p + \cdots + a_0$ *and* $k(z) = b_p z^p + \cdots + b_0$ $(a_p b_p \neq 0)$ *belong to* $\mathfrak{H}(R, S)$. *If* $|a_p| < |b_p|$, *then we have*

(i) $\rho_g = \rho_G = 0$ *and* p *is even;*

(ii) $k(z) = \frac{b_p}{a_p} h(z) + A$, *where* A *is a constant;*

(iii) $\mathfrak{H}(R, S) = \mathfrak{H}_P(R, S)$ *and* $\mathfrak{H}(R, S)$ *consists of just two elements* h *and* k, *that is,* $\mathfrak{H}(R,S) = \{h, k\}$;

(iv) g *satisfies the functional equation*

$$g(\lambda w + A) = B(\lambda w + A - \alpha_1)\, g(w),$$

where $\lambda = \frac{b_p}{a_p}$ *and* α_1 *and* B *constants such that* $\alpha_1 \neq -\frac{A}{\lambda - 1}$, $g(\frac{\alpha_1 - A}{\lambda}) \neq 0$;

(v) g *has just an infinite number of simple zeros,* $\{\alpha_j\}_{j=1}^{\infty}$ *such that*

$$\alpha_{j+1} = \lambda^j \alpha_1 + A(\lambda^j - 1) / (\lambda - 1) \quad (j \geq 1)$$

and $k(z) = \alpha_1 + P(z)^2$, *where* P *is a polynomial of degree* p/2;

(vi) *Examples of this situation indeed occur.*

Theorem 4.9 (Baker [1]). *Assume that* $\rho_g < +\infty$, *there exists two polynomials* h *and* k *of degree* p *belonging to* $\mathfrak{H}(R, S)$ *and the leading coefficients of* h *and* k *are the same in modulus. Then one of the following three cases occurs:*

(i) $k(z) = Lh(z) + M$, *where* L *is a root of unity and* M *is a constant;*

(ii) p *is even and there is a polynomial* r *such that*

$$h(z) = r(z)^2 + A_0 \quad \textit{and} \quad k(z) = \{r(z) + \beta\}^2 + D_0,$$

where A_0, D_0 *and* β *are constants;*

(iii) *the ratio of the leading coefficients of* h *and* k *is a primitive* s-*th root of unity, the* (sp)-*th iterate* Ψ_{ps} *of the expansion* Ψ *of* $k^{-1} \circ h$ *about* ∞ *satisfies* $\Psi_{ps}(z) \equiv z$. *Case* (iii) *can occur only if* $\rho_G \geq 2$.

Further, examples of each of the cases exist.

Now, we assume that P(R) = P(S) = 4. Then from Theorem 3.1 we have

$$F(z)^2G(z) = (e^{H(z)} - \alpha)(e^{H(z)} - \beta), \ H(0) = 0, \ \alpha\beta(\alpha - \beta) \neq 0, \tag{4.4}$$

$$f(w)^2g(w) = (e^{L(w)} - \gamma)(e^{L(w)} - \delta), \ L(0) = 0, \ \gamma\delta(\gamma - \delta) \neq 0, \tag{4.5}$$

where F and f are suitable entire functions, H and L are non-constant entire functions, and α, β, γ and δ are constants. With the above notations we have

Theorem 4.10 (Hiromi-Ozawa [1]). *Suppose that* P(R) = P(S) = 4. *Then there exists an analytic mapping* φ *of* R *into* S *if and only if there exists an entire functions* h(z) *such that either*

$$H(z) = L(h(z)) - L(h(0)), \quad \alpha e^{L(h(0))} = \gamma, \quad \beta e^{L(h(0))} = \delta$$

or

$$H(z) = -L(h(z)) + L(h(0)), \quad \alpha\gamma = e^{L(h(0))}, \quad \beta\delta = e^{L(h(0))}.$$

Theorem 4.11 (Mutō-Niino [1]). *Suppose that* P(R) = P(S) = 4 *and* $\mathfrak{H}(R, S) \neq \phi$. *Then the following hold:*

(I) *If* $\mathfrak{H}_T(R, S) \neq \phi$, *then* $\mathfrak{H}_T(R, S)$ *consist of transcendental entire functions of the same order, the same type and the same class;*

(II) *If* $\mathfrak{H}_P(R, S) \neq \phi$ *and* h *and* k *are two elements of* $\mathfrak{H}_P(R, S)$, *then either*

(i) k(z) = Lh(z) + M, *where* L *is a root of unity and* M *is a constant, or*

(ii) *there is a polynomial* r *such that*

$$h(z) = r(z)^2 + A_0 \text{ and } k(z) = (r(z) + \beta)^2 + D_0,$$

where A_0, D_0 *and* β *are constants.*

We define the order of an ultrahyperelliptic surface R, which is defined by $y^2 = G(z)$, by the order of N(r, 0, G). Then we have

Theorem 4.12 (Ozawa [7]). *Assume that* R *is of finite order and* P(R) = 4, *and* S *is of non-zero finite order. If there is a non-trivial analytic mapping of* R *into* S, *then the order of* S *is a half of an integer.*

We now give the following examples of an analytic mapping of R into S with P(R) = 3, P(S) = 2 and P(R) = 4, P(S) = 2:

Example 1. Let R and S be surfaces defined by

$$y^2 = G(z) := 81e^{4z^2} - 72e^{3z^2} - 2e^{2z^2} - 8e^{z^2} + 1$$

and

$$u^2 = g(w) := w(81e^{4w} - 72e^{3w} - 2e^{2w} - 8e^{w} + 1),$$

respectively. Then P(R) = 3 and P(S) = 2 by Theorems 3.1 and 3.6 (cf. Theorem 3.7). And we have an identity $f(z)^2G(z) = g(h(z))$ with $f(z) = z$ and $h(z) = z^2$. Hence by Theorem 4.1 there is an analytic mapping of R into S whose projection is h.

Example 2. Let R and S be surfaces defined by

$$y^2 = G(z) := e^{e^z} - 1 \quad \text{and} \quad u^2 = g(w) := (e^w - 1)/w,$$

respectively. Then we have P(R) = 4 and P(S) = 2 by Theorems 3.1 and 3.7. And we have $f(z)^2G(z) = g(h(z))$ with $f(z) = e^{-z/2}$ and $h(z) = e^z$. Hence there is an analytic mapping of R into S whose projection is h.

Further we have

Theorem 4.13 (Ozawa [5]). *Let* R *and* S *be surfaces of finite order with* P(R) = 4 *and* P(S) = 3. *Then there is no analytic mapping of* R *into* S.

Now we consider a finite modification $\tilde{R}$ defined by (3.4) of the ultrahyperelliptic surface R defined by (4.1). In general we have

Theorem 4.14 (Ozawa [6]). *If there is a non-trivial analytic mapping of* R *into* $\tilde{R}$, *then it reduces to a conformal mapping of* R *onto* $\tilde{R}$ *whose projection has the form* az + b. *Further if* G *and* $\tilde{G}$ *have the same number of zeros in* $\{|z| < r_0\}$, *then its projection has the form* $e^{2\pi ip/q}z + b$ *with a suitable rational number* p/q.

However, we have the following (cf. Theorem 4.16):

Theorem 4.15 (Niino [2]). *If* P(R) = 4, *then there is no non-trivial analytic mapping of* R *into its finite modification* $\tilde{R}$.

Next we assume P(R) = P(S) = 4. Then G in (4.1) and g in (4.2) satisfy (4.4) and (4.5), respectively. Let $\tilde{R}$ and $\tilde{S}$ be finite modifications of R and S defined by $y^2 = \tilde{G}(z)$ and $u^2 = \tilde{g}(w)$, respectively. Then by the definition of the finite modification $\tilde{G}$ and $\tilde{g}$ have the following form:

$$\tilde{G}(z) = Q(z)G(z) \text{ and } Q(z) = \prod_{j=1}^{\lambda} (z - a_j) \prod_{j=1}^{\mu} (a - b_j)^{-1}, \ \lambda + \mu \geq 1 \qquad (4.6)$$

and

$$\tilde{g}(w) = q(w)g(w) \text{ and } q(w) = \prod_{j=1}^{\sigma} (w - d_j) \prod_{j=1}^{\tau} (w - e_j)^{-1}, \ \sigma + \tau \geq 1. \qquad (4.7)$$

It is clear from Theorem 1.1 and 3.7 that there is no analytic mapping of $\tilde{R}$ into S and no analytic mapping of $\tilde{S}$ into S.

We also have the following still under the condition P(R) = P(S) = 4:

Theorem 4.16 (Niino [2]). *There exists an analytic mapping* φ *of* R *into* $\tilde{S}$ *if and only if* *there exists an entire function* h *and a meromorphic function* f* *satisfying one of the following conditions*

(a) $H(z) = L(h(z) - L(h(0)),$ $\qquad f^*(z)^2 = q(h(z)),$

$\gamma/\alpha = \delta/\beta = e^{L(h(0))}$ *or* $\delta/\alpha = \gamma/\beta = e^{L(h(0))},$

(b) $H(z) = -L(h(z)) + L(h(0)),$ $\qquad f^*(z)^2 = q(h(z)),$

$\alpha\gamma = \beta\delta = e^{L(h(0))}$ *or* $\alpha\delta = \beta\gamma = e^{L(h(0))}$

and further $\sigma + \tau = 1$ *or* 2 *in* (4.7) *when* h *is transcendental and* $\sigma + \tau = 1$ *when* h *is a polynomial.*

Theorem 4.17 (Niino [2]). *There exists an analytic mapping* φ *of* $\tilde{R}$ *into* $\tilde{S}$ *if and only if there exists an entire function* h *and a meromorphic function* f* *satisfying one of the following conditions*

(a) $H(z) = L(h(z)) - L(h(0)),$ $\quad f^*(z)^2 = q(h(z)) / Q(z),$

$\gamma/\alpha = \delta/\beta = e^{L(h(0))}$ *or* $\delta/\alpha = \gamma/\beta = e^{L(h(0))},$

(b) $H(z) = -L(h(z)) + L(h(0)),$ $\quad f^*(z)^2 = q(h(z)) / Q(z),$

$\alpha\gamma = \beta\delta = e^{L(h(0))}$ *or* $\alpha\delta = \beta\gamma = e^{L(h(0))}$

and further $\sigma + \tau = 1$ *or* 2 *in* (4.7) *when* h *is transcendental.*

It follows from Theorem 4.10, 4.16 and 4.17 that

Corollary 4.18 (Niino [2]). *If either there exists an analytic mapping* φ *of* R *into* $\tilde{S}$ *or there exists an analytic mapping* φ *of* $\tilde{R}$ *into* S, *then there exists an analytic mapping of* R *into* S *whose projection is the same as the projection of* φ.

Open problems.

1. Determine all the ultrahyperelliptic surfaces with P(R) = 3.

It is very plausible to conjecture that Theorem 3.6 is also true for arbitrary entire functions H and L.

2. Is there any example of an analytic mapping of an ultrahyperelliptic surface R with P(R)= 4 into another S with P(S) = 3?

This would be negative and depend on the Problem 1.

3. Is Baker's Theorem 4.9 also true without the condition $\rho_G < \infty$?

4. In general does $\mathfrak{H}_T(R, S)$ consist of entire functions of the same order, the same type and the same class without the maximal condition P(R) = P(S) = 4? (cf. Niino [6]).

§ 5. Analytic mappings between two algebroid surfaces.

Let R_n $(n \geq 2)$ and S_m $(m \geq 2)$ be algebroid surfaces formed by elements $p = (z, y)$ and $q = (w, u)$ for each pair (z, y) and (w, u) satisfying the irreducible equations

$$y^n + A_1(z)\, y^{n-1} + \cdots + A_n(z) = 0 \tag{5.1}$$

and

$$u^m + B_1(z)\, u^{m-1} + \cdots + B_m(z) = 0, \tag{5.2}$$

respectively, where $A_1, \cdots, A_n, B_1, \cdots, B_m$ are meromorphic functions in the finite plane. We assume that R_n and S_m have an infinite number of branch points.

Let $\mathfrak{P}_{R_n}$ (resp. $\mathfrak{P}_{S_m}$) be the projection map of R_n (resp. S_m) into z-plane (resp. w-plane) such that $p = (z, y) \to z$ (resp. $q = (w, u) \to w$). Let φ be an analytic mapping of R_n into S_m. Then its projection

$$h(z) := \mathfrak{P}_{S_m} \circ \varphi \circ \mathfrak{P}_{R_n}{}^{-1}(z)$$

is, in general, an n-valued algebroid function of z. However we have

Theorem 5.1 (Hiromi-Mutō [2]). *If there is a non-trivial analytic mapping of* R_n *into* S_m, *then its projection* h *reduces to a* k*-valued algebroid function of* z, *where* k *is a proper divisor of* n.

k in Theorem 5.1 cannot take all proper divisors of n. For example

Theorem 5.2 (Mutō [2]). *Let* R_6 *and* S_2 *be algebroid surfaces defined by irreducible equations* $y^6 = G(z)$ *and* $u^2 = g(w)$, *respectively, where* G *and* g *are entire functions having an infinite number of zeros whose orders are less than* 6 *and* 2, *respectively. If there is an analytic mapping* φ *of* R_6 *into* S_2, *then its projection* h *is either single-valued or three-valued, that is, the case where it is two-valued does not occur.*

An analytic mapping φ of R_n into S_m is called rigid if φ preserves the projection maps, that is,

$$\mathfrak{P}_{S_m}(p)\ \varphi\ \mathfrak{P}_{S_m}\varphi\ (q) \qquad \text{whenever } \mathfrak{P}_{R_n}p = \mathfrak{P}_{R_n}q.$$

It is clear that φ is rigid when and only when its projection h reduces to a single-valued function of z. As a corollary of Theorem 5.1 we have

Theorem 5.3 (Hiromi-Mutō [2]). *Assume that there is an analytic mapping* φ *of* R_n *into* S_m. *If* n *is a prime number, then* φ *is rigid.*

We give two examples of non-rigid analytic mappings (cf. Mutō [2]).

Example 3. Let R_6 and S_2 be algebroid surfaces defined by

$$y^6 = (e^z - 1)\ (e^{e^z - 1} - 1)^3$$

and

$$u^2 = w(e^{w^3} - 1),$$

respectively. Then there exists a non-rigid analytic mapping φ of R_6 into S_2 induced by $h(z) = [e^z - 1]^{1/3}$, that is, $\varphi = \mathfrak{P}_{S_2}{}^{-1} \circ h \circ \mathfrak{P}_{R_4}$, because we have $u(h(z)) = y$.

Example 4. Let R_4 be an algebroid surface defined by

$$y^4 = z^2\ (1 - z^2) \prod_{j=1}^{\infty} ((1 - \tfrac{z^2}{a_n})\ (1 - \tfrac{1-z^2}{a_n}))^2,$$

where $\{a_n\}$ are suitable complex numbers. Then there is analytic mapping φ of R_4 into itself induced by $h^2 + z^2 = 1$, that is, $\varphi = \mathfrak{P}_{R_4}{}^{-1} \circ h \circ \mathfrak{P}_{R_4}$.

We denote by $N(r; B, R_n)$ the counting function for the number of branch points of R_n and $N^*(r; B, R_n)$ the counting function for the number of branch points of R_n whose order of ramification is $n-1$. Then we have a sufficient condition for the rigidity.

Theorem 5.4 (Mutō [2]). *Assume that the inequality*

$$\frac{N^*(r; B, R_n)}{N(r; B, R_n)} \geq \epsilon > 0$$

holds for a set of r *of infinite measure. Then every analytic mapping of* R_n *into* S_m *is rigid whenever it exists.*

As a corollary we have

Corollary 5.5. *If* R_n *is regularly branched, then every analytic mapping of* R_n *into* S_m *is rigid whenever it exists.*

Further we have

Theorem 5.6 (Mutō [2]). *Assume that* S_m *has an infinite number of branch points whose order of ramification are* $m-1$. *If* $n \leq m$, *then every analytic mapping of* R_n *into* S_m *is rigid whenever it exists.*

Now we give non-existence criteria for analytic mappings.

Theorem 5.7 (Mutō [2]). *Let* S_m *be the same as in Theorem* 5.2. *If either* (i) $n < m$ *or* (ii) $n > m$ *and* n *is a prime number, then there is no analytic mapping of* R_n *into* S_m.

Example 4 shows that there is a non-rigid self analytic mapping of R_n. However we have

Theorem 5.8 (Mutō [2]). *If the part of* R_n *which lies over* $|z| > r$ *is connected for all* r, *then every analytic mapping of* R_n *into itself is rigid.*

Theorem 5.9 (Mutō [2]). *If there is a rigid analytic mapping of* R_n *into itself, then it is onto and its projection has the form* $e^{2\pi ip/q}z + b$ *with a suitable rational number* p/q.

Further Ozawa [8] obtained some interesting results on existence of analytic mappings of R_3 into S_3, where R_3 and S_3 are surfaces defined by $y^3 = A(z)\, y + B(z)$ and $u^3 = a(w)\, u + b(w)$ with meromorphic functions A, B, a and b, respectively.

We now consider regularly branched algebroid surfaces R_n and S_m defined by

$$y^n = G(z) \quad \text{and} \quad u^m = g(w) \tag{5.3}$$

with entire functions G and g, respectively. Then many similar results to ones on ultrahyperelliptic surfaces stated in §4 are found in Hiromi-Mutō [2], Mutō [1], [2] and Niino [1], [2], [7], [8], [9]. Typically we state

Theorem 5.10 (cf. Hiromi-Mutō [2] and Niino [7]). *Let* R_n *and* S_m *be regularly branched algebroid surfaces defined by* (5.3) *respectively. If there is an analytic mapping* φ *of* R_n *into* S_m, *then* $n = pm$ *with a positive integer* p *and the projection* h *of* φ *is a single-valued entire function of* z *and satisfies*

$$f(z)^m G(z)^k = g(h(z)), \qquad p \le kp \le n-1, \tag{5.4}$$

where f *is a suitable meromorphic function and* k *is a suitable positive integer which is coprime to* m.

Conversely, if $n = pm$ *with a positive integer* p *and there is an entire function* h *satisfying* (5.4) *with a suitable meromorphic function* f *and a suitable positive integer* k *which is coprime to* m, *then there exists an analytic mapping* φ *of* R_n *into* S_m *whose projection is* h.

§ 6. Relation to factorization of meromorphic functions.

By Theorem 4.1 (cf. Theorem 5.10) the problem of existence of analytic mappings between ultrahyperelliptic surfaces reduces to the problem of the following functional equation including a composite function:

$$f(z)^2G(z) = g(h(z)) \tag{6.1}$$

with two given entire functions G and g, which have no zero other than an infinite number of simple zeros, and two unknown entire functions h and f.

On the other hand we have the following problem: *are there entire functions* g *and* h *such that*

$$F(z) = g(h(z)) \tag{6.2}$$

for a given entire function F? Here we can put $F(z) = f(z)^2G(z)$, where f and G are entire and G has simple zeros only and G is the canonical product of least possible genus over these zeros. Then the equation (6.2) is written such that $f(z)^2G(z) = F(z) = g(h(z))$, which is the same form as (6.2).

Thus the problems of analytic mappings between algebroid surfaces are closely connected with the problems of factorization of meromorphic functions. Some results on analytic mappings are restated as results on factorization of meromorphic functions (cf. Niino [6]). From this point of view we are very interested in factorization of meromorphic functions.

References

Ahlfors, L. V.

[1] The theory of meromorphic curves. Acta Soc. Sci. Fenn. Ser. A, Tom. 3, No. 4 (1941), 31pp.

Aogai, H.

[1] Picard constant of a finitely sheeted covering surfaces. Kōdai Math. Sem. Rep. 25 (1973), pp. 219 - 224.

Baker, I. N.

[1] Analytic mappings between two ultrahyperelliptic surfaces. Aequationes Math. 14 (1976), pp. 461 - 472.

Behnke, H. and Stein, K.

[1] Entwicklung analytischer Funktionen auf Riemannschen Flächen. Math. Ann. 120 (1949), pp. 430 - 461.

Cartan, H.

[1] Sur les zéros des combinaison linéaires de p fonctions holomorphes données. Mathematica (Cluj.) 7 (1933), pp. 5 - 29.

Hayman, W. K.

[1] Meromorphic functions. Clarendon Press, London, 1964.

Hiong, H. L.

[1] Sur les fonctions méromorphes et les fonctions algébroïdes. Extensions d'un théorème de M. R. Nevanlinna. Mém. Sic. Math. 139, Gauthier-Villars, Paris, 1957.

Hiromi, G. and Mutō, H.

[1] On the existence of analytic mappings, I. Kōdai Math. Sem. Rep. 19 (1967), pp. 236 - 244.

[2] On the existence of analytic mappings, II. Kōdai Math. Sem. Rep. 19 (1967), pp. 439 - 450.

Hiromi, G. and Niino, K.

[1] On a characterization of regularly branched three-sheeted covering Riemann surfaces. Kōdai Math. Sem. Rep. 17 (1965), pp. 250 - 260.

Hiromi, G. and Ozawa, M.

[1] On the existence of analytic mappings between two ultrahyperelliptic surfaces. Kōdai Math. Sem. Rep. 17 (1965), pp. 281 - 306.

Mutō, H.

[1] On the existence of analytic mappings. Kōdai Math. Sem. Rep. 18 (1966), pp. 24 - 35.

[2] On analytic mappings among algebroid surfaces. Kōdai Math. Sem. Rep. 21 (1969), pp. 191 - 204.

[3] Analytic mappings between two ultrahyperelliptic surfaces. Kōdai Math. Sem. Rep. 22 (1970), pp. 53 - 60.

[4] On the family of analytic mappings among ultrahyperelliptic surfaces. Kōdai Math. Sem. Rep. 26 (1975), pp. 454 - 458.

Mutō, H. and Niino, K.

[1] A remark on analytic mappings between two ultrahyperelliptic surfaces. Kōdai Math. Sem. Rep. 26 (1974/75), pp. 103 - 107.

Nevanlinna, R.

[1] Zur Theorie der meromorphen Funktionen. Acta Math. 46 (1925), pp. 1 - 99.

[2] Einige Eindeutigkeitssätze in der Theorie der meromorphen Funktionen. Acta Math. 48 (1926), pp. 367 - 391.

[3] Le théorème de Picard-Borel et la théorie des fonctions méromorphes. Gauthier-Villars, Paris, 1929; 2nd edition, Chelsea Pub. Com., New York, 1974.

[4] Eindeutige analytische Funktionen. Springer-Verlag, Berlin-Göttingen-Heidelberg, 1st ed. 1936, 2nd ed. 1653; Analytic functions (English version). Springer-Verlag, New York-Heidelberg-Berlin, 1970.

Niino, K.

[1] On regularly branched three-sheeted covering Riemann surfaces. Kōdai Math. Sem. Rep. 18 (1966), pp. 229 - 250.

[2] On finite modifications of two- or three-sheeted covering open Riemann surfaces. Kōdai Math. Sem. Rep. 19 (1967), pp. 415 - 424.

[3] On the family of analytic mappings between two ultrahyperelliptic surfaces. Kōdai Math. Sem, Rep. 21 (1969), pp. 182 - 190.

[4] On the family of analytic mappings between two ultrahyperelliptic surfaces, II. Kōdai Math. Sem. Rep. 21 (1969), pp. 491 - 495.

[5] On the functional equation $(e^M - \gamma)^k (e^M - \delta)^{m-k} = f^m (e^H - \sigma)^k (e^H - \eta)^{m-k}$. Aequationes Math. 22 (1981), pp. 293 - 301.

[6] On analytic mappings and factorizing meromorphic functions. "Factorization theory of meromorphic functions and related topics", pp. 71 - 82, Lecture Notes in Pure and Appl. Math., 78, Dekker, New York, 1982.

[7] On analytic mappings between two algebroid surfaces. Complex Variables Theory Appl. 2 (1983/84), pp. 283 - 293.

[8] On finite modifications of algebroid surfaces. Kodai Math. J. 8 (1985), pp. 296 - 306.

[9] Analytic mappings between two regularly branched three-sheeted algebroid surfaces. Tôhoku Math. J. 39 (1987), pp. 313 - 328.

[10] Some results on analytic mappings between two ultrahyperelliptic surfaces.

Ozawa, M.

[1] On complex analytic mappings. Kōdai Math. Sem. Rep. **17** (1965), pp. 93 - 102.

[2] On ultrahyperelliptic surfaces. Kōdai Math. Sem. Rep. **17** (1965), pp. 103 - 108.

[3] On complex analytic mappings between two ultrahyperelliptic surfaces. Kōdai Math. Sem. Rep. **17** (1965), pp. 158 - 165.

[4] On the existence of analytic mappings. Kōdai Math. Sem. Rep. **17** (1965), pp. 191 - 197.

[5] On an ultrahyperelliptic surface whose Picard's constant is three. Kōdai Math. Sem. Rep. **19** (1967), pp. 245 - 256.

[6] On a finite modification of an ultrahyperelliptic surface. Kōdai Math. Sem. Rep. **19** (1967), pp. 312 - 316.

[7] A remark on ultrahyperelliptic surfaces. Kōdai Math. Sem. Rep. **19** (1967), pp. 381 - 383.

[8] On analytic mappings among three-sheeted surfaces. Kōdai Math. Sem. Rep. **20** (1968), pp. 146 - 154.

[9] On rigid analytic mappings among surfaces $\{e^W = f(z)\}$. Kōdai Math. Sem. Rep. **20** (1968), pp. 155 - 158.

Rémoundos, G.

[1] Extension aux fonctions algébroïdes multiformes du théorème de M. Picard et de ses généralisations. Mém. Sci. Math. **23**, Gauthier-Villars, Paris, 1927.

Rodin, B. and Sario, L.

[1] Existence of mappings into noncompact Riemann surfaces. J. Analyse Math. 17 (1966), pp. 219 - 224.

Sario, L. and Noshiro, K.

[1] Value distribution theory. D. Van Nostrand Company Inc., Princeton, 1966.

Selberg, H.

[1] Über die Wertverteilung der algebroiden Funktionen. Math. Z. 31 (1930), pp. 709 - 728.

[2] Algebroide Funktionen und Umkehrfunktionen Abelscher Integrale. Avh. Norske Vid. Akad. Oslo 8 (1934), pp. 198 - 220.

Ullrich, E.

[1] Über den Einfluss der Verzweigtheit einer Algebroide auf ihre Wertverteilung. J. Riene Angew. Math **167** (1931), pp. 198 - 220.

Valiron, G.

[1] Sur la dérivée de fonctions algébroïdes. Bull. Soc. Math. France **59** (1931), pp. 17 - 39.

Weyl, H.

[1] Meromorphic functions and analytic curves. Ann. of Math. Studies, No. 12, Princeton Univ. Press, Princeton, 1943.

Weyl, H. and J.

[1] Meromorphic curves. Ann. of. Math. **39** (1938), pp. 516 - 538.

Wittich, H.

[1] Über eine Borelsche Identität. Math. Z. 84 (1964), pp. 233 - 243.

Wu, H. -H.

[1] The equidistribution theory of holomolphic curves. Ann. of Math. Studies, No. 64, Princeton Univ. Press, Princeton; Univ. of Tokyo Press, Tokyo, 1970.

Mitsuru Ozawa
Department of Mathematics
Science University of Tokyo

Kiyoshi Niino
Faculty of Technology
Kanazawa University

INTERPOLATION PROBLEMS OF PICK-NEVANLINNA TYPE

Saburou Saitoh

1. Introduction.

General and abstract versions for interpolation problems of Pick-Nevanlinna type were recently developed by C. H. FitzGerald [7], W. Hengartner and G. Schober [9], D. H. Hamilton [8], F. Beatrous and J. Burbea [3, 5] and F. H. Szafraniec [18, 19] and others. For our back ground, see their papers. However, we have still many delicate problems when the interpolation set E is small, and the famous theorem of Pick-Nevanlinna [12, 11] is still deep from the point of view of the general theory of reproducing kernels. In this paper, from the point of view of the general theory of reproducing kernels, we will deduce general and reasonable versions for the interpolation problem, which are applicable in many problems. The results will be given by two steps, but the author wants to emphasize that these results can be deduced from the general theory of reproducing kernels by Aronszajn [2]. Therefore, the results will be derived from the theory, directly. For interpolation problems of Pick-Nevanlinna, see also [1, 4, 6, 10, 13, 14, 15, 16, 17].

The author wishes to express his sincere thanks to Professors J. Burbea and F. Beatrous for valuable discussions while the author was visiting the University of Pittsburgh.

This paper was previously written as an original paper whose material was used in part in [16, chapter VIII].

2. Positive Definiteness of $K(x_1, x_2)\ (1 - g(x_1)\ \overline{g(x_2)}.)$

Let $K(x_1, x_2)$ be a positive matrix on a set E, and let H_K denote the Hilbert space admitting the reproducing kernel $K(x_1, x_2)$. See [2] for the general properties for reproducing kernels.

We consider an arbitrary complex valued function $g(x)$ on E such that $g(x)$ is not zero identically on E. For $f \in H_K$, we take the function $f^* \in H_K$ which minimizes the norm $\| \hat{f} \|_{H_K}$ among the functions $\hat{f} \in H_K$ satisfying

$$\hat{f}(x) = f(x) \text{ on the subset } E_0 = \{x \in E;\ g(x) \neq 0\} \text{ of } E.$$

Then we obtain

<u>Theorem 2.1 ([3], [14])</u>. The function $K(x_1, x_2)\,(1 - g(x_1)\,\overline{g(x_2)})$ is a positive matrix on E if and only if, for any $f^*(x) \in H_K$

$$f^*(x)g(x) \in H_K \text{ and } \| f^*(x)g(x) \|_{H_K} \leq \| f^*(x) \|_{H_K}. \tag{2.1}$$

In particular, the condition (2.1) can be replaced by the condition that for any $f \in H_K$

$$f(x)g(x) \in H_K \text{ and } \| f(x)\, g(x) \|_{H_K} \leq \| f(x) \|_{H_K}. \tag{2.2}$$

<u>Proof</u>. The condition (2.2) was given by Beatrous-Burbea [3, Proposition 1.8] through [5, Theorem 2.1], and the condition (2.1) was given by the author [14]. The equivalence of (2.1) and (2.2) is also clear. Following the spirit ([2]) of the general theory of reproducing kernels, we shall give the proof of the theorem.

Note that the function $K(x_1, x_2)g(x_1)\,\overline{g(x_2)}$ is the restriction of the product of two reproducing kernels $K(x_1, x_2)$ and $g(y_1)\,\overline{g(y_2)}$ to the diagonal set E of E × E. Hence, in particular, the function $K(x_1, x_2)g(x_1)\overline{g(x_2)}$ is a positive matrix on E, and so there exists the uniquely determined Hilbert space H_{Kg} admitting the reproducing kernel $K(x_1, x_2)\, g(x_1)\, \overline{g(x_2)}$ on E. Let H_g denote the one-dimensional space admitting the reproducing kernel $g(x_1)\, \overline{g(x_2)}$ on E. Then, the space H_{Kg} is composed of all functions

$\{fg; f \in H_K\}$ on E and their norms are given by

$$\| fg \|_{H_{Kg}} = \min \left\{ \| h \|_{H_K} \| g \|_{H_g} \right\}$$
$$= \min \| h \|_{H_K}, \tag{2.3}$$

where the minimum is taken over all functions h satisfying

$$h \in H_K \text{ and } h(x)g(x) = f(x)g(x) \text{ on E.} \tag{2.4}$$

See [2, pp. 357 - 362]. Note here that, in general,

$$\| g \|_{H_g} = 1.$$

Hence, in (2.3) we have the identity

$$\| fg \|_{H_{Kg}} = \| f^* \|_{H_K}. \tag{2.5}$$

On the other hand, by [2, p. 352, Theorem 1 and pp. 382 - 383], the function

$$K(x_1, x_2)\ (1 - g(x_1)\ \overline{g(x_2)})$$

is a positive matrix on E if and only if

$$H_K \supset H_{Kg} \text{ and } \| fg \|_{H_K} \leq \| fg \|_{H_{Kg}}$$
$$\text{for all } fg \in H_{Kg}. \tag{2.6}$$

Hence, from (2.5) and (2.6) we have the condition (2.1). Since $f^*(x)g(x) = f(x)g(x)$ on E, we see that (2.1) is equivalent to (2.6).

3. Extension of $K(x_1, x_2)\ (1 - g(x_1)\ \overline{g(x_2)})$.

We assume that the function $K(x_1, x_2)$ is a positive matrix on E. Here, we assume that the function $K(x_1, x_2)$ is still a positive matrix on a set X such that $E \subset X$. The next problem is:

When is the function $K(x_1, x_2)\ (1 - g\ (x_1)\ \overline{g(x_2)})$ extensible in a positive matrix on X

with an extension of the function g(x)?

Then, in particular, note that the extension $\tilde{g}(x)$ of $g(x)$ is bounded by 1;

$$|\tilde{g}(x)| \le 1 \text{ on } \{ x \in X; K(x,x) \neq 0 \}.$$

The theorem of Pick-Nevanlinna states that:

Let $\hat{K}(z, \bar{u})$ be the Szegö kernel on the unit disc $X = U$ on the complex plane and E be an arbitrary set in U. Then, the function $\hat{K}(z, \bar{u})\,(1 - g(z)\,\overline{g(u)})$ is a positive matrix on E if and only if there exists a bounded analytic function f(z) on U such that

$$f(z) = g(z) \text{ on } E, \; |f(z)| \le 1 \text{ on } U,$$

and then the function

$$\hat{K}(z, \bar{u})\,(1 - f(z)\,\overline{f(u)})$$

is a positive matrix on U.

We will give our general and abstract version for the interpolation problem of Pick-Nevanlinna. In general, however we need an assumption for the size of E. See [3, 8].

Theorem 3.1. We assume that for a function g(x) on E, the function

$$K(x_1, x_2)\,(1 - g(x_1)\,\overline{g(x_2)}) \tag{3.1}$$

is a positive matrix on E.

If there exists a function $f_0 \in H_K$ such that

$$\frac{f}{f_0} \in H_K \text{ for all } f \in H_K, \tag{3.2}$$

then there exists a function $\tilde{g}(x)$ satisfying

$$\tilde{g}(x) = g(x) \text{ on E and } \tilde{g}(x) \in H_K. \tag{3.3}$$

Furthermore, for this function $\tilde{g}(x)$ we assume that, for any f and h $\in H_K$

$$f(x)\tilde{g}(x) = h(x) \text{ on E implies } f(x)\tilde{g}(x) = h(x) \text{ on X}. \tag{3.4}$$

Then, the function

$$K(x_1, x_2)\,(1 - \tilde{g}(x_1)\,\overline{\tilde{g}(x_2)})$$

is a positive matrix on X and for any positive integer n

$$\tilde{g}(x)^n \in H_K. \tag{3.5}$$

Proof. We denote the restriction of $K(x_1, x_2)$ to the set E by $K_E(x_1, x_2)$, and let H_{K_E} be the Hilbert space admitting the reproducing kernel $K_E(x_1, x_2)$. The interrelationship between the two spaces H_K and H_{K_E} is given by [2, p. 351, Theorem]; any member f(x) in H_{K_E} is the restriction to E of a function $\tilde{f}(x) \in H_K$; that is,

$$\tilde{f}(x) = f(x) \text{ on } E \tag{3.6}$$

and, the norm $\| f \|_{H_{K_E}}$ is given by

$$\| f \|_{H_{K_E}} = \min \| \tilde{f} \|_{H_K},$$

where the minimum is taken over all $\tilde{f} \in H_K$ satisfying (3.6). Hence, from Theorem 2.1 and (3.1), we see that, for any $\tilde{f} \in H_K$ and for its restriction $\tilde{f}_E$ to E,

$$\tilde{f}_E(x)\, g\,(x) \in H_{K_E} \tag{3.7}$$

and

$$\| \tilde{f}_E(x) g(x) \|_{H_{K_E}} \leq \| \tilde{f}_E(x) \|_{H_{K_E}}. \tag{3.8}$$

In particular, for a function $f_0 \in H_K$ satisfying (3.2), we see from (3.7) that there exists an H_K function $\tilde{g}(x)$ satisfying (3.3). From the condition (3.4), further we obtain for any $\tilde{f} \in H_K$,

$$\tilde{f}(x)\tilde{g}(x) \in H_K \tag{3.9}$$

and

$$\|\tilde{f}(x)\,\tilde{g}(x)\|_{H_K} = \|\tilde{f}_E(x)\,\tilde{g}(x)\|_{H_{K_E}}$$

$$\leq \|\tilde{f}_E(x)\|_{H_{K_E}} = \|\tilde{f}\|_{H_K}. \tag{3.10}$$

Note here that the condition (3.4) implies that E is a uniqueness set for H_K functions. Hence, by Theorem 2.1 we see that the function $K(x_1, x_2)\,(1 - g(x_1)\,\overline{g(x_2)})$ is a positive matrix on X. Moreover, by setting $\tilde{f} = \tilde{g}$, we see that $\tilde{g}^2 \in H_K$. Hence, we have (3.5) by induction.

Note that the condition (3.2) is equivalent to

$$1 \in H_K.$$

We can use Theorem 3.1 to derive the theorem of Pick-Nevanlinna for a set E such that E is a uniqueness set for the Hardy $H_1(U)$ analytic functions. In Theorem 3.1, we will think that the set X is a bounded domain and H_K norms are realized in terms of a σ finite positive measure $d\mu$ in the form

$$\| f \|_{H_K}^2 = \int_X | f(x) |^2 \, d\mu(x).$$

Then, the condition (3.4) will be satisfied for a uniqueness set E for $L_1(X, d\mu) \cap H_K$ functions in miscellaneous concrete examples. See [3,5].

4. Remarks for the Assumptions.

We will show that the assumption (3.4) is reasonable.

Indeed, we take two functions $f, g \in H_K$, such that

$$f(x)g(x) \text{ is not a member in } H_K, \tag{4.1}$$

and g(x) takes zeros in X. We set

$$E = \{ x \in X:\ g(x) = 0\}. \tag{4.2}$$

Then, from (4.1)

$$f(x)g(x) = 0 \text{ on } E \text{ but } f(x)g(x) \neq 0 \text{ on } X \text{ identically.}$$

Then, of course, the function

$$K_E(x_1, x_2)\,(1 - g(x_1)\,\overline{g(x_2)}) = K_E(x_1, x_2)$$

is a positive matrix on E, but, from (4.1) the function

$$K(x_1, x_2)\,(1 - g(x_1)\,\overline{g(x_2)})$$

is not a positive matrix on X.

The condition (3.4), in particular, implies that the set E is a uniqueness set for H_K functions. But, the converse is, in general, not valid.

Indeed, consider two functions f, g ϵ H_K such that f(x)g(x) is not a member in H_K and the zero point set of fg is a uniqueness set for H_K functions. Then, f(x)g(x) = 0 on E, but f(x)g(x) is not zero identically on X.

Hamilton [8, Theorem 12] proved that for the Szegö kernel $\hat{K}$, for a uniqueness set E for $H_{\hat{K}}$ functions, the theorem of Pick-Nevanlinna is in general valid on a smooth domain in $\mathbb{C}^n$. Szafraniec [19] gave an interesting variation for the Pick-Nevanlinna interpolation problem in an abstract form.

5. Unbounded and Regular Cases.

The assumption (3.2) may be used for bounded regions X's in many different problems. Let X be an unbounded regular region. In order to deduce an extension of the function g(x) to H_K functions, we will be able to use the regularity of $K(x_1, x_2)$ on the boundary ∂X and analyticity of $K(x_1, x_2)$ in X in many concrete examples. In order to show this, we will state one typical example.

Let D be an unbounded regular region surrounded by a finite number of analytic

Jordan curves on the complex plane. Let $K(z, \overline{u})$ be the usual Bergman kernel on D. Then, we have:

Example. For an arbitrary set E of D, when the function

$$K(z, \overline{u})\ (1 - g(z)\overline{g(u)}) \tag{5.1}$$

is a positive matrix on E, then there exists an analytic function $f(z) \in H_K$ whose restriction coincides with g(z) on E.

In order to show this, we use the fundamental Theorem 2.7 in [3]. That is, from (5.1), there exists a function $\mathcal{K}(z, \overline{u})$ such that

$$K(z, \overline{u})\ (1 - g(z)\ \overline{g(u)}) = \mathcal{K}(z, \bar{u}) \text{ on } E \times \overline{E}, \tag{5.2}$$

$$\mathcal{K}(\ \ , \overline{u}) \text{ and } \overline{\mathcal{K}(z, \overline{\ \ })} \in H_K,$$

and

$$\mathcal{K}(z, \overline{u}) \text{ is a positive matrix on D.} \tag{5.3}$$

We thus have

$$|\ g(z)|^2 = 1 - \frac{\mathcal{K}(z, \overline{z})}{K\ (z, \overline{z})} \text{ on E} \tag{5.4}$$

and

$$g(z)\ \overline{g(u)} = 1 - \frac{\mathcal{K}(z, \overline{u})}{K\ (z, \overline{u})} \text{ on } E \times \overline{E}. \tag{5.5}$$

From (5.5), the function g(z) can be extended meromorphically onto D. We let g(z) still denote the extension of g(z) onto D by (5.5). Since $K(z, \overline{z}) \neq 0$ on D, the extension g(z) is locally bounded on D and so, the function g(z) can be extended analytically onto D. For any fixed $u \in D$, the Bergman kernel $K(z, \overline{u})$ is analytic on $D \cup \partial D$. Hence, the extension g(z) belongs to H_K.

In connection with Theorem 3.1, we [15] can construct an example such that:

The set E is a set of uniqueness for H_K functions, the function g is the restriction of a function $\tilde{g}$ in H_K to E, $K(x_1, x_2)(1 - g(x_1)\overline{g(x_2)})$ is a positive matrix on E and the function $K(x_1, x_2)(1 - \tilde{g}(x_1)\overline{\tilde{g}(x_2)})$ is not a positive matrix on X. Furthermore, we can assume that $1 \in H_K$.

This answers an open question of Beatrous-Burbea [3, page 32]. In addition, note that Theorem 1.7 and Theorem 2.1 in [3] are not correct (The authors also admitted their mistakes).

References

[1] Abrahamse, M. B., The Pick interpolation theorem for finitely connected domains, Mich. Math. J. 26 (1979), pp. 195 - 203.

[2] Aronszajn, N., Theory of reproducing kernels, Trans. Amer. Math. Soc. 68 (1950), pp. 337 - 404.

[3] Beatrous, F., and J. Burbea, Reproducing kernels and interpolation of holomorphic functions, Complex Analysis, Functional Analysis and Approximation Theory, Elsevier Science Publishers B. V. (1986), pp. 25 - 46.

[4] Burbea, J., and P. Masani, Banach and Hilbert Spaces of Vector-Valued Functions, their General Theory and Applications to Holomorphy, Pitman Research Notes in Math. 90 (1983), Pitman, London, 1984.

[5] Burbea, J., and F. Beatrous, Positive-definiteness and its applications to interpolation problems for holomorphic functions, Trans. Amer. Math. Soc. 284 (1984), pp. 247 - 270.

[6] Duren, P. L., and P. L. Williams, Interpolation problems in function spaces, J. Funct. Anal. 9 (1972), pp. 75 - 86.

[7] FitzGerald, C. H., Quadratic inequalities and analytic continuation, J. Analyse Math. 31 (1977), pp. 19 - 47.

[8] Hamilton, D. H., Quadratic inequalities and interpolation, J. Analyse Math. 43 (1984), pp. 26 - 50.

[9] Hengartner, W., and G. Schober, Interpolation, continuation and quadratic inequalities, Pacific J. Math. 100 (1982), pp. 139 - 151.

[10] Marshall, D. E., An elementary proof of the Pick-Nevanlinna interpolation theorem, Michigan Math. J. 21 (1974), pp. 219 - 223.

[11] Nevanlinna, R., Über beschränkte Funktionen die in gegebenn Punkten vorgeschreibene Werte annehmen, Ann. Acad. Sci. Fenn. Ser. B. (13), (1919), pp. 25 - 56.

[12] Pick, G., Über der Beschränkungen analytische Funktionen welche durch vorgegebene Funktionswerte bewirkt werden, Math. Ann. 77 (1916), pp. 7 - 23.

[13] Rosenblum, M., and J. Rovnyak, An operator-theoretic approach to theorems of the Pick-Nevanlinna and Loewner types, I. Integral Equation and Operator Theory, 3 (1980), pp. 408 - 436.

[14] Saitoh, S., Applications of the general theory of reproducing kernels, Complex Analysis and Applications '85, Sofia, 1986, pp. 580 - 591.

[15] Saitoh, S., A counter example in the Pick-Nevanlinna interpolation problems. Arch. Math. 51 (1988), pp. 164 - 165.

[16] Saitoh, S., Theory of reproducing kernels and its applications, Pitman Research Notes in Math. 189 (1988), Longman Scientific & Technical, Harlow, Essex, UK.

[17] Snyder, A. K., Multipliers and the Nevanlinna-Pick theorem, Indiana Univ. Math. J. 25 (1976), pp. 1159 - 1170.

[18] Szafraniec, F. H., Interpolation and domination by positive definite kernels, In: Complex Analysis, Fifth Rumanian - Finish Seminar, Part 2. Proc. Bucharest 1981, Lecture Notes in Math. 1014, pp. 291 - 295, Springer-Verlag, 1983.

[19] Szafraniec, F. H., On bounded holomorphic interpolation in several variables, Monatsh. Math. 101 (1986), pp. 59 - 66.

Saburou Saitoh
Department of Mathematics
Faculty of Engineering
Gunma University

DISCRETENESS OF REAL TWO-GENERATOR FREE GROUPS

Hiroki Sato

§0. Introduction.

We set

$$A_1(z) = z/t_1$$

and

$$A_2(z) = \{(\rho - t_2)\, z + \rho(t_2 - 1)\} \,/\, \{(1 - t_2)\, z + (\rho t_2 - 1)\}$$

for $(t_1, t_2, \rho) \in R_I$, where $R_I = \{(t_1, t_2, \rho) \in \mathbf{R}^3 \mid 0 < t_1 < 1, 0 < t_2 < 1, \rho > 0\}$. We consider in which case the group $G(\tau) = \langle A_1(z), A_2(z)\rangle$ generated by $A_1(z)$ and $A_2(z)$ is discrete and/or free. For this question, Purzitsky [3] gave a necessary and sufficient condition for $G(\tau)$ being discrete and free. However it seems that it is difficult to determine immediately by his theorem whether or not $G(\tau)$ represented by $\tau = (t_1,t_2,\rho) \in R_I$ is discrete and free. On the other hand, in the previous paper [4] we represented the shape of the classical Schottky space $R_I\mathfrak{S}_2^0$ of this type (see §1 for the notation) and constructed a fundamental region for the Schottky modular group acting on $R_I\mathfrak{S}_2^0$.

In this paper we give some conditions on $\tau = (t_1, t_2, \rho) \in R_I$ to determine immediately whether or not the group $G(\tau)$ represented by τ is discrete and/or free by using the method in [4], Nielsen transformations and a Jørgensen's inequality (Propositions 4.1, 4.2, 5.1, 6.1, 6.2 and 7.1). The main results are Propositions 5.1 and 7.1. From these propositions we have a result corresponding to the theorem by Purzitsky mentioned above (Theorem 1).

Last we note the following. Marden [2] verified that not every Schottky group is classical. Zarrow [6] claimed to obtain an explicit example of non-classical Schottky group. However Sato [5] pointed out that the Schotty group constructed by Zarrow turns out to be classical. Now the following question arises: In which case is a Schottky group classical? For this question, we know the following result by combining Theorem 1 with Sato [4, Theorem 2]: Every Schottky group of Type I of genus two is classical (see §1 for the definition).

In §1 we will state definitions. In §2 we will consider automorphisms of free groups on two generators. In §3 we will introduce some curves and regions. In §4 through §7 we will study whether or not the groups represented by points on the curves or in the regions introduced in §3 are discrete and/or free. In §8 we will state theorems.

§1. Definitions.

1.1. DEFINITION 1.1 Let C_1, C_{g+1};···; C_g, C_{2g} be a set of $2g$, $g \geq 1$, mutually disjoint Jordan curves on the Riemann sphere which comprise the boundary of a 2g-ply connected region ω. Suppose there are g Möbius transformations $A_1, \cdots, A_g$ which have the property that A_j maps C_j onto C_{g+j} and $A_j(\omega) \cap \omega = \emptyset$, $1 \leq j \leq g$. Then the g necessarily loxodromic transformations A_j generate a *marked Schottky group* $G = \langle A_1, \cdots, A_g \rangle$ of genus g with ω as a fundamental region. In particular, if all C_j ($j = 1, 2, \cdots, 2g$) are circles, then we call $A_1, \cdots, A_g$ a *set of classical generators of G.* A *classical Schottky group* is a Schottky group for which there exists some set of classical generators.

DEFINITION 1.2. We say two marked Schottky groups $G = \langle A_1, \cdots, A_g \rangle$ and $\hat{G} = \langle \hat{A}_1, \cdots, \hat{A}_g \rangle$ to be *equivalent* and denotes the fact by $G \sim \hat{G}$ if there exists a

Möbius transformation T such that $\hat{A}_j = TA_jT^{-1}$ for all j=1, 2, $\cdots$, g. The *Schottky space* of genus *g*, denoted by $\mathfrak{S}_g$, is the set of all equivalence classes of marked Schottky groups of genus $g \geq 1$.

DEFINITION 1.3. The *classical Schottky space*, denoted by $\mathfrak{S}_g^0$, is defined to consist of all elements of $\mathfrak{S}_g$ for which there exists some set of classical generators.

1.2. Let A_j (j=1,2) be loxodromic transformations. Let λ_j ($|\lambda_j| > 1$), p_j and p_{2+j} be the multipliers, the repelling and the attracting fixed points of A_j. Let G = $<A_1, A_2>$ be a marked group such that p_1, p_2, p_3 and p_4 are all distinct. We define t_j by setting $t_j = 1/\lambda_j$ (j=1,2). Thus $t_j \in D^* = \{z|\ 0 < |z| < 1\}$. We determine a Möbius transformation T by $T(p_1) = 0$, $T(p_3) = \infty$ and $T(p_2) = 1$ and define ρ by $\rho = T(p_4)$. Thus $\rho \in C-\{0, 1\}$.

We denote by $\mathfrak{M}_2$ the set of all equivalence classes of marked groups $<A_1, A_2>$ generated by loxodromic transformations A_1 and A_2 whose fixed points p_k (k=1,2,3,4) are all distinct, where an equivalence class is similarly defined as in Definition 1.2. We can define a mapping α of $\mathfrak{M}_2$ into $D^{*2} \times (C-\{0,1\})$ by setting $\alpha([G]) = (t_1, t_2, \rho)$, where [G] denotes the equivalence class of $G = <A_1, A_2>$. Conversely λ_1, λ_2 and p_4 are uniquely determined from a given point $\tau = (t_1,t_2,\rho) \in D^{*2} \times (C-\{0,1\})$ under the normalization condition $p_1 = 0$, $p_3 = \infty$ and $p_2 = 1$; we define λ_j (j=1,2) and p_4 by setting $\lambda_j = 1/t_j$ and $p_4 = \rho$, respectively. We determine Möbius transformations $A_1(\tau,z)$, $A_2(\tau,z)$ from τ as follows: The multiplier, the repelling and the attracting fixed points of $A_j(\tau,z)$ are λ_j, p_j and p_{2+j}, respectively. Thus we obtain a mapping β of $D^{*2} \times (C-\{0,1\})$, into $\mathfrak{M}_2$ by setting $\beta(\tau) = [<A_1(\tau,z), A_2(\tau,z)>]$. Then we note that $\alpha\beta = \beta\alpha = \text{id}$. Therefore we identify $\mathfrak{M}_2$ (resp, $\mathfrak{S}_2$ and $\mathfrak{S}_2^0$) with $\alpha(\mathfrak{M}_2)$ (resp. $\alpha(\mathfrak{S}_2)$

and $\alpha(\mathfrak{S}_2^0)$) and from now on we denote them by the same notation $\mathfrak{M}_2$ (resp, $\mathfrak{S}_2$ and $\mathfrak{S}_2^0$).

1.3. DEFINITION 1.4. A Möbius transformation $A(z) = (az + b) / (cz + d)$ is called a *a real Möbius transformation* if $a, b, c, d \in R$ and $ad - bc \neq 0$. If $G = \langle A_1, A_2, \cdots, A_g \rangle$ is a marked group (resp. marked Schottky group or marked classical Schottky group) such that A_j is a real Möbius transformation for each $j=1,2,\cdots,g$, (that is, for example $\alpha([G]) \in R^3$ if $g = 2$), then we call G a *marked group* (resp. *marked Schottky group* or *marked classical Schottky group*) *of real type.*

In the case of g= 2, there are eight kinds of marked groups (marked Schottky groups, marked classical Schottky groups) of real type (cf. Sato [4]). Here we only consider the following two types:

$$\text{Type I}: \quad t_1 > 0, t_2 > 0 \text{ and } \rho > 0 \tag{1}$$

and

$$\text{Type IV}: \quad t_1 > 0, \ t_2 > 0 \text{ and } \rho < 0. \tag{2}$$

For each k = I, IV, we denote by $R_k\mathfrak{M}_2$ the set of all equivalence classes of marked groups $\langle A_1, A_2 \rangle$ of Type k such that all fixed points of A_1 and A_2 are distinct.

DEFINITION 1.5. For each k = I, IV, the intersection of $R_k\mathfrak{M}_2$ with $\mathfrak{S}_2$ (resp. $\mathfrak{S}_2^0$) is called the *real Schottky space* (resp. the *real classical Schottky space*) *of Type k*, and is denoted by $R_k\mathfrak{S}_2$ (resp. $R_k\mathfrak{S}_2^0$).

§2. Automorphisms of a free group on two generators.

In this section we will state some results about Nielsen transformations in Sato [4, §§2,3 and 5]. Let $G = \langle A_1, A_2 \rangle$ be a free group on two generators.

Theorem A (Neumann). *The group Φ_2 of automorphisms of G has the following* *presentation:*

$$\Phi_2 = \langle N_1, N_2, N_3 \mid (N_2N_1N_2N_3)^2 = 1,\ N_1N_3N_1N_3 = N_3N_1N_3N_1,$$
$$N_3^{-1}N_2N_3N_2N_1N_3N_1N_2N_1 = 1\rangle,$$

where N_1: $(A_1,A_2)\mapsto(A_1,A_2^{-1})$, N_2: $(A_1,A_2)\mapsto(A_2,A_1)$ *and* N_3:$(A_1,A_2)\mapsto(A_1,A_1A_2)$.

We call the mappings N_1, N_2 and N_3 in Theorem A *Nielsen transformations.* We can regard N_j (j=1,2,3) as automorphisms of the space of all equivalence classes of marked free groups on two generators (see Sato [4,§2]).

Let (t_1, t_2, ρ) be the point in $\mathfrak{M}_2$ corresponding to a marked group $G = \langle A_1,A_2\rangle$. Let $(t_1(j), t_2(j), \rho(j))$ be the images of (t_1, t_2, ρ) under the mapping N_j (j=1,2,3). Let p and q be two solutions of the equation

$$t_1(1 - t_2)z^2 - (\rho - t_2 - \rho t_1t_2 + t_1)\, z + \rho(1 - t_2) = 0.$$

We set $X = \rho - t_2 - \rho t_1t_2 + t_1$ and $Y = \rho - t_2 + \rho t_1t_2 - t_1$. Then we have the following.

Lemma 2.1 (Lemma 2.1 in [4]).

(1) $t_1(1) = t_1$, $t_2(1) = t_2$ *and* $\rho(1) = 1/\rho$.

(2) $t_1(2) = t_2$, $t_2(2) = t_1$ *and* $\rho(2) = \rho$.

(3) $t_1(3) = t_1$, $t_2(3)^{1/2} + 1/t_2(3)^{1/2} = Y/t_1^{1/2}t_2^{1/2}(\rho - 1)$,

and if $\rho > 0$, *then*

$$\rho(3)^{1/2}+1/\rho(3)^{1/2} = \begin{cases} X/(\rho^{1/2}t_1^{1/2}(1-t_2)) & \text{if } p > 0 \text{ and } q > 0 \\ -X/(\rho^{1/2}t_1^{1/2}(1-t_2)) & \text{if } p < 0 \text{ and } q < 0, \end{cases}$$

if $\rho < 0$, *then*

$$(-\rho(3))^{1/2} + 1/(-\rho(3))^{1/2}$$

$$= (X^2 - 4t_1\rho(1 - t_2))^{1/2} / (-\rho)^{1/2}t_1^{1/2}(1 - t_2).$$

We easily see the following.

Lemma 2.2. *Let* $G = <A_1,A_2>$ *be a marked two-generator group. Then for each* $j = 1,2,3$, $N_j(g)$, *the image of G under the mapping* N_j, *is the same group G except marking.*

Corollary. *Let* $G = <A_1,A_2>$ *be a marked two-generator group. Then* $\phi(G)$, *the image of G under a mapping* $\phi\in\Phi_2$, *is the same group G except marking.*

DEFINITION 2.1. Let ϕ_1 and ϕ_2 be automorphisms of $G = <A_1,A_2>$. We say ϕ_1 and ϕ_2 are *equivalent* if $\phi_1(G)$ is equivalent to $\phi_2(G)$, and denote this by $\phi_1\sim\phi_2$ (see Definition 1.2).

We note that if $\phi_1\in\Phi_2$ is equivalent to $\phi_2\in\Phi_2$, then $\phi_1([G]) = \phi_2([G])$ for $[G]\in\mathfrak{M}_2$, that is $\phi_1(\tau) = \phi_2(\tau)$ for $\tau = (t_1,t_2,\rho)\in\mathfrak{M}_2$.

Lemma 2.3. $N_3N_1N_3N_1\sim 1$.

Proof. This comes from straightforward calculations, q.e.d.

§3. Curves and regions.

3.1. In this section we will introduce some curves and regions in R_I, where $R_I = \{(t_1, t_2, \rho) \mid 0 < t_1 < 1, 0 < t_2 < 1, \rho > 0\}$. Throughout this section we fix t_1 with $0 < t_1 < 1$.

DEFINITION 3.1. For each positive integer $n = 1, 2,\cdots$, and for fixed t_1 with $0 < t_1 < 1$, curves $\alpha(t_1;n)$, $\beta(t_1;n)$ and $\gamma(t_1;n)$ are defined as follows:

$$(1) \quad \alpha(t_1;n) \ : \ \rho^{1/2} = (1 + t_1^{n/2}t_2^{1/2}) / (t_1^{n/2} + t_2^{1/2}) \ \ (0 < t_2 < t_{2,n}(t_1)^{1/2}),$$

where

$$t_{2,n}(t_1)^{1/2} = \frac{t_1^{(n-1)/2}\{(1 - t_1^n)^{1/2} - t_1^{1/2}(1 - t_1^{n-1})^{1/2}\}}{(1 - t_1^{n-1})^{1/2} + (1 - t_1^n)^{1/2}}.$$

(2) $\beta(t_1;n)\ :\ \rho^{1/2} = (1 - t_1^{n/2} t_2^{1/2}) \,/\, (t_1^{n/2} - t_2^{1/2})\quad (0 < t_2 < t_1^n).$

(3) $\gamma(t_1;n)\ :\ \rho^{1/2} = (1 - t_1^{n/2} t_2^{1/2}) \,/\, (t_2^{1/2} - t_1^{n/2})\quad (t_1^n < t_2 < 1).$

For each negative integer $n = -1,-2,\cdots$, we define $\alpha(t_1;n)$, $\beta(t_1;n)$ and $\gamma(t_1;n)$ by replacing ρ with $1/\rho$ in the equations defining the curves $\alpha(t_1;-n)$, $\beta(t_1;-n)$ and $\gamma(t_1;-n)$, respectively.

For each $n = \pm1, \pm2,\cdots$, we denote the equations defining the curves $\alpha(t_1;n)$, $\beta(t_1;n)$ and $\gamma(t_1;n)$ by $\rho = \alpha_n(t_1,t_2)$, $\rho = \beta_n(t_1,t_2)$ and $\rho = \gamma_n(t_1,t_2)$, respectively.

3.2. Next we will definite some regions $D_j(t_1;n)$ $(j=1,2,3)$ in the set R_I for each $n = \pm1,\pm2,\cdots$, and for fixed t_1 with $0 < t_1 < 1$. We set

$$\epsilon(n) = \begin{cases} 1 & \text{if} \quad n = 1,\ 2,\cdots \\ -1 & \text{if} \quad n = -1,-2,\cdots \\ 0 & \text{if} \quad n = 0 \end{cases}$$

DEFINITION 3.2.

(1) $D_1(t_1;n)\ (n \geq 1)\ :\ \beta_{n-1}(t_1,t_2) < \rho < \alpha_n(t_1,t_2)\ ,\ (0 < t_2 < t_{2,n}(t_1)).$

$D_1(t_1;n)\ (n \leq -1)\ :\ \alpha_n(t_1,t_2) < \rho < \beta_{n+1}(t_1,t_2)\ ,\ (0 < t_2 < t_{2,n}(t_1)),$

where $\beta_0(t_1,t_2) = 1\ (0 < t_2 < 1)$ and $t_{2,n}(t_1)\ (n \geq 1)$ are as defined in Definition 3.1 and $t_{2,n}(t_1)\ (n \leq -1)$ are

$$t_{2,n}(t_1)^{1/2} = \frac{(1 - t_1^{-n})^{1/2} - (1 - t_1^{-(n+1)})^{1/2}}{t_1^{-(n+1)/2}\{t_1^{1/2}(1 - t_1^{-(n+1)})^{1/2} + (1 - t_1^{-n})^{1/2}\}}.$$

(2) $D_2(t_1;n)$: The region bounded by four curves $\alpha(t_1;n)$, $\beta(t_1;\epsilon(n)(|n| - 1))$, $\gamma(t_1;n)$ and $\beta(t_1;n)$, where $\beta(t_1;0)$ means the point $(t_1,1,1)$.

(3) $D_3(t_1;n)$: The region bounded by two curves $\beta(t_1;\epsilon(n)(|n| - 1))$ and $\gamma(t_1;n)$ in $t_1^{|n|} < t_2 < t_1^{|n|-1}$, where $\beta(t_1;0)$ means the point $(t_1,1,1)$, namely

$\rho > \beta_{\epsilon(n-1)(|n|-1)}(t_1,t_2)$ and $\rho > \gamma_n(t_1, t_2)$ if $n \geq 1$;

$\rho > \beta_{\epsilon(n+1)\ (|n|-1)}(t_1,t_2)$ and $\rho < \gamma_n(t_1,t_2)$ if $n \leq -1$.

We set $D_j = \bigcup_{t_1} \bigcup_{n} \{D_j(t_1;n) \mid 0 < t_1 < 1\ ,\ n \neq 0\}$.

3.3. Let N_1 and N_3 be the Nielsen transformations defined in §2. We have the following lemmas from Lemma 2.1 by straightforward calculations (cf. Sato [4]).

Lemma 3.1. *For each* $n=\pm1,\pm2,\cdots$, *and for fixed* t_1 *with* $0 < t_1 < 1$,

(1) $N_3^{\epsilon(n)}(\alpha(t_1;n)) = \alpha(t_1;\epsilon(n))\ (|n| + 1)$,

(2) $N_3^{\epsilon(n)}(\beta(t_1;n)) = \beta(t_1;\epsilon(n))\ (|n| + 1))$,

(3) $N_3^{\epsilon(n)}(\gamma(t_1;n)) = \gamma(t_1;\epsilon(n))\ (|n| + 1))$,

(4) $N_1(\alpha(t_1;n)) = \alpha(t_1;-n)$,

(5) $N_1(\beta(t_1;n)) = \beta(t_1;-n)$,

(6) $N_1(\gamma(t_1;n)) = \gamma(t_1;-n)$.

Lemma 3.2. *For each* $n = \pm1,\pm2,\cdots$ *and for fixed* t_1 *with* $0 < t_1 < 1$,

(1) $N_3^{\epsilon(n)}(D_j(t_1;n)) = D_j(t_1;\epsilon(n)\ (|n| + 1))$ (j=1,2,3)

(2) $N_1(D_j(t_1;n)) = D_j(t_1;-n)$ (j=1,2,3).

§4. Groups in $D_1(t_1;n)$ and $D_2(t_1;n)$.

Let $\tau = (t_1,t_2, \rho) \in R_I$. We denote by $G(\tau)$ the group corresponding to τ.

Proposition 4.1 *Let* $\tau \in D_1(t_1;n)$ $(n=\pm1,\pm2,\cdots)$. *Then* $G(\tau)$ *is a classical Schottky group.*

Proof. By Lemma 3.2, it suffices to show this proposition in the case of $\tau \in D_1(t_1;1)$. For $\tau \in D_1(t_1;1)$, $G(\tau)$ is a classical Schottky group by Theorem 2 in Sato [4], q.e.d.

Throughout this section we set

$$A_1 = \frac{1}{t_1^{1/2}} \begin{pmatrix} 1 & 0 \\ 0 & t_1 \end{pmatrix} \qquad (0 \le t_1 \le 1)$$

and

$$A_2 = \frac{1}{t_1^{1/2}(\rho - 1)} \begin{pmatrix} \rho - t_2 & \rho(t_2 - 1) \\ 1 - t_2 & \rho t_2 - 1 \end{pmatrix} \qquad (0 \le t_2 \le 1,\ \rho \ge 0).$$

Then

$$A_1^{-1}A_2 = \frac{1}{t_1^{1/2}t_2^{1/2}(\rho - 1)} \begin{pmatrix} t_1(\rho - t_2) & t_1\rho(t_2 - 1) \\ 1 - t_2 & \rho t_2 - 1 \end{pmatrix}.$$

We denote by $\sigma(A)$ the trace of $A \in SL(2,C)$. Then

$$\sigma(A_1^{-1}A_2) = \{t_1(\rho - t_2) + \rho t_2 - 1\} / t_1^{1/2}t_2^{1/2}(\rho - 1).$$

Proposition 4.2. *Let* $\tau = (t_1,t_2,\rho) \in D_2(t_1;n)$ $(n = \pm 1, \pm 2, \cdots)$. *Then* $G(\tau) = \langle A_1,A_2 \rangle$ *is not a free group.*

Proof. By Lemma 3.2, it suffices to show this proposition in the case of $\tau \in D_2(t_1;1)$. We divide $D_2(t_1;1)$ into the following three parts: (i) $\alpha_1(t_1,t_2) < \rho < \beta_1(t_1,t_2)$ $(0 < t_2 < t_1)$; (ii) $t_2 = t_1$, $\rho > \alpha_1(t_1,t_2)$; (iii) $\alpha_1(t_1,t_2) < \rho < \gamma_1(t_1,t_2)$ $(t_1 < t_2 < 1)$.

First we note that

$$(*) \qquad \sigma(A_1^{-1}A_2) > -2$$

from $\rho > \alpha_1(t_1,t_2)$.

Case (i). We have $\sigma(A_1^{-1}A_2) < 2$ from $\rho < \beta_1(t_1,t_2)$ by direct calculations. Hence combining this with the above fact (*), we have that $A_1^{-1}A_2$ is elliptic and $G(\tau)$ is not free.

Case (ii). By substituting $t_2 = t_1$ in the equation of $\sigma(A_1^{-1}A_2)$, we have $\sigma(A_1^{-1}A_2) = (-t_1^2 - 1 + 2\rho t_1) / (t_1(\rho - 1)$. Hence $\sigma(A_1^{-1}A_2) < 2$. Hence combining this with the above fact (*), we have $A_1^{-1}A_2$ is elliptic and $G(\tau)$ is not free.

Case (iii). We have $\sigma(A_1^{-1}A_2) < 2$ from $\theta > \gamma_1(t_1,t_2)$. By the same method as in (1), we have that $A_1^{-1}A_2$ is elliptic and $G(\tau)$ is not free, q.e.d.

§5. Groups in $D_3(t_1;n)$.

Let $A = \begin{pmatrix} a & b \\ c & d \end{pmatrix}$, $ad - bc = 1$. We denote the norm of A by $\|A\| = (|a|^2 + |b|^2 + |c|^2 + |d|^2)^{1/2}$. We denote by I the identity matrix.

Lemma 5.1. (Beardon [1, p.105]). *If* $G = \langle A_1,A_2 \rangle$ *is a discrete group, then*

$$\max(\|A_1 - I\|, \|A_2 - I\|) > 0.146.$$

Corollary. *If*

$$\max(\|A_1 - I\|, \|A_2 - I\|) \le 0.146 ,$$

then the group generated by A_1 *and* A_2 *is not discrete.*

Lemma 5.2. *Let* N_2 *and* N_3 *be the Nielsen transformations defined in* §2. *Let* $\tau = (t_1,t_2,\rho) \in D_3(t_1;1)$. *Set* $N_2(\tau) = \tau^* = (t_1^*, t_2^*, \rho^*)$. *Then*

(1) $t_1 < t_1^* = t_2$, $t_2^* = t_1 < t_2$, $t_2^* < t_1^*$ *and* $\rho^* = \rho$.

(2) *If* $\tau^* \in D_3(t_1^*;m)$ $(m \ge 1)$, *then*

(i) $N_3^{-(m-1)}(\tau^*) = (t_{11}, t_{21}, \rho_1) \in D_3(t_1^*;1)$, *and*

(ii) $t_{11} = t_1^*$, $t_{21} > t_2^*$, $t_{11} < t_{21}$ *and* $\rho_1 < \rho^*$,

Proof. This lemma comes from Lemma 2,1 by straightforward calculations, q.e.d.

Lemma 5.3. *Let* $\tau = (t_1, t_2, \rho) \in D_3(t_1;1)$. *Suppose that*

$$\phi_k \phi_{k-1} \cdots \phi_1(\tau) = \tau_k = (t_{1,k}, t_{2,k}, \rho_k) \in D_3(t_{1,k};1)$$

for all $k = 1,2,\cdots$, *where* $\phi_j = N_3^{-m_j} N_2$. *Then either all sequences* $\{t_{1,k}\}$, $\{t_{2,k}\}$ *and* $\{\rho_k\}$ *tend to one as* $k \to \infty$, *or* $G(\tau)$ *is not a discrete group.*

Proof. By Lemma 5.2, $t_1 < t_{1,1} < t_{1,2} < \cdots < t_{1,k} < \cdots \leq 1$; $t_2 < t_{2,1} < t_{2,2} < \cdots < t_{2,k} < \cdots \leq 1$ and $\rho > \rho_1 > \rho_2 > \cdots > \rho_k > \cdots \geq 1$. Then there exist limits of the sequences $\{t_{1,k}\}$, $\{t_{2,k}\}$ and $\{\rho_k\}$. We set $t_{1,0} = \lim_{k\to\infty} t_{1,k}$, $t_{2,0} = \lim_{k\to\infty} t_{2,k}$ and $\rho_0 = \lim_{k\to\infty} \rho_k$.

First we suppose that $\rho_0 = 1$. Since $\tau_k \in D_3(t_1;1)$, we have

$$(\rho_k^{1/2} t_k^{1/2} + 1) \,/\, (\rho_k^{1/2} + t_k^{1/2}) < t_{2,k}^{1/2} < 1.$$

Hence

$$(t_{1,0} + 1) \,/\, (1 + t_{1,0}) \leq t_{2,0}^{1/2} \leq 1.$$

Thus $t_{2,0} = 1$. Since $t_{1,k} = t_{2,k-1}$ by Lemma 5.2, we have

$$t_{1,0} = \lim_{k\to\infty} t_{1,k} \, , = \lim_{k\to\infty} t_{2,k-1} = t_{2,0} = 1.$$

Next we suppose that $\rho_0 > 1$. If $t_{10} \neq 1$, then $t_{2,0} = t_{1,0} \neq 1$ from $t_{2,k-1} = t_{1,k}$ (n=1,2,$\cdots$). Hence using

$$\rho_2^{1/2} > (1 - t_{1,k}^{1/2} t_{2,k}^{1/2}) \,/\, (t_{2,k}^{1/2} - t_{1,k}^{1/2}),$$

we have $\rho_0 = +\infty$, which contradicts $\rho_0 < \rho$. Therefore $t_{1,0} = 1$, and so $t_{2,0} = 1$.

Let $G(\tau_k) = \langle A_{1,k}, A_{2,k} \rangle$. Then

$$A_{1,k} = \frac{1}{t_{1,k}^{1/2}} \begin{pmatrix} 1 & 0 \\ 0 & t_{1,k} \end{pmatrix}$$

and

$$A_{2,k} = \frac{1}{t_{2,k}^{1/2}(\rho_k - 1)} \begin{pmatrix} \rho_k - t_{2,k} & \rho_k(t_{2,k} - 1) \\ 1 - t_{2,k} & \rho_k t_{2,k} - 1 \end{pmatrix}$$

We have

$$\|A_{1,k} - I\|^2 = (1 - t_{1,k}^{1/2})^2(1 + 1/t_1)$$

and

$$\|A_{2,k} - I\|^2 = H_k/t_{2,k}(\rho_k - 1)^2,$$

where

$$H_k = \{\rho_k - t_{2,k} - t_{2,k}^{1/2}(\rho_k - 1)\}^2 + \rho_k^2(t_{2,k} - 1)^2$$

$$+ (1 - t_{2,k})^2 + \{\rho_k t_{2,k} - 1 - t_{2,k}^{1/2}(\rho_k - 1)\}^2.$$

Since $\rho_0 > 1$ and $t_{10} - t_{20} = 1$, we have

$$\lim_{k\to\infty} \|A_{1,k} - I\|^2 = 0 \text{ and } \lim_{k\to\infty} \|A_{2,k} - I\|^2 = 0.$$

Thus from corollary to Lemma 5.1, $G(\tau_k)$ is not a discrete group for all sufficiently large k. Hence by Lemma 2.2, $G(\tau)$ is not a discrete group, q.e.d.

Lemma 5.4. *Fix* t_1 *with sufficiently close to one and fix* ρ_0 *with* $1 < \rho_0$. *Let* $\tau_1 = (t_1, t_{2,1}, \rho_0)$ *be the intersection point of the curve* $\gamma(t_1;1)$ *with the line* $\rho = \rho_0$, *and let* $\tau_2 = (t_1, t_{2,2}, \rho_0)$ *be a point on the line* $\rho = \rho_0$ *in* $D_3(t_1;1)$. *Let* $G(\tau_1) = \langle A_1(\tau_1), A_2(\tau_1)\rangle$ *and* $G(\tau_2) = \langle A_1(\tau_2), A_2(\tau_2)\rangle$ *be the groups corresponding to* τ_1 *and* τ_2, *respectively. Then*

$$\|A_2(\tau_2) - I\| < \|A_2(\tau_1) - I\|.$$

Proof. Note that

$$\|A_2(\tau_j) - I\|^2 = \{\rho_0^2(t_{2,j}-1)^2 + (1 - t_{2,j})^2 + H_{1,j}^2 + H_{2,j}^2\} \,/\, t_{2,j}(\rho_0 - 1)^2$$

for j = 1,2, where

$$H_{1,j} = \rho_0 - t_{2,j} - t_{2,j}^{1/2}(\rho_0 - 1)$$

and

$$H_{2,j} = \rho_0 t_{2,j} - 1 - t_{2,j}^{1/2}(\rho_0 - 1).$$

By the assumption $0 < t_{2,1} < t_{2,2} < 1$, we have that

$$1/t_{2,1}(\rho_0 - 1)^2 > 1/t_{2,2}(\rho_0 - 1)^2,$$

$$(1 - t_{2,1})^2 > (1 - t_{2,2})^2$$

and

$$H_{1,1} - H_{1,2} = (t_{2,2}^{1/2} - t_{2,1}^{1/2})\,(t_{2,2}^{1/2} + t_{2,1}^{1/2} + \rho_0 - 1) > 0.$$

Furthermore we have $H_{2,1} < 0$, $H_{2,2} < 0$ and

$$H_{2,1} - H_{2,2} = \left(t_{2,2}^{1/2} - t_{2,1}^{1/2}\right)\{\rho_0(t_{2,2}^{1/2} + t_{2,1}^{1/2}) - \rho_0 + 1\} > 0,$$

since $t_{2,1}$ and $t_{2,2}$ are both sufficiently close to one. Hence $\|A_2(\tau_1) - I\| >$ $\|A_2(\tau_2) - I\|$, q.e.d.

<u>Lemma 5.5</u>. *Let* $\tau = (t_1, t_2, \rho) \in \gamma(t_1;1)$ *and* $G(\tau) = \langle A_1(\tau), A_2(\tau)\rangle$ *be the group corresponding to* τ. *Then* $\|A_2(\tau) - I\| \to 0$ *as* $t_1 \to 1$.

<u>Proof</u>. Since $\tau \in \gamma(t_1;1)$, $\rho^{1/2} = (1 - t_1^{1/2} t_2^{1/2}) / (t_2^{1/2} - t_1^{1/2})$.

Thus we have

$$\rho - 1 = (1 - t_1)\,(1 - t_2) / (t_2^{1/2} - t_1^{1/2})^2,$$

$$\{\rho - t_2 - t_2^{1/2}(\rho - 1)\}^2 = (1 - t_2^{1/2})^2\,(\rho + t_2^{1/2})^2$$

and

$$\{\rho t_2 - 1 - t_2^{1/2}(\rho - 1)\}^2 = (1 - t_2^{1/2})^2\,(1 + \rho t_2^{1/2})^2.$$

Hence

$$\|A_2(\tau) - I\|^2 = \frac{H}{t_2(1 + t_2^{1/2})^2}\left(\frac{t_2^{1/2} - t_1^{1/2}}{1 - t_1^{1/2}}\right)^2\left(\frac{t_2^{1/2} - t_1^{1/2}}{1 + t_1^{1/2}}\right)^2,$$

where

$$H = (\rho + t_2^{1/2})^2 + (\rho^2 + 1)\,(1 + t_2^{1/2})^2 + (1 + \rho t_2^{1/2})^2.$$

Since $t_2^{1/2} - t_1^{1/2} < 1 - t_1^{1/2}$, we have

$$\frac{1}{t_2(1 + t_2^{1/2})^2}\left(\frac{t_2^{1/2} - t_1^{1/2}}{1 - t_1^{1/2}}\right)^2 \left(\frac{t_2^{1/2} - t_1^{1/2}}{1 + t_1^{1/2}}\right)^2$$

$$< \frac{(t_2^{1/2} - t_1^{1/2})^2}{t_2(1 + t_1^{1/2})^2(1 + t_2^{1/2})^2} \quad .$$

If t_1 tends to one, then t_2 and ρ also tend to one, since $t_1 < t_2$ and $\tau = (t_1, t_2, \rho) \in \gamma(t_1;1)$. Hence there exists a positive number M such that $H < M$ for t_1 sufficiently close to one. Thus given arbitrarily small $\epsilon > 0$,

$$\|A_2(\tau) - I\|^2 < M(t_2^{1/2} - t_1^{1/2})^2 \,/\, t_2(1 + t_1^{1/2})^2(1 + t_2^{1/2})^2$$
$$< 2M(t_2^{1/2} - t_1^{1/2})^2 < \epsilon$$

for t_1 sufficiently close to one, q.e.d.

We have the following from Lemmas 5.4 and 5.5.

Corollary 1. *Let* $\tau = (t_1, t_2, \rho) \in D_3(t_1;1)$ *and* $G(\tau) = \langle A_1(\tau), A_2(\tau)\rangle$ *be the group corresponding to* τ. *Then* $\|A_2(\tau) - I\|^2 \to 0$ *as* $t_1 \to 1$.

From corollary to Lemma 5.1 and Corollary 1, we have the following.

Corollary 2. *Let* $\tau = (t_1, t_2, \rho) \in D_3(t_1;1)$ *and* $G(\tau) = \langle A_1(\tau), A_2(\tau)\rangle$ *be the group corresponding to* τ. *Then* $G(\tau)$ *is not a discrete group for* t_1 *sufficiently close to one.*

Let $\tau = (t_1, t_2 \rho) \in D_3(t_1;1)$. If $N_2(\tau) = (t_1^*, t_2^*, \rho^*) \in D_3(t_1;j)$, then $N_3^{-(j-1)} N_2(\tau) \in D_3\,(t_1^*;1)$, where N_2 and N_3 are the Nielsen transformations defined in §2. We denote by Cl(D) the closure of set D relative to R_I.

Proposition 5.1 *Let* $\tau = (t_1,t_2,\rho) \in D_3(t_1;n)$ *and* $G(\tau)$ *be the group corresponding to* τ. *Then either the following* (i) *or* (ii) *holds.*

(i) *There exists a sequence* $\{m(1), m(2), \cdots, m(j)\}$ *of positive integers such that* $N_2\phi_{m(j)}\cdots\phi_{m(2)}\phi_{m(1)}(\tau) \in Cl(D_1) \cup Cl(D_2)$ *if* $n \geq 1$ *and* $N_2\phi_{m(j)}\cdots\phi_{m(2)}\phi_{m(1)}(N_1(\tau)) \notin Cl(D_1) \cup Cl(D_2)$ *if* $n \leq -1$, *where* $\phi_{n(k)} = N_3^{-(m(k)-1)}N_2$ $(k=1,2,\cdots, j)$.

(ii) $G(\tau)$ *is not a discrete group.*

Proof. Suppose that $N_2\phi_{m(j)}\cdots\phi_{m(2)}\phi_{m(1)}(\tau) \notin Cl(D_1) \cup Cl(D_2)$ if $n \geq 1$ and $N_2\phi_{m(j)}\cdots\phi_{m(2)}\phi_{m(1)}(N_1(\tau)) \notin Cl(D_1) \cup Cl(D_2)$ if $n \leq -1$ for any sequence $\{m(1), m(2), \cdots, m(j)\}$ $(j=1,2,\cdots\cdots)$. Then from Lemma 5.3 and Corollary 2 to Lemma 5.5, we see that $G(\tau)$ is not a discrete group, q.e.d.

§6. Groups on $\alpha(t_1;n)$ and $\gamma(t_1;n)$.

In this section we will consider discreteness of groups corresponding to points of $\alpha(t_1;n)$ and $\gamma(t_1;n)$.

Lemma 6.1 *Let* $\tau = (t_1,t_2,\rho) \in \alpha(t_1;n)$ *and* $G(\tau) = <A_1(\tau), A_2(\tau)>$ *be the group corresponding to* τ. *Then* $G(\tau)$ *is a free and discontinuous group, and* $A_1(\tau)^{-1}A_2(\tau)$ *is a parabolic transformation.*

Proof. See Sato [4, Proposition 3.1], q.e.d.

Let $P_n(\alpha,\beta) = (t_1,t_{2,n}(\alpha,\beta), \rho_n(\alpha,\beta))$ be the intersection point of $\alpha(t_1;n)$ and $\beta(t_1;n-\epsilon(n))$. From Lemmas 2.2 and 6.1 we have the following.

Proposition 6.1. *Let* $\tau = (t_1,t_2,\rho) \in \alpha(t_1;n)$ *and* $G(\tau) = <A_1(\tau), A_2(\tau)>$ *be the group corresponding to* τ. *Then* $G(\tau)$ *is a free and discontinuous group, and* $A_1(\tau)^{-n}A_2(\tau)$ *is a parabolic transformation.*

Proposition 6.2. *Let* $\tau = (t_1,t_2,\rho) \in \gamma(t_1;n) \cap \partial D_3$ *and* $G(\tau) = <A_1(\tau), A_2(\tau)>$ *be the group corresponding to* τ, *where* ∂D_3 *denotes the boundary of* D_3. *Then* $A_1(\tau)^{-n}A_2(\tau)$ *is a parabolic transformation, and either the following* (i) *or* (ii) *holds.*

(i) *There exists a sequence* $\{m(1), m(2),\cdots, m(j)\}$ *of positive integers such that*

$$N_2\phi_{m(j)}\cdots \phi_{m(2)}\phi_{m(1)}(\tau) \in Cl(D_1) \cup Cl(D_2) - \gamma(t_1;1) \text{ if } n \geq 1 \text{ and}$$

$$N_2\phi_{m(j)}\cdots\phi_{m(2)}\phi_{m(1)}(N_1(\tau)) \in Cl(D_1) \cup Cl(D_2) - \gamma(t_1;1) \text{ if } n \leq -1,$$

where $\phi_{m(k)}$ $(k=1,2,\cdots, j)$ *are the elements of* Φ_2 *as in Proposition* 5.1

(ii) $G(\tau)$ *is not a discrete group.*

Proof. This Proposition comes from Corollary to Lemma 5.1 and Lemmas 5.3 and 5.5, q.e.d.

§7. Groups on $\beta(t_1;1)$.

7.1. Let $\tau = (t_1,t_2,\rho) \in \beta(t_1;n)$. Set

$$A_1 = \frac{1}{t_1^{1/2}}\begin{pmatrix} 1 & 0 \\ 0 & t_1 \end{pmatrix}$$

and

$$A_2 = \frac{1}{t_2^{1/2}(\rho - 1)}\begin{pmatrix} \rho - t_2 & \rho(t_2 - 1) \\ 1 - t_2 & \rho t_2 - 1 \end{pmatrix}$$

Then $G = <A_1,A_2>$ is the group corresponding to τ. We set

$$T = \begin{pmatrix} 0 & -1 \\ 1 & -\rho^{1/2}t_1^{1/2} \end{pmatrix}$$

$A_1^* = TA_1T^{-1}$ and $A_2^* = TA_1^{-1}A_2T^{-1}$. Since $\tau\in\beta(t_1;1)$, that is

$$\rho^{1/2} = (1 - t_1^{1/2} t_2^{1/2}) / (t_1^{1/2} - t_2^{1/2}),$$

we have

$$A_1^* = \begin{pmatrix} t_1^{1/2} & 0 \\ -c & t_1^{-1/2} \end{pmatrix} \quad \text{and} \quad A_2^* = \begin{pmatrix} 1 & -h \\ 0 & 1 \end{pmatrix}$$

where $c = (1 - t_1^{1/2} t_2^{1/2})(1 - t_1)/(t_1^{1/2} - t_2^{1/2})$ and $h = (t_1^{1/2} - t_2^{1/2})^2 / t_1^{1/2} t_2^{1/2} (1 - t_1)$.

We set $u_n = nt_1 - 2t_1^{1/2} + n$ and $v_n = nt_1 + 2t_1^{1/2} + n$, $U_n = \{u_n - (u_n^2 - 4(n - 1)^2 t_1)^{1/2}\} / 2(n-1)t_1^{1/2}$ and $V_n = \{v_n - (v_n^2 - 4(n-1)^2 t_1)^{1/2}\} / 2(n-1)t_1^{1/2}$ for $n=2,3,\cdots$. We divide the interval $0 < t_2 < t_1$ as follows:

$$0 < V_2^2 < U_2^2 < U_3^2 < U_4^2 < \cdots < U_{2n-1}^2 < U_{2n}^2 < U_{2n+1}^2 < \cdots < t_1.$$

Set

$$Z = \{1 + t_1^{1/2} + t_1 - ((1 + t_1^{1/2} + t_1)^2 - 4t_1)^{1/2}\} / 2t_1^{1/2}. \tag{*}$$

<u>Lemma 7. 1 (Jørgensen's Inequality [1, p. 105]).</u> *Let* $G = \langle A,B \rangle$ *be a non-elementary Möbius transformation group. If*

$$|\sigma^2(A) - 4| + |\sigma(ABA^{-1}B^{-1}) - 2| < 1, \tag{**}$$

then G is not a discrete group.

<u>Lemma 7.2</u>. *Let* $\tau = (t_1, t_2, \rho) \in \beta(t_1;1)$ *and* $G(\tau) = \langle A_1(\tau), A_2(\tau) \rangle$ *be the group corresponding to* τ. *If* $Z < t_2^{1/2} < t_1^{1/2}$, *then* $G(\tau)$ *is not a discrete group, where Z is as defined in the above.*

<u>Proof</u>. We may set $A_1(\tau) = A_1$ and $A_2(\tau) = A_2$, where A_1 and A_2 are as in the beginning of this section. We substitute $A_1^{-1}A_2$ and A_1 for *A* and *B*, respectively in the inequality (**) in Lemma 7.1. Since $A_1^{-1}A_2$ is a parabolic transformation, $\sigma^2(A) = 4$. By straightforward calculations, we have

$$\sigma(ABA^{-1}B^{-1}) - 2 = (t_1^{1/2} - t_2^{1/2})^2(1 - t_1^{1/2}t_2^{1/2})^2/t_1t_2.$$

Thus the inequality (**) turns out

$$(t_1^{1/2} - t_2^{1/2})^2(1 - t_1^{1/2}t_2^{1/2})^2 < t_1t_2,$$

and so $t_1^{1/2} > Z$. By Lemma 7.1, $\langle A,B\rangle = \langle A_1^{-1}A_2,A_1\rangle$ is not discrete. Since we obtain $\langle A_1,A_2\rangle$ from $\langle A_1^{-1}A_2,A_2\rangle$ by applying a finite number of Nielsen transformations, $G(\tau)$ is not a discrete group if $t_2^{1/2} > Z$ by corollary to Lemma 2.2, q.e.d.

Lemma 7.3. *Let* $\tau = (t_1,t_2,\rho) \in \beta(t_1;1)$ *and* $G(\tau) = \langle A_1(\tau), A_2(\tau)\rangle$ *be the group corresponding to* τ. *Then*

(1) $V_n < U_n$ *for* $n = 2,3,4,\cdots$.

(2) *If* $V_{n+1} < U_n$, *then* $V_{n+1+k} < U_{n+k}$ *for* $k = 0,1,2,\cdots$.

(3) $U_\ell \leq Z < U_{\ell+1}$ *if and only if* $\sigma(A_1(\tau)) - 2 \leq \ell < \sigma(A_1(\tau)) - 1$.

(4) $U_m < V_{m+1}$ *if and only if* $m < (\sigma(A_1(\tau)) + 2)/4$.

Proof. This lemma comes from straightforward calculations, q.e.d.

7.2. Proposition 7.1. *Let* $\tau = (t_1,t_2,\rho) \in \beta(t_1;1)$ *and* $G(\tau = \langle A_1(\tau), A_2(\tau)\rangle$ *be the group corresponding* τ. *Let Z be as defined in* §7.1. *Let* ℓ *be the integer with* $\sigma(A_1(\tau)) - 2 \leq \ell < \sigma(A_1(\tau)) - 1$ *and let m be the greatest integer with* $m \leq (\sigma(A_1(\tau) + 2)/4$. *Then* $A_1(\tau)^{-1}A_2(\tau)$ *is a parabolic transformation. Furthermore*

(1) if $0 < t_2 \leq V_2^2$, *then* $G(\tau)$ *is free and discontinuous group.*

(2) If $V_j^2 < t_2 < U_j^2$ ($j= 2,3,\cdots,m$), then $G(\tau)$ *is not a free group.*

(3) *If* $U_j^2 \leq t_2 \leq V_{j+1}^2$ ($j = 2,3,\cdots,m$) , *then* $G(\tau)$ *is a free and discontinuous group.*

(4) *If* $V_{m+1}^2 < t_2 < t_1$, *then* $G(\tau)$ *is not a free group.*

(5) *If* $Z^2 < t_2 < t_1$ *, then* $G(\tau)$ *is not a discrete group.*

Proof. See Sato [4, Proposition 3.3] for the proofs of parabolicity of $A_1(\tau)^{-1}A_2(\tau)$ and (1).

(2) We divide these intervals into the following two classes (i) and (ii) : (i) $V_{2j}^2 < t_2 < U_{2j}^2$ (j=1,2,···, [m/2]), where [n] denotes the greatest integer with $[n] \leq n$; (ii) $V_{2j+1}^2 < t_2 < U_{2j+1}^2$(j = 1,2,···, [(m−1) /2]).

Case of class (i). We may set $A_1(\tau) = A_1$ and $A_2(\tau) = A_2$ where A_1 and A_2 are as in the beginning of §7.1. Let T, $A_1^*$and A_2^* be the matrices defined in §7.1. Then we easily see that

$$\begin{aligned}&\sigma(A_2^{*j-1}A_1^{*-1}A_2^{*j})\\ &= t_1^{1/2}+1/t_1^{1/2}-\{(2j-1)\,(1-t_1^{1/2}t_2^{1/2})\,(t_1^{1/2}-t_2^{1/2})\}/t_1^{1/2}t_2^{1/2}.\end{aligned}$$

Since $V_{2j}^2 < t_2 < U_{2j}^2$, we have $-2 < \sigma(A_2^{*j-1}A_1^{*-1}A_2^{*j}) < 2$ by straightforward calculations. Hence $A_2^{*j-1}A_1^*A_2^{*j}$ is elliptic. Thus $G^*(\tau) = \langle A_1^*,A_2^*\rangle$ is not free and so $G(\tau) = \langle A_1(\tau), A_2(\tau)\rangle$ is not free.

Case of class (ii). We easily see that

$$\sigma(A_2^{*j}A_1^{*-1}A_2^{*j})=t_1^{1/2} + 1/t_1^{1/2}- \{2j(1-t_1^{1/2}t_2^{1/2})\,(t_1^{1/2}-t_2^{1/2})\}/t_1^{1/2}t_2^{1/2}$$

From the condition $V_{2j+1}^2 < t_2 < U_{2j+1}^2$, we have $-2 < \sigma(A_2^{*j}A_1^{*-1}A_2^{*j}) < 2$ by straightforward calculations. Hence $A_2^{*j}A_1^{*-1}A_2^{*j}$ is elliptic. Thus $G^*(\tau)$ is not free and $G(\tau)$ is not free.

(3) Let A_1^*, A_2^* and $G^*(\tau)$ be as in (2). For j = 2k or j = 2k + 1 ($2 \leq j < m$), we set $A_{1j}^* = A_2^{*-k}A_1^*A_2^{*-k}$, $A_{2j}^* = A_2^*$ and $G_j^*(\tau) = \langle A_{1j}^*, A_{2j}^*\rangle$. Then

$$A_{1j}^* = \begin{pmatrix} t_1^{1/2} - khc & kht_1^{1/2} - k^2h^2c + (kh/t_1^{1/2}) \\ -c & -khc + (1/t_1^{1/2}) \end{pmatrix}$$

where c and h are as in §7.1. Let $C_{1j}^* = I\,(A_{1j}^*)$ and $C_{3j}^* = I\,(A_{1j}^{*-1})$ be the isometric circles of A_{1j}^* and A_{1j}^{*-1}, respectively. Then we have

$$C_{1j}^* \;:\; |z + kh - (1/ct_1^{1/2})\,| = 1/c$$

and

$$C_3^* \;:\; |z - kh + (t_1^{1/2}/c)| = 1/c.$$

For j = 2k, the assumption $t_2^{1/2} \leq V_{j+1} = V_{2k+1}$ is equivalent to

$$(kh - t_1^{1/2}/c - 1/c) - (-kh + 1/ct_1^{1/2} + 1/c) \geq 0,$$

which means that C_{1j}^* and C_{3j}^* are mutually exterior except a tangent point if there is. Furthermore the assumption $U_j = U_{2k} \leq t_2^{1/2}$ is equivalent to

$$(kh - t_1^{1/2}/c + 1/c) - (-kh + 1/ct_1^{1/2} - 1/c) \leq h,$$

which means that C_{1j}^* and C_{3j}^* lie between two vertical lines $C_{2j}^* : \{z \in C \mid z = x_0\}$ and $C_{4j}^* : \{z \in C \mid z = x_0 - h\}$ for some real number x_0. Hence $\omega_j = \bigcap_{i=1}^{4} \text{Ext}\,(C_{ij}^*)$ is a fundamental region for $G_j^*(\tau)$, where Ext (C_{ij}^*) denotes the exterior of C_{ij}^* for each i=1,2,3,4 and Ext $(C_{2j}^*) = \{z \in C \mid \text{Re } z < x_0\}$, Ext $(C_{4j}^*) = \{z \in C \mid \text{Re } z > x_0 - h\}$. Thus $G_j^*(\tau)$ is a free and discontinuous group and so is $G(\tau)$.

For j = 2k + 1, the assumption $U_j = U_{2k+1} \leq t_2^{1/2}$, and $t_2^{1/2} \leq V_{j+1} = V_{2k+2}$ are equivalent to

$$(-kh + 1/ct_1^{1/2} - 1/c) - (kh - t_1^{1/2}/c + 1/c) \geq 0$$

and

$$(-kh + 1/ct_1^{1/2} + 1/c) - (kh - t_1^{1/2}/c - 1/c) \leq h\,,$$

respectively. By the same method as in the above, we obtain that C_{1j}^* , C_{3j}^* , C_{2j}^* and

C^*_{4j} comprise a fundamental region for $G^*_j(\tau)$. Thus $G^*_j(\tau)$ is a free and discontinuous group and so is $G(\tau)$.

(4) Let $G^*(\tau)$ be as in (2). Since $V_{m+2} < U_{m+1}$, we have $V_{m+2+j} < U_{m+1+j}$ for $j = 0,1,2,\cdots$ by Lemma 7.3 (2). By the same method as in the proof of (2) we see that $G^*(\tau)$ is not a free group for t_2 with $V^2_{m+j} < t_2 < U^2_{m+j}$ and for t_2 with $V^2_{m+j+1} < t_2 < U^2_{m+j+l}$ $(j = 1,2,\cdots)$. Since $V_{m+j+1} < U_{m+j}$, we see that $G^*(\tau)$ is not a free group for t_2 with $V^2_{m+j} < t_2 < U^2_{m+j+1}(j = 1,2,\cdots)$. Hence $G^*(\tau)$ and $G(\tau)$ are not free groups for t_2 with $V^2_{m+1} < t_2 < t_1$.

(5) This case comes from Lemma 7.2, q.e.d.

Corollary. *Let* $\tau = (t_1,t_2,\rho) \in \beta(t_1;n)$. *According as* $N_3^{-(n-1)}(\tau)$ $(n \geq 1)$ *and* $N_3^{n+1}(N_1(\tau))$ $(n \leq -1)$ *are contained in case* (j) *in Proposition* 7.1 *for each* $j = 1,2,3,4,5$, *the result corresponding to the case holds.*

§8. Theorems.

8.1. Theorem 1. *Set* $\tilde{F}_I([\Phi_2]) = \{(t_1,t_2,\rho) \mid 1 < \rho \leq (1+t_1^{1/2}t_2) / (t_1^{1/2}+t_2)$, $0 < t_2 \leq t_1$, $0 < t_1 < 1\}$. *If* $G=<A_1,A_2>$ *is a free, purely loxodromic discrete group with* $\tau(G) = (t_1,t_2,\rho) \in R_I$, *then there exists an element* $\phi \in \Phi_2$ *with* $\phi(\tau(G)) \in \tilde{F}_I([\phi_2])$, *where* $\tau(G)$ *is the point corresponding to G defined in* §1.

Proof. We set $D_j(n) = \bigcup_{0<t_1<1} D_j(t_1;n)$ for j=1,2,3 and for each integer $n \neq 0$, and set $D_j = \bigcup_{n\neq 0} D_j(n)$ (cf. §3.2). Since *G* contains no parabolic transformations, $\tau(G) \in D_1 \cup D_2 \cup D_3$. Hence $\tau(G) \in D_j(n)$ for some *j* and for some *n*. Then $\phi_1(\tau(G)) \in D_j(1)$ for some $\phi_1 \in \Phi_2$ by Lemma 3.2. Since *G* is free and discrete, $\phi_1(\tau(G)) \in D_1(1) \cup D_3(1)$ from Propositions 4.1, 4.2 and 5.1.

Suppose $\phi_1(\tau(G)) \in D_3(1)$. Since G is discrete, $\phi_2\phi_1(\tau(G)) \in Cl(D_1) \cup Cl\,(D_2)$

for some $\phi_2 \in \Phi_2$ from Proposition 5.1. Since G is discrete and purely loxodromic, $\phi_2\phi_1(\tau(G)) \in D_1$, that is $\phi_2\phi_1(\tau(G)) \in D_1(m)$ for some integer *m* from Propositions 4.1, 4.2, 6.1, 6.2 and 7.1. Hence $\phi_3\phi_2\phi_1(\tau(G)) \in D_1(1)$ for some $\phi_3 \in \Phi_2$ by Lemma 3.2. By applying N_2 and/or N_3N_1, if necessary, to $\phi_3\phi_2\phi_1(\tau(G))$ (resp. $\phi_1(\tau(G))$) if $\phi_1(\tau(G)) \in D_3(1)$ (resp. $\phi_1(\tau(G)) \in D_1(1)$), we have the desired result, q.e.d.

Corollary 1. $R_I\mathfrak{S}_2^0 = R_I\mathfrak{S}_2$, *namely every Schottky group of Type* I *is classical.*

Proof. We will only prove that $R_I\mathfrak{S}_2^0 \supset R_I\mathfrak{S}_2$. Suppose $\tau \in R_I\mathfrak{S}_2$. Let $G(\tau)$ be the group corresponding to τ. Since $G(\tau)$ is free, purely loxodromic and discontinuous, $G(\tau)$ satisfies the assumption of Theorem 1. Hence $\phi(\tau) \in \tilde{F}_I([\Phi_2])$ for some $\phi \in \Phi_2$. Since $\tilde{F}_I([\Phi_2]) \subset D_1(1)$, we see that $G(\tau)$ is a classical Schottky group from Proposition 4.1. Hence $\tau \in R_I\mathfrak{S}_2^0 = R_I\mathfrak{S}_2$, q.e.d.

From Theorem 1, Sato [4, Theorem 4] and Purzitsky [2, Theorem 3], we have the following.

Corollary 2. *Every Fuchsian Schottky group of genus two is classical.*

8.2. Lemma 8.1. *Let* $\tau \in \beta(t_1;n)$ *and* $G(\tau)$ *be the group corresponding to* τ. *If* $G(\tau)$ *is a free and discrete group, then* τ *is a boundary point of* $R_I\mathfrak{S}_2^0$.

Proof. Since $G(\tau)$ is free and discrete, τ is of cases (1) or (3) in Proposition 7.1. If τ is of case (1), τ is a boundary point of $D_1(t_1;2)$ for some t_1. Hence τ is a boundary point of $R_I\mathfrak{S}_2^0$ by Proposition 4.1. Next suppose that τ is of case (3). By slightly deforming the fundamental region for $G_j^*(\tau)$ constructed in the proof of Proposition 7.1 (3), we easily see that there exists a sequence of classical Schottky groups tending to $G_j^*(\tau)$. Hence in this case τ is also a boundary point of $R_I\mathfrak{S}_2^0$, q.e.d.

Theorem 2. *Let* $G = \langle A_1, A_2 \rangle$ *is a free and discrete group with* $\tau(G) \in R_I$, *then* $\tau(G) \in$

$Cl\,(R_I\mathfrak{S}_2^0)$, *where* $\tau(G)$ *is the point corresponding to G.*

Proof. First we see from Theorem 1 that if G is a free, purely loxodromic discrete group, then $\phi(\tau) \in \tilde{F}_I([\Phi_2])$ for some $\phi \in \Phi_2$. Since $\tilde{F}_I([\Phi_2]) \subset R_I\mathfrak{S}_2^0$, $\tau \in R_I\mathfrak{S}_2^0$.

Next if *G* is a free and discrete group containing parabolic transformations, then

$$\tau \in \bigcup_{0<t_1<1} \bigcup_{n\neq 0} (\alpha(t_1;n) \cup \beta(t_1;n) \cup \gamma(t_1;n)).$$

If $\tau \in \bigcup_{0<t_1<1} \bigcup_{n\neq 0} (\alpha(t_1;n)$, then τ is a boundary point of $R_I\mathfrak{S}_2^0$ by Propositions 4.1 and 6.1. If $\tau \in \bigcup_{0<t_1<1} \bigcup_{n\neq 0} \beta(t_1;n)$ and *G* is a free and discrete group, then τ is a boundary point of $R_I\mathfrak{S}_2^0$ by Lemma 8.1. If $\tau \in \bigcup_{0<t_1<1} \bigcup_{n\neq 0} \gamma(t_1;n)$ and *G* is a free and discrete group, then $\phi(\tau) \in Cl\,(D_1) \cup Cl\,(D_2) - \gamma(t_1;1)$ for some $\phi \in \Phi_2$ by Proposition 6.2. Since G contains parabolic transformations, $\tau \in \alpha(t_1;n) \cup \beta(t_1;n)$ for some t_1 and for some integer *n*. By Lemma 3.1, $\phi(\tau) \in \alpha(t_1;1) \cup \beta(t_1;1)$ for some $\phi \in \Phi_2$. By the above fact, we see that τ is a boundary point of $R_I\mathfrak{S}_2^0$, q.e.d.

Corollary. *Let* τ *be as in Theorem* 2. *Then there exists an element* $\phi \in \Phi_2$ *with* $\phi(\tau) \in Cl\,(N_3N_1\,(\tilde{F}_I([\Phi_2])))$.

Proof. This comes from Theorem 2 and Sato [4, Theorem 2], q.e.d.

Acknowledgements. The work was partially supported by the Grants-in-Aid Scientific and Co-operative Research, the Ministry of Education, Science and Culture, Japan.

References

[1] Beardon, A. F., The Geometry of Discrete Groups, Springer Verlag, New York, Heidelberg, Berlin, 1983.

[2] Marden, A., Schottky groups and circles, in Contributions to Analysis (L. V. Ahlfors,ed.) pp. 273 - 278, Academic Press, New York-London, 1974.

[3] Purzitsky, N., Real two-dimensional representation of two-generator free groups, Math.Z. 127 (1972), pp. 95 - 104.

[4] Sato, H., Classical Schottky groups of real type of genus two, I, Tôhoku Math. J. 40 (1988), pp. 51 - 75.

[5] Sato, H., On a paper of Zarrow, Duke Math. J. 56 (1988), pp. 205 - 209.

[6] Zarrow, R., Classical and non-classical Schottky groups, Duke Math. J. 42 (1975), pp. 717 - 724.

Department of Mathematics
Faculty of Science
Shizuoka University

CONFORMAL EMBEDDINGS OF AN OPEN RIEMANN SURFACE INTO CLOSED SURFACES OF THE SAME GENUS

Masakazu Shiba

Dedicated to Professor Kôtaro Oikawa on his sixtieth birthday

Introduction

Every open Riemann surface of finite genus can be conformally embedded into a closed Riemann surface of the same genus. Such a closed surface is classically called a compact continuation [of the given open surface] of the same genus. Generally there are various compact continuations of a given open surface. The aim of this paper is to study the space of compact continuations of an open Riemann surface by considering their (normalized) period matrices. This means that all surfaces have to be considered together with their canonical homology bases (modulo dividing cycles for open surfaces) and that the space of compact continuations is grasped in the Torelli space or the Siegel upper half space.

In the former paper [10] we have paid particular attention to the diagonal entries of the period matrices and shown that the diagonal actually characterizes a special subspace of compact continuations, the space of canonical hydrodynamic continuations. We have also obtained a generalization of Schiffer's span (AD-span) to surfaces of positive genera.

We will now consider the nondiagonal entries as well and prove that the period matrices of compact continuations of a given open Riemann surface is bounded in the Siegel upper half space. This is a generalization of Heins' theorem ([3]).

For the completeness we state some results in [10] and [11], with which the present work is closely related. Theorems 2 and 5 give generalization and refinement of the preceding paper [10], and Theorem 6 gives information upon the behavior of the period matrices of hydrodynamic continuations. Finally, Heins' theorem is generalized in Theorems 7 and 8.

§ 1. Throughout this paper (R, χ) denotes a pair of an open Riemann surface R of finite genus g (≥ 1) and a fixed canonical homology basis $\chi = \{a_j, b_j\}_{j=1}^{g}$ of R modulo dividing cycles.

A realization of (R, χ) is, by definition, a triple (R', χ', I') of a closed Riemann surface R' of genus g, a canonical homology basis $\chi' = \{a_j', b_j'\}_{j=1}^{g}$ of R', and a conformal embedding I' of R into R' such that $I'(a_j)$ and $I'(b_j)$ are respectively homologous to a_j' and b_j' ($j = 1, 2, ..., g$). Two realizations (R', χ', I') and (R'', χ'', I'') of (R, χ) are said to be <u>equivalent</u>, if there is a conformal mapping f of R' onto R'' with $f \circ I' = I''$ (on R) and each equivalence class is called a <u>compact continuation</u> of (R, χ). The compact continuation determined by the realization (R', χ', I') is denoted by $[R', \chi', I']$, and the space of compact continuations of (R, χ) by $C(R, \chi)$.

Since two equivalent realizations of (R, χ) obviously define the same "Torelli surface" (i.e., the same point in the Torelli space $\mathfrak{T}_g$ of genus g, and have the same normalized period matrix, we can speak of the <u>(normalized) period matrix of a compact continuation</u> of (R, χ). We denote by $\mathfrak{S}_g$ the Siegel upper half space of genus g and define τ: $C(R, \chi) \to \mathfrak{S}_g$ by setting $\tau[R', \chi', I']$ to be the normalized period matrix of $[R', \chi', I']$, that is, of the Torelli surface (R', χ').

§ 2. Let F be an S_t- function on R, $-1 < t \leq 1$ (see [8] and [11]). This means that F is a (single- or multiple-valued) meromorphic function on R such that $\mathrm{Im}(e^{-1/2\pi i t} dF)$ is distinguished in the sense of Ahlfors. When we need no explicit reference to t, we simply call F an S-function. The name S-function comes from the german word "Strömungsfunktion". It should be noted, however, that "Strömungsfunktion" does not mean the stream function in fluid dynamics but the complex velocity potential function (of a stationary flow of an ideal fluid on R). Cf. [12].

Theorem 1 ([8], [11]). For any non-constant S_t- function F on R there exists a realization (R^F, χ^F, I^F) which has the following properties.

(I) The (Lebesgue) area of $R^F \backslash I^F(R)$ vanishes.

(II) The pull-back of $d\tilde{F}$ via I^F coincides with dF (on R).

(III) $\tilde{F}$ is single-valued and holomorphic on (a neighborhood of) $R^F \backslash I^F(R)$, and $\mathrm{Im}(e^{-1/2\pi i t}\tilde{F})$ takes on a constant value on each component $R^F \backslash I^F(R)$.

We call (R^F, χ^F, I^F) a hydrodynamic realization and $[R^F, \chi^F, I^F]$ a hydrodynamic continuation of (R, χ) with respect to F. In fact, $e^{-1/2\pi i t}\tilde{F}$ and R^F have a remarkable physical meaning. To see this, we simplify the matter and identify $I^F(R)$ with R itself, and consider $\tilde{F}$ as a flow on R^F. Then the total energy of the flow $\tilde{F}$ on $\tilde{R}$ is the same as that of the flow F on R, and each ideal boundary component of R is realized on R^F as a (generally branched) arc on the stream lines of $e^{-1/2\pi i t}\tilde{F}$.

Let $C(R, \chi; F)$ be the space of hydrodynamic continuations of (R, χ) with respect to F. We know ([11]) that $C(R, \chi; F)$ does often contain more than one point.

§ 3. For any $A = (\alpha_1, \alpha_2, ..., \alpha_g) \in \mathbb{C}^g$ and t, $-1, < t \leq 1$, there exists an S_t- function Φ_A^t on R with

$$\int_{a_j} d\Phi_A^t = \alpha_j, \quad j = 1, 2, ..., g.$$

Furthermore, Φ_A^t is uniquely determined up to an additive constant. For the details, see e.g., [8].

We write Φ_k^t for $\Phi_{A_k}^t$, where $A_k = (\delta_{k1}, \delta_{k2}, ..., \delta_{kg})$, k = 1, 2, ..., g. The system $\{\Phi_k^t\}_{k=1}^{g}$ is called a canonical basis for the holomorphic S_t- functions on (R, χ). It is not difficult to see that if $A = (\alpha_1, \alpha_2, ..., \alpha_g) \in \mathbb{C}^g$ then

$$d\Phi_A^t = \sum_{k=1}^{g} \{\text{Re } \alpha_k\} \, d\Phi_k^t + (\text{Im } \alpha_k) \, d\Psi_k^t\},$$

where

$$d\Psi_k^t := d\Phi_k^s, \qquad s \equiv t + 1 \pmod 2), -1 < s \leq 1.$$

The corresponding results for closed surfaces are well known (cf. e.g., [1] or [12]). In particular, for any $[R', \chi', I'] \in C(R, \chi)$, there exists a unique holomorphic differential φ'_A on R' such that

$$\int_{I'(a_j)} \varphi'_A = \alpha_j, \; j = 1, 2, ..., g.$$

The g complex numbers

$$\beta'_j(A) = \int_{I'(b_j)} \varphi'_A \;, j = 1, 2, ..., g$$

are also uniquely determined. Thus, we can define, for any complex g-vector $A = (\alpha_1, \alpha_2, ..., \alpha_g)$ and a real number $t \in (-1, 1]$, the mapping

$$\Pi_A^t\colon \; C(R, \chi) \to \mathbb{R}$$

by setting

$$\Pi_A^t \, [R', \chi', I'] = \text{Im}(e^{-\pi i t} \sum_{j=1}^{g} \alpha_j \beta'_j(A)).$$

By applying the same argument as in [10], we can prove

<u>Theorem 2</u>. For any <u>complex</u> g-vector A and $t \in (-1, 1]$, Π_A^t assumes its minimum value exactly on the space $C(R, \chi; \Phi_A^t)$:

$$\Pi_A^t[R', \chi', I'] \geq \Pi_A^t[\tilde{R}, \tilde{\chi}, \tilde{I}]$$

for all $([R', \chi', I'], [\tilde{R}, \tilde{\chi}, \tilde{I}]) \in C(R, \chi) \times C(R, \chi; \Phi_A^t)$ and the equality holds if and only if $[R', \chi', I'] \in C(R, \chi; \Phi_A^t)$.

§ 4. Now, we set for any $t \in (-1, 1]$

$$\tau^t_{kj} = \int_{b_j} d\Phi^t_k, \quad j, k = 1, 2, \ldots, g,$$

and let

$$\tau_\ell[\tilde{R}, \tilde{\chi}, \tilde{I}]$$

denote the ℓ-th row of the period matrix $\tau[\tilde{R}, \tilde{\chi}, \tilde{I}]$ of $[\tilde{R}, \tilde{\chi}, \tilde{I}] \in C(R, \chi; \Phi^t_k)$. Then we have

$$\tau_k[\tilde{R}, \tilde{\chi}, \tilde{I}] = (\tau^t_{k1}, \tau^t_{k2}, \ldots, \tau^t_{kg}),$$

but the corresponding equalities for $\ell \neq k$ do not always hold. In other words, the matrix

$$T_t := (\tau^t_{jk})_{j,\, k = 1, 2, \ldots, g}$$

is not necessarily the period matrix of $(\tilde{R}, \tilde{\chi})$.

We set furthermore for $j, k = 1, 2, \ldots, g$

$$\tau^*_{kj} = \tfrac{1}{2}(\tau^1_{kj} + \tau^0_{kj}),$$

$$\rho_{kj} = \tfrac{1}{2}\,|\,\tau^1_{kj} - \tau^0_{kj}|$$

and

$$t_{kj} = 1 - \tfrac{1}{\pi}\arg(\tau^1_{kj} - \tau^0_{kj}) \in (-1, 1].$$

Then we know

Theorem 3 ([10]).

(I) For each j and k $(1 \leq j, k \leq g)$,

$$\tau^t_{kj} = \tau^*_{kj} + \rho_{kj} e^{(t - t_{kj})\pi i}, \quad t \in (-1, 1]. \tag{1}$$

(II) The diagonal entries τ'_{kk} of the normalized period matrix of $[R', \chi', I'] \in C(R, \chi)$ satisfy the inequalities

$$\mathrm{Im}\ \tau^0_{kk} \leq \mathrm{Im}\ \tau'_{kk} \leq \tau^1_{kk}.$$

The left (resp. right) equality holds if and only if $[R', \chi', I']$ is a (canonical) hydrodynamic continuation with respect to Φ^0_k (resp. Φ^1_k).

(III) $t_{kk} = \frac{1}{2}$ for $k = 1, 2, ..., g$.

§ 5. The following theorem shows that the space $C(R, \chi; \Phi^t_k)$ of canonical hydrodynamic continuations is fully characterized by the diagonal entries of the period matrices.

Theorem 4 ([10]). Let Δ: $C(R, \chi) \to \mathbb{C}^g$ be defined by

$$\Delta[R', \chi', I'] = (\tau'_{11}, \tau'_{22}, ..., \tau'_{gg})$$

with $\tau'_{kk} = \tau_{kk}[R', \chi', I']$, $k = 1, 2, ..., g$. Then, $\Delta(C(R, \chi))$ is contained in the closed polydisk

$$P = P(R, \chi) := \prod_{k=1}^{g} \{|\tau_{kk} - \tau^*_{kk}| \leq \rho_{\kappa\kappa}\}.$$

Furthermore $\Delta[R', \chi', I'] \in \partial P$, the topological boundary of P, if and only if $[R', \chi', I']$ belongs to $C(R, \chi; \Phi^t_k)$ for some $(t, k) \in (-1, 1] \times \{1, 2, ..., g\}$.

We can show that Δ: $C(R, \chi) \to P$ is not always surjective; even the distinguished boundary of P may not be covered by $\Delta(C(R, \chi))$.

§ 6. We are now concerned with non-diagonal entries of the normalized period matrix $\tau[R', \chi', I'] \in \mathfrak{S}_g$ of $[R', \chi', I'] \in C(R, \chi)$. We have already shown in [9] and [10] that the consideration of the diagonal entries also bring about a generalization of the notion of Schiffer's span. Here we generalize this result a little more and prove the following theorem, which states that any value ρ_{jk} (j, k = 1, 2, ..., g) serves as a generalized span. This generalized span may be called AD-span. Cf. [7].

Theorem 5. The following eight conditions are equivalent to each other.

(a) $\rho_{kk} = 0$ for some k $(1 \le k \le g)$.

(b) $\rho_{jj} = 0$ for every $j = 1, 2, ..., g$.

(c) $\rho_{jk} = 0$ for all j, k $(1 \le j, k \le g)$.

(d) P is degenerate.

(e) $C(R, \chi)$ consists of a single point.

(f) R is of class O_{AD}.

(g) R is uniquely prolongable in the sense of Oikawa ([6]).

(h) R is uniquely prolongable in the sense of Nevanlinna ([4]).

In fact, we have already proved in [10] that all the statements, except (c) and (d), are equivalent to one another, and the rest of the proof follows directly from the other part of the theorem and Theorem 3.

§ 7. The following proposition will be often used in the sequel.

Proposition. Let s, t ϵ (−1, 1] and let j, k be two integers with $1 \le j, k \le g$. Then

$$\operatorname{Im}\left(e^{-\frac{1}{2}(s+t)\pi i}\left(\tau^{s}_{jk} - \tau^{t}_{kj}\right)\right) = 0 \tag{2}$$

Proof. We apply the generalized Riemann's bilinear relation (cf. e.g., [10]) to

$<e^{-\frac{1}{2}\pi it}\, d\Phi^{t}_{k},\ e^{-\frac{1}{2}\pi is}\, d\Phi^{s}_{j}>$ in two ways.

On one hand, this inner product is equal to

$$2 \cos \tfrac{1}{2}\pi t \cos \tfrac{1}{2}\pi s \operatorname{Im} \tau_{jk}^{s} + 2 \sin \tfrac{1}{2}\pi t \sin \tfrac{1}{2}\pi s \operatorname{Im} \tau_{jk}^{t}$$

$$- 2 \cos \tfrac{1}{2}\pi t \sin \tfrac{1}{2}\pi s \operatorname{Re} (\tau_{jk}^{s} - \tau_{kj}^{t}),$$

and, on the other hand, it is also equal to

$$< e^{-\frac{1}{2}\pi is} d\Phi_j^s, e^{-\frac{1}{2}\pi it} d\Phi_k^t >$$

$$= 2 \cos \tfrac{1}{2}\pi s \cos \tfrac{1}{2}\pi t \operatorname{Im} \tau_{kj}^{t} + 2 \sin \tfrac{1}{2}\pi s \sin \tfrac{1}{2}\pi t \operatorname{Im} \tau_{jk}^{s}$$

$$- 2 \cos \tfrac{1}{2}\pi s \sin \tfrac{1}{2}\pi t \operatorname{Re} (\tau_{kj}^{t} - \tau_{jk}^{s}).$$

Now the desired equality immediately follows by simple computations.

On setting $t = s$ in equation (2), we have

Corollary. $\operatorname{Im}(e^{-\pi it}T_t)$ is a symmetric matrix for every $t \in (-1, 1]$.

We are now ready to prove

Theorem 6. For $j, k = 1, 2, ..., g$,

$$\tau_{jk}^{*} = \tau_{kj}^{*}, \tag{3}$$

$$\rho_{jk} = \rho_{kj}, \tag{4}$$

and, (unless $\rho_{jk} = \rho_{kj} = 0$)

$$t_{jk} + t_{kj} = \pm 1. \tag{5}$$

For the proof, consider the particular case $t = s$ in the preceding proposition (or the above corollary):

$$\operatorname{Im} (e^{-\pi it}(\tau_{jk}^{t} - \tau_{kj}^{t})) = 0. \tag{6}$$

Substitution of (1) into (6) yields

$$\operatorname{Im}\,(e^{-\pi it}(\tau^*_{jk} - \tau^*_{kj}) + \rho_{jk}e^{-\pi it_{jk}} - \rho_{kj}e^{-\pi it_{kj}}) = 0.$$

Since t can assume an arbitrary value in $(-1, 1]$, we get equation (3). We also have

$$\rho_{jk}\sin(t_{jk}\pi) = \rho_{kj}\sin(t_{kj}\pi). \tag{7}$$

This equation can be also obtained by letting $s = t = \frac{1}{2}$ in the proposition.

Next, we consider another special case $s = 0$, $t = 1$ (or $s = 1$, $t = 0$) in Proposition. Then we have

$$\rho_{jk}\cos(t_{jk}\pi) = -\rho_{kj}\cos(t_{kj}\pi). \tag{8}$$

Combination of (7) and (8) yields the remaining conclusions of the theorem.

The above reasoning does not exclude the special case $j = k$, so that we have from equation (5) the following corollary, which is a weaker result than Theorem 3, (III).

Corollary. $t_{kk} = \pm\frac{1}{2}$, $k = 1, 2, \ldots, g$.

§ 8. From Theorem 2 we can prove the following theorem; for the details of the proof see [10].

Theorem 7. The period matrix $(\tau'_{jk})_{j,\,k\,=\,1,\,2,\,\ldots,\,g}$ of $[R', \chi', I'] \in C(R, \chi)$ satisfy

$$\sum_{j,k=1}^{g} \operatorname{Im}\,[e^{-\pi it}\,(\tau'_{jk} - \tau^*_{jk})\,\alpha_j\alpha_k - \rho_{jk}e^{-\pi it_{jk}}\overline{\alpha}_j\alpha_k] \geq 0$$

for any complex g-vector $A = (\alpha_1, \alpha_2, \ldots, \alpha_g) \neq 0$. The equality holds for $[R', \chi', I'] \in C(R, \chi; \Phi^t_A)$.

We shall now prove the following theorem.

Theorem 8. Let $(\tau'_{jk})_{j,\,k\,=\,1,\,2,\,\ldots,\,g}$ be the period matrix of $[R', \chi', I'] \in C(R, \chi)$, then

$$(\operatorname{Im}\,[e^{-\pi it}\,(\tau'_{jk} - \tau^*_{jk}) - \rho_{jk}e^{-\pi it_{jk}}])_{j,\,k\,=\,1,\,2,\,\ldots,\,g}.$$

is a positive semi-definite matrix for any $t \in (-1, 1]$.

Proof. We first note that

$$(\mathrm{Im}\, [e^{-\pi it} (\tau'_{jk} - \tau^*_{jk}) - \rho_{jk} e^{-\pi it_{jk}}])_{j,\, k = 1, 2, \ldots, g}.$$

is symmetric. This follows immediately from Corollary to Proposition and the well-known symmetry of the matrix $T' = (\tau'_{jk})_{j,\, k = 1, 2, \ldots, g}$.

Next, let $A = (\alpha_1, \alpha_2, \ldots, \alpha_g)$ be a real g-vector and consider the S_t-function Φ^t_A. As we have shown in section 3, $d\Phi^t_A = \sum_{j=1}^{g} \alpha_j\, d\Phi^t_j$, so that, we have, for $k = 1, 2, \ldots, g$,

$$\beta^t_k(A) = \int_{b_k} d\Phi^t_A = \sum_{j=1}^{g} \alpha_j \int_{b_k} d\Phi^t_j = \sum_{j=1}^{g} \alpha_j\, \tau^t_{jk}.$$

Similarly we have $k = 1, 2, \ldots, g$

$$\beta'_k(A) = \sum_{j=1}^{g} \alpha_j\, \tau'_{jk}.$$

It follows from Theorem 2 (or Theorem 7) that

$$\mathrm{Im}\, (e^{-\pi it} \sum_{j=1}^{g} \alpha_j \sum_{k=1}^{g} \alpha_k (\tau'_{jk} - \tau^t_{jk})) \geq 0,$$

or

$$\sum_{j,\, k=1}^{g} \mathrm{Im}\, (e^{-\pi it} (\tau'_{jk} - \tau^t_{jk}))\, \alpha_j \alpha_k \geq 0. \qquad (9)$$

Substituting (1) into the last inequality, we have

$$\sum_{j,\, k=1}^{g} \mathrm{Im}\, (e^{-\pi it} (\tau'_{jk} - \tau^*_{jk}) - \rho_{jk} e^{-\pi it_{jk}})\, \alpha_j \alpha_k \geq 0$$

This completes the proof of the theorem.

For two g x g matrices T_1, T_2 with the symmetric imaginary parts, we shall write

$$T_1 \succ T_2$$

if $\mathrm{Im}\, (T_1 - T_2)$ is positive semi-definite. Then " $\succ$ " defines a quasi-ordering in $\mathfrak{S}_g$. Furthermore, we set $Q = \{T' = (\tau'_{jk}) \in \mathfrak{S}_g \mid e^{-\pi it}\, T' \succ e^{-\pi it} T_t$ for all $t \in (-1, 1]\}$. Obviously Q is a convex set. From inequality (9) we can restate Theorem 8 as follows:

Theorem 8'. $\tau(C(R, \chi)) \subset Q$.

We note that Q only gives a bound for the set $\tau(C(R, \chi))$ in general, since, as was noted earlier, the matrix T_t is not always the period matrix of a compact continuation of (R, χ). Cf. the remark after Theorem 2.

From Theorem 8' (or Theorem 8) we can easily conclude the following theorem, which generalizes Heins' result ([3]).

Theorem 9. τ: $C(R, \chi) \to \mathfrak{S}_g$ is bounded. That is, the $\frac{1}{2}g(g+1)$ functions

$$\tau_{jk}\colon\ C(R, \chi) \to \mathbb{C},\quad \tau_{jk}\,[R', \chi', I'] = \tau'_{jk},$$

are bounded on $C(R, \chi)$.

References

[1] Farkas, H. M., and I. Kra, Riemann surfaces, Springer, New York-Heidelberg-Berlin, 1980, 337pp.

[2] Grötzsch, H., Die Werte des Doppelverhältnisses bei schlichter konformer Abbildung, Sitzungsber. Preuss. Akad. Wiss. Berlin (1933), pp. 501 - 515.

[3] Heins, M., A problem concerning the continuation of Riemann surfaces, in Contributions to the theory of Riemann surfaces, ed. by L. V. Ahlfors et al. Princeton Univ. Press, Princeton, 1953, pp. 55 - 62.

[4] Mori, A., A remark on the prolongation of an open Riemann surface of finite genus, J. Math. Soc. Japan 4 (1952), pp. 27 - 30.

[5] Oikawa, K., On the prolongation of an open Riemann surface of finite genus, Kōdai Math. Sem Rep. 9 (1957), pp. 34 - 41.

[6] Oikawa, K., On the uniqueness of the prolongation of an open Riemann surface of finite genus, Proc. Amer. Math. Soc. 11 (1960), pp. 785 - 787.

[7] Sario, L., and K. Oikawa, Capacity functions, Springer, Berlin-Heidelberg-New York, 1969, 361pp.

[8] Shiba, M., The Riemann-Hurwitz relation, parallel slit covering map, and continuation of an open Riemann surface of finite genus, Hiroshima Math. J. 14 (1984), pp. 371 - 399.

[9] Shiba, M., The moduli of compact continuations of an open Riemann surface of genus one, Trans. Amer. Math. Soc. 301 (1987), pp. 371 - 399.

[10] Shiba, M., The period matrices of compact continuations of an open Riemann surface of finite genus, in " Holomorphic functions and Moduli", vol. I. ed. by D. Drasin, Springer, 1988, pp. 237-246.

[11] Shiba, M., and K. Shibata, Hydrodynamic continuations of an open Riemann surface of finite genus, Complex Variables Theory Appl. 8 (1987), pp. 205 - 211.

[12] Springer, G., Introduction to Riemann surfaces, Addison-Wesley, Reading, 1957, 309pp. (Reprint: Chelsea, New York, 1981).

Department of Mathematics
Hiroshima University

DIRICHLET PRINCIPLE AND THE ANALYTIC PROLONGATION THEOREM ONTO ANALYTIC BOUNDARY

Keiichi Shibata

Dedicated to Professor Yûsaku Komatu on his 75th birthday

0. Introduction

0.1. Of the three cases classified in the statement of Riemann's mapping theorem that concerns the conformal maps of simply connected plane regions, the hyperbolic case is of course the most important, whose original proof was based on the so-called Dirichlet Principle in an immediate way.

Given a pair of conformally equivalent regions with regular analytic arcs on their boundaries, the function associating their interiors one-to-one conformally with one another, can be continued analytically, as is well known, up to the like one-to-one conformal homeomorphism between the boundary arcs.

0.2. The underlying idea in this paper is as follows: Under a stronger assumption that the boundary analyticity prevails over the whole contour, the Dirichlet Principle enables us to convert the implication in the above propagation of analyticity from interior to boundary, i.e., we can elucidate how the boundary analyticity gives rise to the interior analyticity by annihilating the first variation of Dirichlet's functional in an appropriate space.

In case in which one of those analytic contours lies in a higher dimensional space $\mathbb{R}^n$ ($n \geq 3$), the observations above find their analogy in the analytic prolongation of minimal surface beyond the analytic boundary. In what follows we will argue mainly in terms of minimal surfaces rather than of plane conformal mappings except the final section § 6, on account of almost the same circumstances in both settings $n = 2$ and $n \geq 3$.

To be precise, we provide at first in § 1 a criterion concerning the interrelationship between the regular analyticity of the boundary and the interior harmonicity in an isothermal parameter, putting aside temporarily the existence proof — a necessary and sufficient condition for a minimal surface S_0 to span a system of ℓ regular analytic closed curves $\gamma = \{\gamma^{(1)}, \ldots, \gamma^{(\ell)}\}$ ($1 \leq \ell < +\infty$) with reference to the structures of the one- and two-dimensional submanifolds in question which are separately of analytic natures. Here the closed curves may not necessarily be simple.

According to this criterion, the contour γ of any minimal surface remains regular analytic in the surface parameter itself, only if γ can be regular analytic with regard to at least one curve parameter, which suggests us to apply Dirichlet Principle to the aggregate of smooth surfaces (i.e., vector-valued functions) with boundary value that makes it regular analytic. In this way an alternative for the existence proof will be worked out in § 2. §§ 3 - 4 deal with a similar criterion in reference to the Dirichlet's energy integral, utilizing new variational formulae for this functional.

0.3. In this context our circle of ideas has achieved derivation of a new uniqueness result as a by-product:

Uniqueness Theorem. Every single closed contour, which is regular analytic, bounds a unique orientable minimal surface to within the conformal automorphisms of the parameter domain.

Our method thus offers on the other hand a new insight into the conformal mapping theory of a simply connected region, ramifying over $\mathbb{C}$ with analytic boundary, or, of bordered Riemann subsurfaces, which draws our interest per se.

1. Minimal surfaces spanning a regular analytic contour

1.1. The main body of this section shall be preceded by an exposition of some notations, terminology and definitions frequently used throughout the paper.

$\mathbb{R}^n$ ($2 \leq n \in \mathbb{Z}$): the totality of the real n-vector $x = (x^1, ..., x^n)$; $\mathbb{C} = \mathbb{R}^2$; $w = u + \sqrt{-1}v = e^{\sqrt{-1}\theta}$ ($u, v \in \mathbb{R}$), $\theta \in \mathbb{R}$.

$B^{(j)}$ ($j = 1, 2, ..., \ell$) denotes disjoint closed disks lying in the complex w-plane, the first one $B^{(1)}$ of them being the unit disk, in particular, and the others (in case $\ell > 1$) located in the interior of $B^{(1)}$ (which is abbreviated as Int B).

$R(b) = \{w \in |1 - |w|| < b\}$ ($0 < b < 1$): a concentric annular region;

When circumstances require to distinguish the independent variable w or $\zeta \in \mathbb{C}$ in §§ 2 - 4, the unit disk B is specified by the preciser symbols B_w of B_ζ respectively.

It is not so out of date to make mention about a rigorous treatment on the regular analytic curve, not an arc but the regular analytic closed curve immersed in $\mathbb{R}^n$ ($n > 1$), because it is seldom encountered in available mathematical books or papers. Emphasis is laid here upon our attitude of mind that we aim at setting up a unified treatment of the regular analytic closed curve which is not only an embedding but also an immersion of the circumference into $\mathbb{C} = \mathbb{R}^2$ or $\mathbb{R}^n$ ($n > 2$).

Definition 1. A closed curve γ is regular analytic with its parameter circle ∂B if and only if its suitable parametrization $\gamma(w)$ satisfies the requirements: (i) each component of the n-vector $\gamma(w)$ extends to a complex function single-valued and holomorphic in a neighbourhood of ∂B; (ii) $d\gamma(w)/dw \neq 0$, where it is defined.

Remark 1. The above characteristic feature of the regular analytic curve rests unaltered only under one-to-one conformal transformations of the parameter circle.

Remark 2. The definition of an analytic arc is described sometimes (in particular, by differential geometers) in terms of Taylor expandability in its arc-length parameter, which turns out to be essentially equivalent to ours. But Definition 1 is better adapted for our purpose dealing with the closed curves.

Under an orientable surface $S = S(w)$ of disk type we understand an equivalence class modulo C^2– automorphisms of Int B, whose representative is a C^2– mapping of Int B into $\mathbb{R}^n$. S(w) is said to span (or, to be bounded by) the system of regular analytic closed curves $\gamma^{(j)}(w)$ $(j = 1, 2, \ldots, \ell)$, if it is continuous up to B and if some orientation-preserving homeomorphism τ of $\partial B^{(j)}$ satisfies the relation $\tau \circ S(w) \mid_{w \in B_{(j)}} = \gamma^{(j)}(w)$. In case S(w) is of rank 2 everywhere in the interior of B, it is called an immersed surface.

<u>Definition 2</u>. The parametrized surface S(w) on B becomes a minimal surface, if it is harmonic with respect to an isothermal parameter, i.e.

$$\nabla^2 S(w) = 0, \tag{1}$$

$$|\partial S/\partial u|^2 - |\partial S/\partial v|^2 = (\partial S/\partial u)\,(\partial S/\partial v) = 0 \tag{2}$$

The orbit of the submanifolds such as γ, S etc. appearing in the observations so far, i.e., the bare point sets in $\mathbb{R}^n$ free from any parametrizations, will be named <u>locus</u> (or, <u>carrier</u>) of them and be denoted by the symbols Loc γ, Loc S etc. respectively.

<u>1.2</u>. We are now prepared to state and prove the first of our criteria:

<u>Theorem 1</u>. In order that a minimal surface $S_0 = S_0(w)$ parametrized on B might span the finite system $\gamma(w)$ of regular analytic closed curves it is necessary and sufficient that the complex analytic structure for the boundary curve γ is the unique continuation of the analytic structure which renders the harmonic surface $S_0(w)$ its own conformal structure.

<u>Remark 3</u>. Although an analogue to Theorem 1 may be found of course in H. Lewy [6], it should be noted that we allow the contour γ to have some multiple points, perhaps infinite in number, and that the method of proof differs from it.

Proof of Theorem 1. The sufficiency follows immediately from Definition 1 and one has only to prove the necessity.

Since the local validity of the necessity proof implies the global one, it loses no generality to confine ourselves to the special case $\ell = 1$, i.e., $B = B^{(1)}$.

Suppose a minimal surface $S_0(w)$ spans a regular analytic closed curve $\gamma(w)$, the former of which is parametrized on B and the latter on $R(a_0)$ for some positive constant a_0 (< 1) respectively.

For an arbitrary point $w = e^{\sqrt{-1}\theta}$ on ∂B we deform the homeomorphism $S_0^{-1} \circ \gamma\,(e^{\sqrt{-1}\theta})$ between ∂B up to $\phi(\theta)$, so that the resulting bijection $\phi(\theta)$ may be quasisymmetric. Then $\phi(\theta)$ is extensible up to a quasiconformal diffeomorphism $\tilde{w} = \phi(w)$ of Int B onto Int $B = \{\tilde{w} \mid |\tilde{w}| < 1\}$. In a similar manner the correspondence $\gamma^{-1} \circ S_0 \circ \phi^{-1}\,(\tilde{w})$ between the unit circumferences $|\tilde{w}| = 1$ and $|w| = 1$ can be extended quasiconformally up to a diffeomorphism $w = \tilde{\phi}\,(\tilde{w})$ between the exteriors of those unit disks. Welding of the two contiguous regions Int B and $(R(a_0) \setminus (B \cap R(a_0))$ along their common boundary ∂B by means of the quasisymmetric identification $\theta \to \phi(\theta)$ produces a simply connected Riemann region Δ lying over $\mathbb{C}$ and hence permits us to consider the point set in $\mathbb{C}^3$

$$F = \text{Loc}\ S_0 \circ \phi^{-1}(w)|_{w \in B}\ \cup\ \text{Loc}\ \gamma\,(w)\ |\ _{w \in R(a_0) \setminus (B \cap R(a_0))}$$

to be parametrized by a quasiconformal mapping of Δ.

There exists a $1:1$ quasiconformal mapping $\hat{w} = \chi(\tilde{w})$ of Δ onto a simply connected region in the $\hat{w}$-plane, such that $\hat{w}$ is an isothermal (or, equivalently, conformal) parameter for F. More precisely, the mapping

$$\hat{w} \to \text{Loc}\ \gamma(w)\ |\ _{w \in R(a_0) \setminus (B \cap R(a_0))}$$

is holomorphic both inside and outside of the welded track $\chi \circ \phi\ (\partial B)$, while the variable $\hat{w} \in \chi \circ \hat{\phi}(\text{Int } B)$ remains isothermal for the harmonic surface $S_0(w)$. Thus we have proved that the boundary curve γ is regular analytic with respect to the isothermal parameter of minimal surface S_0 spanning γ. q.e.d.

2. The absolute minimum of the Dirichlet functional

2.1. Variationally characterized, the minimal surface with a sufficiently smooth boundary is a critical point of the area functional, which may be formulated in the setting of fixed boundary value problem owing to its invariance character under any diffeomorphic changes of parameter. As to the Dirichlet's energy functional, however, being invariant only by means of conformal parameter change, the vanishing point of its first variation in a fixed boundary value setting is the unique solution to the so-called Dirichlet problem for the Laplacian operator expressible in the Poisson integral formula, where the critical value coincides, of course, with the minimum. Hence, if one will take a minimal surface spanning a given rectifiable Jordan curve for solution to a variational problem of Dirichlet's energy functional, it must inevitably bear the feature, so to say, of a movable boundary value problem, or, a conditionally free boundary value problem.

The surface of least energy (least area) may, among all the minimal surfaces, have long since been the most intensively studied well, partly because the temporarily prescribed boundary value determines a harmonic vector uniquely, which is just of least energy.

The well known Douglas-Courant's existence proof on minimal surfaces for a given rectifiable contour was, roughly speaking, based on the following idea: to solve the Dirichlet problem with the boundary data of all monotone mappings that takes a parameter curve to the locus of the contour and minimize the Dirichlet's energy functional within the class of such harmonic mappings. The resulting extremal map actually fulfills the conditions of minimal surface and the boundary correspondence is one-to-one.

2.2. Theorem 1 in the preceding section suggests, however, a condition to be satisfied by the Dirichlet integral of a harmonic map, spanning the regular analytic contour γ, of a fixed source manifold, in order that it may be a minimal surface. Indeed, we notice at a first glance of a simple example that the case $\ell \geq 2$ must be out of question under these circumstances in view of the so-called 'Modulsatz' and that we have only to be concerned with the harmonic extension of the regular analytic mapping that defines the contour, or, equivalently, with the class of harmonic maps of Int $B^{(1)}$ into $\mathbb{R}^n$ admitting all equivalent representations of the regular analytic closed curve γ as their boundary values. But previous to starting this procedure we will have something to do with, it is what the equivalent representations of γ should mean.

There is a positive constant a (< 1), which we intend to fix once and for all, such that the definition domain of our regular analytic closed curve $\gamma(w)$ comprises the concentric annulus R(a).

Let $\mathfrak{F}$ denote the family consisting of all conformal mappings of the region R(a) onto a doubly connected region with the inner boundary component $|\zeta| = 1$ in the ζ-plane. If one chooses an arbitrary f of $\mathfrak{F}$, then $\gamma \circ f^{-1}(\zeta)$ will be another parametrization of the same contour γ defined in the parameter region f(R(a)). For brevity we commit ourselves hereafter to understand under the symbol $\partial^{-1} \circ f(\partial B)$ the plane region bounded by the Jordan curve with the locus $f(\partial B)$. We now start carrying out the existence proof of the minimal surface spanning γ along our present lines of argument.

Proposition 1. The family $\mathfrak{F}$ is normal in R(a) and is compact in the topology of uniform convergence on an arbitrary compact subregion of R(a).

Proof. The first part of the assertion is immediate from Montel's theorem. The Weierstrass double series theorem and a theorem of Hurwitz guarantee that any cluster element of $\mathfrak{F}$ is a holomorphic injection of R(a) taking every value such that $|\zeta| - 1 \ (> 0)$ is sufficiently small.

Proposition 2. Suppose a sequence $\{f_n(w)\}_{n = 1, 2, ...}$ of $\mathfrak{F}$ converges to a limit $f_*(w) \in \mathfrak{F}$ compact-uniformly in R(a). Then to any compact subregion κ of the region $f_*(R(a))$ there corresponds an $n_0 \in \mathbb{Z}^+$ such that $\{f_n^{-1}(\zeta)\}_{n = n_0, n_0 + 1, ...}$ converges to $f_*^{-1}(\zeta)$ uniformly on κ.

Proof. If we choose a sufficiently large $n_0 \in \mathbb{Z}^+$ for the given compact set κ, the interior of all $f_n(R(a))$ comprises κ so far as $n \geq n_0$. Montel's theorem produces the normality of $\{f_n^{-1}(\zeta)\}_{n = n_0, n_0 + 1, ...}$ on κ. It contains a subsequence $\{f_{n_k}^{-1}(\zeta)\}_{k = 1, 2, ...}$ convergent to some limit $g(\zeta)$ uniformly on κ. Taking an arbitrary point ζ_* on κ, we set $w_* = f_*^{-1}(\zeta_*)$. We have

$$\zeta_* = \lim_{k \to \infty} f_{n_k} \circ f_{n_k}^{-1} (\zeta_*) = f_* \circ g (\zeta_*).$$

Since f_* was injective, it follows that

$$g (\zeta_*) = w_* = f_*^{-1} (\zeta_*).$$

In view of arbitrariness of $\zeta_* \in \kappa$ we get $g(\zeta) = f_*^{-1}(\zeta)$, which shows the uniform convergence $\lim_{k \to \infty} = f_{n_k}^{-1}(\zeta) = f_*^{-1}(\zeta)$ on κ. We have thus verified any subsequence of $\{f_n^{-1}(\zeta)\}_{n = n_0, n_0+1, ...}$ converges to $f_*^{-1}(\zeta)$, so far as it converges uniformly on κ at all.

If the original sequence $\{f_n^{-1}(\zeta)\}_{n = n_0, n_0 + 1, ...}$ does not converge to $f_*^{-1}(\zeta)$ uniformly on κ, a point sequence $\{\zeta_n\}_{n = n_0, n_0 + 1, ...}$ exists, satisfying

$$| f_n^{-1}(\zeta_n) - f_*^{-1}(\zeta_n) | \geq c$$

for some constant $c > 0$. We may assume without losing the generality that

$\{\zeta_n\}_{n = n_0, n_0+1, \dots}$ converges to some $\zeta_0 \in \kappa$. In virtue of the first part of this proof there is a sequence of indices $\{n_k\}_{k = 1, 2, \dots} \in \mathbb{Z}^+$ such that

$$| f_{n_k}^{-1} (\zeta_{n_k}) - f_*^{-1}(\zeta_0) | \geq c.$$

This is a contradiction. q.e.d.

2.3. For an arbitrary element f of $\mathfrak{F}$ the solution of the Dirichlet problem with boundary value $\gamma \circ f^{-1}(\zeta)$ $(\zeta \in f(\partial B))$ can be regarded as the unique harmonic extension of the boundary function $\gamma \circ f^{-1}(\zeta) |_{f(\partial B)}$ into the interior region $\partial^{-1} \circ f(\partial B)$ and may be denoted by $(\mathrm{Ext}_H \gamma \circ [f^{-1}(\zeta)]_{f(\partial B)})(\zeta)$, or briefly by

$$\mathrm{Ext}_H \gamma \circ f^{-1}(\zeta),$$

which is of course one of its piecewise C^2-extensions. Although the prescription of a boundary value $\gamma \circ f^{-1}(\zeta)$ $(\zeta \in f(\partial B), f \in \mathfrak{F})$ never determines its smooth extensions uniquely, we still insist to denote it by a similar symbol

$$\mathrm{Ext}_S \gamma \circ f^{-1}(\zeta),$$

by analogy. We admit from the outset that the Dirichlet's energy integral

$$E [| d\mathrm{Ext}_S \gamma \circ f^{-1}(\zeta) |^2],$$

of it taken over the whole region $\partial^{-1} \circ f(\partial B)$ $(f \in \mathfrak{F})$ of definition is always finite without restricting the generality. Note that the parameter domain of the surface to be studied was not a fixed disk B but the variable region $\partial^{-1} \circ f(\partial B)$ for ever $f \in \mathfrak{F}$. In this context let us remark

Proposition 3. Suppose the parameter w of a C^2-surface $\mathfrak{C}(w)$ is changed to $\tilde{w}$ under a K-quasiconformal transformation $w \to \tilde{w} = \psi(w)$. If we set $\tilde{\mathfrak{C}}(\tilde{w}) = \mathfrak{C} \circ \psi^{-1}(\tilde{w})$, then

$$E[\,|\,d\,\tilde{\mathfrak{C}}(\tilde{w})|^2] \le KE\,[\,|\,d\mathfrak{C}(w)\,|^2].$$

Proof. Note that

$$E[\,|\,d\mathfrak{C}(w)\,|^2] = \frac{\sqrt{-1}}{4}\iint (g_{11} + g_{22})\,dw \wedge d\overline{w}$$

with the metric tensors g_{ij} (i, j = 1, 2) of the surface $\mathfrak{C}(w)$ considered. Since $\partial\tilde{\mathfrak{C}}/\partial\tilde{w} = (\partial\mathfrak{C}/\partial w)\,(\partial w/\partial\tilde{w}) + (\partial\mathfrak{C}/\partial\overline{w})(\partial\overline{w}/\partial\tilde{w})$, it follows that

$$\left|\frac{\partial\tilde{\mathfrak{C}}}{\partial\tilde{w}}\right|^2 = \left|\frac{\partial\mathfrak{C}}{\partial w}\right|^2 |\,p + \overline{q}\,|^2 \le \left|\frac{\partial\mathfrak{C}}{\partial w}\right|^2 \left(|p| + |q|\right)^2,$$

where $p = \partial w/\partial\tilde{w}$, $q = \partial w/\partial\overline{\tilde{w}}$. If we write $J = |p|^2 - |q|^2$, then

$$E[|d\tilde{\mathfrak{C}}(\tilde{w})\,|^2| = \frac{\sqrt{-1}}{4}\iint (\tilde{g}_{11} + \tilde{g}_{22})\,d\tilde{w} \wedge d\overline{\tilde{w}}$$

$$\le \frac{\sqrt{-1}}{4}\iint (g_{11} + g_{22})\,\frac{(|p| + |q|)^2}{J}\,dw \wedge d\overline{w}$$

$$\le K\,\frac{\sqrt{-1}}{4}\iint (g_{11} + g_{22})\,dw \wedge d\overline{w}. \quad \text{q.e.d.}$$

Corollary. The Dirichlet integral is invariant under any 1 : 1 conformal transformation of its independent variable.

Proposition 4. The infimum of Dirichlet's functional $E[|\,d\mathrm{Ext}_H\,\gamma \circ f^{-1}(\zeta)|^2]$ on $\mathfrak{F}$ is attained by some $f_* \in \mathfrak{F}$.

Proof. Let $\{f_n(w)\}_{n = 1, 2, \dots}$ be a minimizing sequence for this extremum problem. By virtue of Propositions 1, 2 it loses no generality to assume the original sequence $\{f_n(w)\}_{n = 1, 2, \dots}$ converges uniformly on an arbitrarily fixed compact subregion Clo $R(a')$ of $R(a)$ to some $f_*(w) \in \mathfrak{F}$. Setting $B' = \{w|\,|w| < 1 + a'\}$, we map the unit disk Int B_* conformally onto the region $\partial^{-1} \circ f_n(\partial B')$ (resp. $\partial^{-1} \circ f_*(\partial B')$) by means of the biholomorphic function $\phi_n(z)$ (resp. $\phi_*(z)$) normalized so as to satisfy

$f_n(1) = \phi_n(0)$, $\phi'_n(0) > 0$ (resp. $f_*(1) = \phi_*(0)$, $\phi_*(0) > 0$). Since $f_n(w)$ tends to $f_*(w)$ uniformly on $\partial B'$ as $n \to \infty$, the sequence $\{\phi_n(z)\}_{n = 1, 2, \dots}$ converges to $\phi_*(z)$ uniformly on any compact subregion of Int B_z. There is a circle $C = \{z|\ |z| = r\}$ $(0 < r < 1)$ such that $\phi_n(C)$ is comprised in the interior of $f_*(R(a'))$ for all index n from some n_0 on. Since $(\mathrm{Ext}_H\ \gamma \circ f_n^{-1}) \circ \phi_n(z)$ converges to $(\mathrm{Ext}_H\ \gamma \circ f_*^{-1}) \circ \phi_*(z)$ uniformly on C (Proposition 2), the Dirichlet integral E'_n of the former harmonic vector tends to the one E'_* of the latter as $n \to \infty$, so far as the integration domain is restricted to $\partial^{-1}C$.

But $\partial f_*^{-1}(\zeta)/\partial\zeta$ is bounded on $f_*(R(a'))$, whose area can be made as small as one pleases only if one takes a' sufficiently small. Hence for any given $\epsilon > 0$ there is a $\delta > 0$ such that $E[|d\mathrm{Ext}_H\ \gamma \circ f_*^{-1}(\zeta)|^2] - (\epsilon/2) < E'_*$ for any $a' < \delta$. Therefore we have

$$E\,[\,|\,d\mathrm{Ext}_H\ \gamma \circ f_*^{-1}(\zeta)\,|^2 - \epsilon < \lim_{n\to\infty} E\,[\,|\,d\mathrm{Ext}_H\ \gamma \circ f_n^{-1}(\zeta)^2]$$

and the proof is completed.

In the next section it will be shown that $\mathrm{Ext}_H\ \gamma \circ f^{-1}(\zeta)$ thus obtained furnishes surely a harmonic surface sought after.

3. Variational formulae on Dirichlet's functional

3.1. As suggested by Courant [4] already in his classical approach implicitly, the use of quasiconformal transformations of surface parameters lends itself to the purpose of calculating the first variation of the Dirichlet functional.

For a parametrization of a smooth oriented surface $\mathfrak{C}(w)$ ($w \in$ Int B) in general, a simple calculations shows that

$$\left(\frac{\partial \mathfrak{C}}{\partial w}\right)^2 = \frac{1}{4}\left(\frac{\partial \mathfrak{C}}{\partial u} - \sqrt{-1}\,\frac{\partial \mathfrak{C}}{\partial v}\right)^2 = \frac{1}{4}\left(\left|\frac{\partial \mathfrak{C}}{\partial u}\right|^2 + \left|\frac{\partial \mathfrak{C}}{\partial v}\right|^2 - 2\sqrt{-1}\,\frac{\partial \mathfrak{C}}{\partial u}\,\frac{\partial \mathfrak{C}}{\partial v}\right)$$
$$= \frac{1}{4}\left(g_{11} - g_{22} - 2\sqrt{-1}\,g_{12}\right),$$

where $g_{ij} = g_{ji}$ (i, j, = 1, 2) denote the inner products of the base tangent vectors on

$\mathfrak{C}(w)$. Ignoring the constant factor 1/4, we may call this component $g_{11} - g_{22} - 2\sqrt{-1}\, g_{12}$ of a covariant tensor under the name square complex derivative of $\mathfrak{C}(w)$.

Proposition 5. Let $\mathfrak{C} = \mathfrak{C}(w)$ be the C^2-parametrization of a surface. If $\mathfrak{C}(w)$ is harmonic, then its square complex derivative is holomorphic. Conversely, under the condition $(\partial\mathfrak{C}/\partial w)^2$ is holomorphic, $\mathfrak{C}(w)$ is harmonic unless it reduces to a constant vector.

Proof. The assertion follows immediately from the identity

$$\frac{\partial}{\partial \overline{w}}\left(\frac{\partial \mathfrak{C}}{\partial w}\right)^2 = 2\,\frac{\partial \mathfrak{C}}{\partial w}\left(\frac{\partial^2 \mathfrak{C}}{\partial \overline{w}\,\partial w}\right).$$

A quasiconformal deformation of a surface parameter induces a similar deformation of the surface itself. Specifically, an isothermally parametrized C^2- surface, after a quasiconformal change of parameter, comes to have the metric tensors g_{ij} (i, j = 1, 2) satisfying no more the isothermality identities $g_{11} - g_{22} = g_{12} = 0$ in general, which motivates introduction of a quantity $\mathfrak{Q}$ (≥ 1), dilatation-quotient, by means of the relation

$$g_{11} + g_{22} + \sqrt{(g_{11} - g_{22})^2 + 4g_{12}^2}$$

$$= \mathfrak{Q}^2\left(g_{11} + g_{22} - \sqrt{(g_{11} - g_{22})^2 + 4g_{12}^2}\,\right)$$

on the analogy of the plane quasiconformal mappings. If $\mathfrak{Q} \leq K$ with some finite constant K, we commit ourselves to call it to be a K-quasiconformally parametrized surface $\mathfrak{C} = \mathfrak{C}(w)$. Note that we have

$$g_{11} + g_{22} = (\mathfrak{Q} + (1/\mathfrak{Q}))\sqrt{g_{11}g_{22} - g_{12}^2}. \tag{4}$$

More precisely, as the pull-back of the line-element

$$ds^2 = |d\mathfrak{C}(w)|^2 = \tfrac{1}{2}(g_{11} + g_{22})|dw|^2 + \tfrac{1}{2}\mathrm{Re}\,\{(g_{11} - g_{22} - 2\sqrt{-1}g_{12})dw^2\}$$

shows, the infinitesimal ellipse in the Int B_w to be sent to the infinitesimal circle on the surface $\mathfrak{C}(w)$ has the major and minor axes $\sqrt{g_{11} + g_{22} \pm \sqrt{(g_{11} - g_{22})^2 + 4g_{12}^2}}|dw|$ respectively, and the direction of the former points to arg dw = $(-1/2)\mathrm{Tan}^{-1}(-2g_{12}/(g_{11} - g_{22}))$ at every point, where $\mathfrak{C}(w)$ is not conformal.

Proposition 6. Of all K-quasiconformally parametrized C^1-surfaces with one and the same locus, their Dirichlet energy integrals are minimized if and only if the parametrization is isothermal.

Proof. The estimate

$$\frac{\sqrt{-1}}{4}\iint (g_{11} + g_{22})\, dw\wedge d\overline{w} = \frac{1}{2}\iint (\mathfrak{Q}(\mathfrak{C}) + \frac{1}{\mathfrak{Q}(\mathfrak{C})})\, d\sigma(\mathfrak{C})$$
$$\geq \iint d\sigma(\mathfrak{C}) = \text{Area of Loc } \mathfrak{C}$$

holds with the equality sign, only if $\mathfrak{Q}(\mathfrak{C}) = 1$. q.e.d.

3.2. Let κ denote a closed disk comprised in Int B_ζ. With this spot in eye we are able to derive a variety of parameter changes as well as variational formulae associated with them. In the first place we mention:

General Q. C. Change of Parameter. Let $h(\zeta)$ be an arbitrary function class $C^2[\mathbb{C}, \mathbb{C}]$, which behaves biholomorphic off κ. If a constant $\varepsilon \in \mathbb{C}$ is sufficiently small, the transformation

$$\zeta \to \tilde{\zeta} = H(\zeta) = \zeta + \epsilon h(\zeta)$$

is quasiconformal in the whole region $\partial^{-1} \circ f(\partial B)$ at every choice of $f \in \mathfrak{F}$, and the latter variable $\tilde{\zeta}$ provides another parameter than the former ζ for the surface $\mathrm{Ext}_S\, \gamma \circ f^{-1}(\zeta)$.

Theorem 2. The first variation of Dirichlet's functional $E[\,|d\mathrm{Ext}_S \gamma \circ f^{-1}(\zeta)\,|^2]$ with respect to General Q. C. Change of Parameter equals

$$\mathrm{Re}\,\{\epsilon \iint (g_{11} - g_{22} - 2\sqrt{-1}\, g_{12})\, h_{\overline{\zeta}} d\zeta\wedge d\overline{\zeta}\}, \tag{5}$$

where g_{ij} (g_{ji}; i, j = 1, 2) are the tensor components of the first fundamental form of the surface $\mathrm{Ext}_S\ \gamma \circ f^{-1}(\zeta)$ considered.

Proof. We write $\mathfrak{C}(\zeta)$ in place of the full expression $\mathrm{Ext}_S\ \gamma \circ f^{-1}(\zeta)$ for brevity and set J for Jacobian of the transformation $\zeta \to \tilde{\zeta}$. Then making use of the identities $\partial\zeta/\partial\tilde{\zeta} = -(\partial\tilde{\zeta}/\partial\bar{\zeta})/J$, $\partial\zeta/\partial\bar{\tilde{\zeta}} = \overline{(\partial\tilde{\zeta}/\partial\zeta)}/J$, our computation proceeds as follows:

$$\frac{2}{\sqrt{-1}}\, E[\,|d\mathfrak{C}\circ H^{-1}(\tilde{\zeta})\,|^2] = \iint \left| \frac{\partial\mathfrak{C}}{\partial\zeta}\frac{\partial\zeta}{\partial\tilde{\zeta}} + \frac{\partial\mathfrak{C}}{\partial\bar{\zeta}}\frac{\partial\bar{\zeta}}{\partial\tilde{\zeta}} \right|^2 J\, d\zeta\wedge d\bar{\zeta}$$

$$= \iint \frac{1}{J} \left| \frac{\partial\mathfrak{C}}{\partial\zeta} \right|^2 \left[\, |1 + \epsilon h_\zeta - \epsilon h_{\bar{\zeta}}|^2 - 2\mathrm{Re}\, \{(\frac{\partial\mathfrak{C}}{\partial\zeta})^2 (1 + \epsilon h_\zeta \epsilon h_{\bar{\zeta}}\} \right] d\zeta\wedge d\bar{\zeta}.$$

Hence

$$\frac{2}{\sqrt{-1}}\, (E[\,|\,d\mathfrak{C}\circ H^{-1}(\tilde{\zeta})\,|^2] - E[\,|d\mathfrak{C}(\zeta)\,|^2])$$
$$= \iint \left|\frac{\partial\mathfrak{C}}{\partial\zeta}\right|^2 (1 + O(\epsilon^2))\, d\zeta \wedge d\bar{\zeta} - \iint \left|\frac{\partial\mathfrak{C}}{\partial\zeta}\right|^2 d\zeta\wedge d\bar{\zeta}$$
$$-2\,\mathrm{Re}\,\{\epsilon \iint (g_{11} - g_{22} - 2\sqrt{-1}g_{12})h_{\bar{\zeta}}\, d\zeta\wedge d\zeta\} + O(\epsilon^2)$$

q.e.d.

Corollary. The vanishing of the first variation of Dirichlet's functional with respect to General Q.C. Change of Parameter is equivalent to

$$\iint (g_{11} - g_{22} - 2\sqrt{-1}\, g_{12})h_{\bar{\zeta}}\, d\zeta\wedge d\bar{\zeta} = 0. \tag{6}$$

Proof. The assertion follows from the arbitrariness of arg ϵ.

Special Q. C. Change of Parameter (I). It is allowable to choose, in particular, so that $h(\zeta) = 0$ off κ.

Corollary 1 (Variational Formula I). For any $h(\zeta) \in C^2[\mathbb{C}, \mathbb{C}]$ supported by κ, the vanishing of the first variation of $E[\,|d\mathrm{Ext}_S\ \gamma\circ f^{-1}(\zeta)\,|^2]$ with respect to Special Q.C. Change of parameter (I) is equivalent to

$$\iint (g_{11} - g_{22} - 2\sqrt{-1}\, g_{12})_{\bar{\zeta}}\, h(\zeta)\, d\zeta\wedge d\bar{\zeta} = 0$$

(cf. Gerstenhaber - Rauch [5]).

Proof. A simple application of Green's theorem to (6) yields the conclusion desired.

Special $\mathbb{Q}$. C. Change of Parameter (II). If we set

$$h(\zeta) = \frac{1}{2\sqrt{-1}\pi} \iint \frac{q(z)}{z - \zeta} dz \wedge d\bar{z} \tag{7}$$

for an arbitrary $q(\zeta) \in C^{\infty}[\mathbb{C}, \mathbb{C}]$ supported by κ, then $h(\zeta)$ belongs to $C^2[\mathbb{C}, \mathbb{C}]$ and is holomorphic off κ. It remains to be checked, however, whether

$$\tilde{\zeta} = \zeta + \epsilon h(\zeta)$$

furnishes a quasiconformal homomorphism of $\mathbb{C}$. Since $\partial h/\partial\zeta$ tends uniformly to zero as $\zeta \longrightarrow \infty$, it is bounded (Ahlfors [1], pp. 87 - 88). From the total differential $dH = (1 + h_{\zeta})d\zeta + h_{\bar{\zeta}}d\bar{\zeta}$ it follows for an arbitrary pairs of points $\zeta_1 \neq \zeta_2$ that

$$H(\zeta_2) - H(\zeta_1) = \zeta_2 - \zeta_1 + \int_{\zeta_1}^{\zeta_2} (h_{\zeta}d\zeta + h_{\bar{\zeta}}d\bar{\zeta})$$

and that

$$|\, H(\zeta_2) - H(\zeta_1) \,| \leq |\, \zeta_2 - \zeta_1| - \int_{\zeta_1}^{\zeta_2} (|\, h_{\zeta} \,| + |q(\zeta)\,|\,)\, |d\zeta|.$$

Hence one concludes that $H(\zeta)$ provides a quasiconformal homemomorphism of $\mathbb{C}$, only if $|\epsilon|$ is sufficiently small. q. e.d.

Corollary 2 (Variational Formula II). If the first variation of Dirichlet's functional with respect to Special Q. C. Change of Parameter (II) vanish, then we must have

$$\iint (g_{11} - g_{22} - 2\sqrt{-1}\, g_{12})\, q\, (\zeta)\, d\zeta \wedge d\bar{\zeta} = 0$$

for every function $q(\zeta)$ supported in the interior of B_{ζ}.

Proof. The conclusions follows immediately by substituting the relation $h_{\bar{\zeta}} = q$, occurring through differentiation of (7), into (6). q.e.d.

Variational Formula II finds its immediate application in presenting an alternative proof to the following known fact:

Proposition 7. The harmonic surface $\mathrm{Ext}_H\ \gamma \circ f_*^{-1}(\zeta)$, that minimizes the Dirichlet energy functional within the family $\{\mathrm{Ext}_H\ \gamma \circ f_*^{-1}(\zeta) \mid f \in \mathfrak{F}\}$, is a minimal surface spanning the given regular analytic closed curve γ.

Proof. We have only to show that the parameter ζ is isothermal on the region of definition $\partial^{-1} \circ f_*(\partial B)$. Since the surface in question is harmonic, its metric tensors satisfy $(g_{11} - g_{22} - 2\sqrt{-1}\, g_{12})_{\bar\zeta} = 0$ everywhere on $\partial^{-1} \circ f_*(\partial B)$. Assume on the contrary that Int B_ζ contains a point ζ_0, at which the holomorphic function $g_{11} - g_{22} - 2\sqrt{-1} g_{12}$ ceases to vanish. Then there exists a $q(\zeta)$, of class C^∞ [C, C] supported by a closed disk κ comprised in $B \setminus (R(a) \cup B)$, such that

$$\iint (g_{11} - g_{22} - 2\sqrt{-1}\, g_{12})\, q(\zeta)\, d\zeta \wedge d\bar\zeta \neq 0. \tag{8}$$

In view of the absolute minimality of $E[\,|d\mathrm{Ext}_H\ \gamma \circ f^{-1}(\zeta)\,|^2]$ with respect to every quasiconformal deformation of the parameter in $\partial^{-1} \circ f_*(\partial B)$, we thus see that (8) contradicts Variational Formula II. Hence $g_{11} - g_{22} - 2\sqrt{-1} g_{12} = 0$ throughout $\partial^{-1} \circ f_*(\partial B)$ by the unicity theorem of homolomorphic functions. q.e.d.

4. Quasiconformal extension of the boundary parametrization

4.1. According to what the classical calculus of variations tells us, the Euler-Lagrange equation for the Dirichlet functional with admissible C^2-function U and a point-wise prescribed boundary value is $\nabla^2 U = 0$, while the boundary value problem for the Laplacian is uniquely solved by means of the Poisson integral. How must then the matters stand in case of movable boundary value problem? It is the purpose of this section to answer such questions in a necessary and sufficient condition, putting aside the existence proof for a moment. Our interest centres around clarifying the class admitted to concurrence.

4.2. At the end of the preceding section we have effected on the isothermal parameter ζ of the minimal surface $\mathrm{Ext}_H\ \gamma \circ f_*^{-1}(\zeta)$ a quasiconformal deformation

$$\zeta \to \tilde{\zeta} = H(\zeta) = \zeta + \epsilon h(\zeta). \tag{9}$$

The parametrized surface

$$(\mathrm{Ext}_H\ \gamma \circ_*^{-1}) \circ H^{-1}(\tilde{\zeta}), \tag{10}$$

that comes out in this way, can be regarded as a quasiconformal extension of the regular analytic boundary parametrization $\gamma \circ f_*^{-1}(\zeta)$ ($\zeta \in f_*$ (R(a)). So we are free to write Ext_Q $\gamma \circ [f_*^{-1}(\zeta)]|_{\zeta \epsilon f_*(\partial B_w)}$ for the expression (10), or

$$\mathrm{Ext}_Q\ \gamma \circ f_*^{-1}(\zeta)$$

in an abridged symbol.

On the other hand we know the existence of many smooth surfaces, whose locus actually differs from Loc $[\mathrm{Ext}_H\ \gamma \circ f_*^{-1}(\zeta)\ ]$ and which remain yet to span the regular analytic contour γ. To see this, employing again the abbreviation $\mathcal{C}(\zeta)$ instead of the notation $\mathrm{Ext}_H \gamma \circ f_*^{-1}(\zeta)$, we denote by $\nu(\mathcal{C}(\zeta))$ the unit normal vector field to Loc $\mathcal{C}(\zeta)$ and let $p(\zeta)$ be an arbitrary function of class C^∞ $[\mathbb{C}, \mathbb{R}]$ supported on B. Then if a real constant ϵ is sufficiently small, the newly parametrized surface

$$\mathcal{C}_*(\zeta) = \mathcal{C}(\zeta) + \epsilon p(\zeta)\nu(\mathcal{C}(\zeta)) \tag{11}$$

is K'-quasiconformal, to the effect that the dilatation-quotient $\mathcal{Q}$ of $\mathcal{C}_*(\zeta)$ ($\zeta \in$ B) in the sense of (3) is bounded above by K'. And what is better, all such surfaces sufficiently near to $\mathcal{C}(\zeta)$ in some Sobolev topology are written in the form (11) conversely (cf. Böhme-Tomi [3]).

Motivated by these circumstances and identifying the surface $\mathcal{C}(\zeta)$ with its conformally equivalent parameter region B, we arrive at the following definition for quasiconformal maps of B into $\mathbb{R}^n$ ($n \geq 3$) in analogy to the case $n = 2$ and for quasiconformal neighbourhoods of the minimal surface:

Definition 3. Given a constant $K(>1)$, the $(K-1)$-quasiconformal neighbourhood of $\mathfrak{C}(\zeta)$ is the composition of all K'-quasiconformal deformation of the parameter ζ of type (3) and of all K''-quasiconformal distortion of Loc $\mathfrak{C}(\zeta)$ of type (11), such that $K'K'' \leq K$.

Thus we have proved

Proposition 8. There is an $\epsilon_0 > 0$, such that the deleted $|\epsilon|$-quasiconformal neighbourhood of the minimal surface $\mathfrak{C}(\zeta)$ is non-void for every $|\epsilon| < \epsilon_0$.

4.3. We will prove

Proposition 9. Let V be the ϵ-quasiconformal neighbourhood of the minimal surface $\mathrm{Ext}_H\, \gamma \circ f_*^{-1}(\zeta)$. Assume that V contains any element $\mathfrak{C}_*(\zeta)$ whatever, minimizing the Dirichlet functional $E[\,|\,d\mathrm{Ext}_Q\, \gamma \circ f_*^{-1}(\zeta)|^2]$ in V. Then $\mathfrak{C}_*(\zeta)$ is a minimal surface.

Proof. Suppose, contrary to the assertion, that the extreme element $\mathfrak{C}_*(\zeta)$ in V is no conformal mapping in the variable ζ. Under such circumstances we set

$$\mu(\zeta) = \frac{\sqrt{(g_{11} - g_{22})^2 + 4g_{12}^2}}{g_{11} + g_{22}} \exp \frac{1}{2\sqrt{-1}} \mathrm{Tan}^{-1}(2g_{12}/(g_{11} - g_{22}))$$

at every non-conformal point ζ of $\partial^{-1} \circ f_*\,(\partial B_w)$, while $\mu(\zeta) = 0$ elsewhere, to define a quasiconformal homeomorphism $z = \chi(\zeta)$ of $\mathbb{C}$ satisfying the Beltrami differential equation

$$\partial z/\partial\overline{\zeta} = \mu\,(\zeta) \bullet \partial z/\partial\zeta.$$

Through the K-quasiconformal parameter change $\zeta \rightarrow z$ the surface $\mathfrak{C}_*(\zeta)$ is deformed to another surface $\mathfrak{C}_* \circ \chi^{-1}(z)$ with the same locus as $\mathfrak{C}_*(\zeta)$, which belongs to V. Proposition 6 then produces the strict inequality

$$E[\,|d\mathfrak{C}_{*}\circ\chi^{-1}(z)|^2 < E[\,|d\mathfrak{C}_{*}(\zeta)|^2], \tag{12}$$

since we have supposed there is at least one point $\zeta_0 \in \mathrm{Int}\, B_\zeta$ satisfying $|\mu(\zeta_0)| > 0$. But this is a contradiction on account of the minimality of the right-hand side in (12).

Owing to the conformality just verified the deformation of the surface $\mathfrak{C}_{*}(\zeta)$ occurring in application of Special Change of Parameter falls again in V, so far as ϵ is sufficiently small. The minimality of the energy functional $E[\,|\,d \bullet |^2]$ at $\epsilon = 0$ permits us to apply Variational Formula I and we conclude with the aid of arbitrariness of $h(\zeta)$ that the square complex derivative of $\mathfrak{C}_{*}(\zeta)$ is holomorphic. It follows from Proposition 5 that $\mathfrak{C}_{*}(\zeta)$ itself must be harmonic in the isothermal parameter ζ, i.e., a minimal surface. q.e.d.

<u>4.4.</u> The main purpose of this section is to prove

<u>Theorem 3</u>. If the Dirichlet's functional defined on the space of all quasiconformal extension $\mathrm{Ext}_Q\, \gamma \circ f^{-1}(\zeta)|_{\zeta \in f(\partial B_w)}$ $(f \in \mathfrak{F})$ of the regular analytic contour γ has a critical point at all, then every such point is a minimal surface spanning γ. Conversely, every minimal surface spanning γ gives some critical value to the Dirichlet's functional within the space of all quasiconformal extensions of the regular analytic contour γ.

<u>Proof</u>. (Necessity). Suppose there is some $\mathrm{Ext}_Q\, \gamma \circ f^{-1}(\zeta)$ $(f_{*} \in \mathfrak{F})$ such that $E[\,|d\mathrm{Ext}_Q\, \gamma \circ f_{*}^{-1}(\zeta)|^2]$ is stationary in the space $\{\mathrm{Ext}_Q\, \gamma \circ f_{*}^{-1}(\zeta) \mid f \in \mathfrak{F}\}$. Since the transform $\mathrm{Ext}_Q\, \gamma \circ f_{*}^{-1}(\zeta) \circ H^{-1}(\tilde{\zeta})$ under any $\tilde{\zeta} = H(\zeta)$ belonging to Special Q.C. Change of Parameter (I) again enters the same space, Corollary 1 applies and the criticalness assumption yields the orthogonality

$$\iint (g_{11} - g_{22} - 2\sqrt{-1}g_{12})\overline{\zeta}\, h(\zeta)\, d\zeta \wedge \overline{\zeta} = 0$$

with an arbitrary function $h \in C^2[\mathbb{C}, \mathbb{C}]$ supported on a compact set $\kappa \subset \mathrm{Int}\, B$. In view of arbitrariness of κ we have the harmonicity of the surface $\mathrm{Ext}_Q\ \gamma\circ f^{-1}(\zeta)$ in Int B_ζ, which of course is holomorphic in the interior of the annular region $f_*(R(a))$ by definition. On the other hand $g_{11} - g_{22} - 2\sqrt{-1}g_{12}$ must be of class C^2 in the whole region $\partial^{-1}\circ f_*(\partial B_W)$, which can accordingly be welded homomorphically across the circumference ∂B_ζ (Painlevé's theorem).

Next suppose the open disk Int B_ζ contains a point, say ζ_0, at which the holomorphic function $g_{11} - g_{22} - 2\sqrt{-1}g_{12}$ ceases to vanish. Then there exists a disk κ comprised in Int B_ζ as well as a function $q(\zeta) \in C^2[\mathbb{C}, \mathbb{C}]$ supported by κ such that

$$\iint (g_{11} - g_{22} - 2\sqrt{-1}g_{12})q(\zeta)\, d\zeta \wedge d\bar{\zeta} \neq 0.$$

The last inequality contradicts obviously Variation Formula II. The holomorphic function $g_{11} - g_{22} - 2\sqrt{-1}g_{12}$ in $\partial^{-1}\circ f_*(\partial B_W)$ must vanish everywhere in Int $B_{\zeta'}$, accordingly in the whole region $\partial^{-1}\circ f_*(\partial B_W)$, where it behaves holomorphic. In other words the surface $\mathrm{Ext}_Q\ \gamma\circ f_*^{-1}(\zeta)$ is harmonic with respect to the isothermal parameter ζ.

<u>(Sufficiency.)</u> Assume that a quasiconformal extension $\mathcal{C}_*(\zeta) = \mathrm{Ext}_Q\ \gamma\circ f_*^{-1}(\zeta)$ is a minimal surface. It is possible to assign to this surface its ϵ-quasiconformal neighbourhood U consisting of all surfaces

$$\mathcal{C}_*(\zeta) + \lambda p(\zeta) \bullet \nu\big(\mathcal{C}_*(\zeta)\big)$$

of type (11) with sufficiently small $\lambda > 0$. Under the condition that U contains some element $\mathcal{C}(\zeta)$ minimizing the Dirichlet functional within U whatever, it must be conformal and harmonic in the parameter ζ (Proposition 9). Hence $\mathcal{C}(\zeta)$ coincides with $\mathcal{C}_*(\zeta)$ by virtue of the unicity of solution in the Dirichlet's problem. It amounts to say that the minimum problem in U is solved uniquely by $\mathcal{C}_*(\zeta)$ itself. Namely, any minimal surface give the Dirichlet functional a locally least value, still more a stationary value. q.e.d.

5. Uniqueness theorem

5.1. There seems still to subsist a good number of questions left unsolved in the classical Plateau problem, even in the simplest setting of a single Jordan contour lying in $\mathbb{R}^n$; especially, our knowledge lacks in counting the number of those solutions.

In accordance with the conventional idea that of all harmonic vectors with prescribed correspondences between ∂B and Loc γ an appropriate choice furnishes us with the desired minimal surface, we have demonstrated in § 4 the existence of a minimal surface with the regular analytic boundary γ, minimizing the Dirichlet functional in a narrower class of harmonic vectors than the usual ones. One would fully be aware that there was some unnaturalness in choosing the family $\mathfrak{F}$ of boundary parametrizations originating from the postulate for its normality and compactness. We are now prepared to remove such artifice and ambiguity.

Theorem 4. Let S be the oriented minimal surface of disk type spanning the given regular analytic closed curve γ. Then S is determined uniquely by γ up to one-to-one conformal automorphisms of parameter domains.

Proof. Suppose, contrary to the assertion, that the regular analytic contour γ bounds two minimal surfaces, say S and $\tilde{S}$, with the properties questioned.

Let $\gamma(w)$ denote as earlier the defining function on R(a) of the given regular analytic contour γ, with whose restriction $\gamma(\omega)$ to ∂B_W the integral transform

$$\mathfrak{Z}(w) = \frac{1}{\sqrt{-1}\pi} \int_{\partial\beta} \frac{\tilde{\gamma}(\omega)}{\omega - w}\, d\omega$$

of Cauchy type can be defined. Then we have $S(w) = \mathrm{Re}\,\{\mathfrak{Z}(w)\}$ for all $w \in B$ (Theorem 1). Even though $\tilde{S}(w)$ may be parametrized not on B but on a closed Jordan region Clo $\partial^{-1}$of(∂B) with some $f \in \mathfrak{F}$ in general (Theorem 2), we may assume from the outset (by making use of an auxiliary conformal transformation if necessary) that this parameter domain is B_w, owing to the conformal invariance of minimality of surfaces and to the regular analyticity of f(∂B). Therefore we may still write

$$\tilde{S}(w) = \mathrm{Re}\ \{\frac{1}{\sqrt{-1}\pi} \int_{\partial\beta} \frac{\tilde{\gamma}(\omega)}{\omega - w}\, d\omega\} \tag{12}$$

with another parametrization $\tilde{\gamma}(\omega)$ of the same boundary curve γ.

Meanwhile, the two boundary parametrizations $\gamma(\omega)$ and $\tilde{\gamma}(\omega)$ are standing under some biholomorphic relationship in a neighbourhood of ∂B. Hence referring to the 'Invariance of Analytic Relation' we can conclude that S(w) and $\tilde{S}(w)$, the direct analytic prolongations of $\gamma(\omega)$ and $\tilde{\gamma}(\omega)$ into Int B respectively, are also analytically connected with one another. In other words $\tilde{S}$ is achieved from S via some 1 : 1 conformal transformation of the parameter w. q.e.d.

5.2. Among various open questions on minimal submanifolds it is asked to settle in what cases their concrete representations can be described (cf. e.g., Almgren [2]). They think it will be a very difficult problem in general to write down the defining equation of a minimal surface by means of the given data. We should remark that the equation (12) answers this question although in a special case.

6. The Riemann mapping theorem in a modified form

6.1. In this section we fix our eyes upon a complex single-valued holomorphic function z = f(w) in some neighbourhood of ∂B such that $f'(w) \neq 0$. The projection of the image curve f(∂B) on the z-plane ℂ is a regular analytic closed curve with multiple points in general.

Propositions 4, 7 permit us to find a minimal surface which spans f(∂B) and possesses the parameter disk B (and which may shares the symbol f(w) with the boundary curve itself owing to Theorem 1). The coordinate w is isothermal for this harmonic map f(w), which accordingly supplies a conformal mapping of Int B onto a region with projection $\partial^{-1}$of(∂B). Except for the case that $f(\partial B)$ is a simple curve, where the above result reduces to a known theorem in the classical theory of functions, the image region is a covering surface over $\mathbb{C}$ which involves at least one branch point (Argument Principle). Thus we have

Theorem 5. Suppose given a regular analytic closed curve γ on the complex z-plane parametrized on the circle ∂B. Then there exists a holomorphic map f of B into $\mathbb{C}$, whose restriction to ∂B agrees with a regular analytic parametrization of γ. Under the assumption that γ has the rotation index m (ϵ $\mathbb{N}$), the image region f(w) reveals m − 1 branch points: f(w) is unique modulo conformal transformations of the isothermal parameter.

Example. Let the regular analytic closed curve $\gamma = \gamma(w)$ be parametrized on $|w| = 1$ by $\gamma(w) = w^2$. Then the minimal surface spanning γ is parametrized in $|w| < 1$ by the mapping

$$S_\alpha(w) = \frac{w^2 - \alpha}{1 - \bar{\alpha}w^2}$$

with an arbitrary complex constant $\alpha \in \text{Int } B_w$. S_α possesses the branch point of order 1 over the point $z = -\alpha$.

The Plateau problem with this γ has infinitely many solution surfaces $\left\{S_\alpha(w)\right\}_{|\alpha| < 1}$, all of which can be achieved by the 1 : 1 conformal transformation

$$w^2 \to \frac{w^2 + \alpha}{1 + \overline{\alpha}w^2},$$

starting at $S_0(w)$.

6.2. Now let us consider the following situation:

R is a finite bordered Riemann surface of positive genus and γ one of its boundary components;

f is a single-valued meromorphic function on R satisfying the condition that Im f = c (= constant) on Loc γ.

In our previous paper [7] we have introduced the notion of hydrodynamic continuation of R in the following manner: namely, identifying every appropriate pair of points of the slit f(Loc γ) with one and the same coordinate z and pulling it back by f^{-1} we have continued R into a closed Riemann surface $\tilde{R}$ of the same genus. Thus the meromorphic function f on R has also been continued to a like meromorphic function $\tilde{f}$ on $\tilde{R}$.

Now we shall take occasion here to interpret the non-uniqueness of the hydrodynamic continuation $(\tilde{R}, \tilde{f})$ of (R, f), applying the present circle of ideas. Without losing the generality in argument we may only deal with the continuation over a single boundary component γ.

Take an end U of R, i.e., a doubly connected subregion of R, one of whose boundary components agrees with γ. Since f is holomorphic up to γ, the differential df is well defined on the closure of U. It enables us to choose U from the outset, so that the multivalency of f remains constant everywhere on U and that the other parametrized component of γ' of U than γ is a regular analytic dividing cycle of R. To be precise, γ' shall be oriented so as to be homologous to γ.

Let U be mapped by a biholomorphic function ϕ conformally into the complex w-plane $\mathbb{C}_W$. Let $\tilde{U}$ denote a simply connected subregion of $\tilde{R}$ bounded by the regular analytic closed curve $\tilde{f}^{-1} \circ f(\gamma')$ and let $\tilde{U}$ be mapped into $\mathbb{C}_W$ by a biholomorphic function $\tilde{\phi}$. It is possible to choose the first coordinate map ϕ so as to satisfy $\phi(\gamma') =$

$\tilde{\phi}(\partial\tilde{U})$. Then the hydrodynamic continuation $\tilde{f}$ of f is, restricted to U, regarded as a minimal surface parametrized on Clo $[\partial^{-1}\circ\phi(\gamma')]$ spanning $f(\gamma')$.

6.3. We are now prepared to specify all the hydrodynamic continuations of (R, f) across γ with the aid of the theorems already obtained:

Theorem 6. Let ϕ be a one-to-one conformal map of an appropriate end of R into $\mathbb{C}$, whose ideal boundary is γ. Let Σ denote the minimal surface with parameter domain Clo $\partial^{-1}\circ\phi(\gamma')$ spanning $f(\gamma')$. Then Σ characterizes the hydrodynamic closing-up of R along γ with respect to the S-function f (cf. [8]).

Proof. We have already observed above that prescription of an arbitrary continuation $\tilde{f}$ of the S-function f implies the existence of some conformal map ϕ U $\to$ $\mathbb{C}_W$. The range of Clo $[f(p) \mid_{p\epsilon U}]$ is a region bounded by $f(\gamma')$ ramifying over $\mathbb{C}_W$, namely, a minimal surface Σ spanning the contour $f(\gamma')$, whose parameter region can be $\partial^{-1}\circ\phi(\gamma')$. The representation formula

$$\Sigma = \Sigma(w) = f\circ\phi^{-1}(w) = \frac{1}{2\sqrt{-1}\pi}\int_{\phi(\gamma')}\frac{f\circ\phi^{-1}(\omega)}{\omega - w}d\omega$$

holds.

Conversely, and 1 : 1 conformal map $\phi : U \to \mathbb{C}_W$ determines a plane Jordan region bounded by the regular analytic curve $\phi(\gamma')$ and the conformal mapping of $\phi(\gamma')$ onto $f(\gamma')$ gives rise to a boundary parametrization of some minimal surface of disk type. Meanwhile, the regular analytic contour $f(\gamma')$ can bound an essentially unique minimal surface $\Sigma(w)$ parametrized on $\partial^{-1}\circ\phi(\gamma')$, which furnishes its energy integral with the minimal value, still more the critical value (Theorems 4, 3). Therefore $\Sigma = \Sigma(w)$, pulled back by ϕ^{-1} to U, must realize the horizontal slit mapping of U on account of Dirichlet Principle, i.e., gives hydrodynamic closing-up $\tilde{f}$ of the S-function f beyond γ.

Once $f|_U$ has been continued up to an $\tilde{f}$ on $\phi^{-1} \circ \Sigma^{-1}$, we are able to get $\tilde{R}$ by welding the simply connected 'cap' $\phi^{-1} \circ \Sigma^{-1}$ with R \ Clo U by means of the biholomorphic transformation $\tilde{f}^{-1} \circ f(p)$ along γ'. Thus $\tilde{f}$ is also continued to a single-valued meromorphic S-function on the whole surface $\tilde{R}$. q.e.d.

References

[1] Ahlfors, L. V., Lectures on quasiconformal mappings, Van Nostrand, 1966, 146 pp.

[2] Almgren, F. J. Jr., Geometric measure theory and elliptic variational problems, C.I.M.E., III Circo 1972, Edizioni Cremonese, Roma 1973, pp. 33 - 117.

[3] Böhme, R., and F. Tomi, Zur Struktur der Lösungsmenge des Plateau problems, Math. Z. 133 (1973), pp. 1 - 29.

[4] Courant, R., Dirichlet's principle, conformal mapping, and minimal surfaces, Interscience Publishers, 1950 330 + xiii pp.

[5] Gerstenhaber, M., and H. E. Rauch, On extremal quasiconformal mappings I, Proc. Nat. Acad. Sci. U.S.A. 40 (1954), pp. 808 - 812.

[6] Lewy, H., On the boundary behaviour of minimal surfaces, Proc. Nat. Acad. Sci. U.S.A. 37 (1951), pp. 103 - 110.

[7] Shiba, M., and K. Shibata, Singular hydrodynamical continuations of finite Riemann surfaces, J. Math. Kyoto Univ. 25 (1985), pp. 745 - 755.

[8] Shiba, M., and K. Shibata, Hydrodynamic continuations of an open Riemann surface of finite genus, Complex Variables 8 (1987), pp. 205 - 211.

Faculty of Science
Okayama University of Science

ON A DISTORTION THEOREM OF THE LOGARITHMIC CAPACITY

Nobuyuki Suita

1. *Introduction*. Let A be a compact set on the unit circle $|z| = 1$. We denote by Cap A the logarithmic capacity of A. Let f(z) be a univalent function in the outside of the unit circle $D = \{|z| > 1\}$ with expansion near the point at infinity

$$f(z) = z + a_0 + a_1/z + \cdots. \tag{1}$$

The following distortion theorem was given by Pommerenke [2] except an exact equality statement.

Theorem 1. *It holds that*

$$\text{Cap } f(A) \geq (\text{Cap } A)^2. \tag{2}$$

Equality in (2) *for a set* A *of positive capacity occurs if and only if* f(z) *maps* D *onto a domain* Δ *such that the Green's function* g(w) *of the connected component of* $f(A)^c$ *containing the point at infinity with pole at infinity satisfies* $dg^* = 0$ *along the other boundary part of* Δ *than* f(A).

Here f(A) denotes the cluster set of f at A, that is the set of all accumulation points of f(z) as z tends to ζ, $\zeta \in A$.

The inequality (2) was first proved by Schiffer for an circular arc A [3]. He used a variational method for the proof. Pommerenke extended it for an arbitrary set by making use of a distortion theorem of Golusin. Both Schiffer and Pommerenke stated that the inequality is sharp and that there are many extremal functions.

In the present paper we shall give an alternative proof of the inequality (2) by the method of extremal metrics. This method gives us the stated characterization of extremal functions.

2. *Extremal metrics.* We define a defining sequence $\{A_n\}_{n=1}^{\infty}$ of a compact set A on the unit circle. Let A_n be an open set consisting of a finite number of mutually disjoint simply-connected subregions of D which satisfies

i) each component of A_n is bounded by an arc on the unit circle containing a subset of A in its interior and by a piece wise analytic arc in D expect end points,

ii) $A_n \supset A_{n+1}$, n = 1, 2, $\cdots$ and

iii) $\bigcap_{n=1}^{\infty} \overline{A_n} = A$, where $\overline{A_n}$ is taken as the closure on the Riemann sphere.

Then the cluster set f(A) coincides with $\bigcap_{n=1}^{\infty} \overline{f(A_n)}$. Let $\Gamma(f(A))$ be the family of locally rectifiable closed curves separating f(A) from the point at infinity. $\Gamma_R(f(A))$ denotes the subfamily of $\Gamma(f(A))$ contained in $|w| < R$. Let $M(\Gamma_R(f(A)))$ denote the module of $\Gamma_R(f(A))$. Then we have

$$\log(1/\text{Cap } f(A)) = \lim_{R\to\infty} (2\pi M(\Gamma_R(f(A))) - \log R). \tag{3}$$

The right-hand side is a monotone increasing limit. Let ρ_R be the extremal metric in the module problem for $\Gamma_R(f(A))$. For $R' > R$, from a variational inequality [4] for extremal metrics we have

$$\| \rho_{R'} - \rho_R \|^2_{\Omega_R} \leq \| \rho_{R'} \|^2_{\Omega_R} - \| \rho_R \|^2,$$

where Ω_R denotes subregion of $\Omega = f(A)^c$ contained in $|w| < R$. Since $(2\pi|w|)^{-1}$ is extremal for the module problem of the curve family separating $|w| = R$ from $|w| = R'$, we have

$$\| \rho_{R'} - (2\pi|w|)^{-1} \|^2_{R<|w|<R'} \leq \| \rho_{R'} \|^2_{R<|w|<R'} - \frac{1}{2\pi} \log(R'/R)$$

and we obtain

$$\| \rho_{R'} - \rho_R \|^2_{\Omega_R} \leq M(\Gamma_{R'}(f(A))) - (\log R')/2\pi - (M(\Gamma_R(f(A))) - (\log R)/2\pi)$$

which implies uniform convergence of ρ_R on every compact set of Ω as $R \to \infty$. It is easy to see that $\rho_R(w)$ is equal to a constant multiple of the gradient of the harmonic measure of $|w| = R$ in Ω_R. The constant is determined so that the constant multiple of the harmonic measure has a flux equal to 1 along $|w| = R$.

Let g(w) be the Green's function of Ω with pole at the point at infinity. Then ρ_R tends to a metric $\rho_R = |\operatorname{grad} g(w)|/2\pi$, since the sequence of constant multiples of harmonic measures converges to the Green's function g(w) uniformly on every compact subset of Ω. ρ_0 satisfies

$$\log(1/\operatorname{Cap} A) = \lim_{R\to\infty} (2\pi\|\rho_0\|^2 - \log R) \tag{4}$$

and possesses the following extremal property. Let ρ be an admissible metric for the curve family $\Gamma(f(A))$ i.e. $\int_\gamma \rho|dw| \geq 1$ for every curve in $\Gamma(f(A))$ and let ρ be expressed as the gradient of a harmonic function in a neighborhood of the point at the infinity. Then the reduced logarithmic area $a(\rho) = \lim_{R\to\infty} (\|\rho\|^2 - (2\pi)^{-1}\log R)$ is defined. We obtain from the variational inequality

$$\| \rho - \rho_0 \|^2 \leq a(\rho) - a(\rho_0). \tag{5}$$

We call ρ_0 an *extremal metric* for $\Gamma(f(A))$.

3. *Proof of Theorem 1*. In order to prove the theorem we consider a family $\Gamma(A)$ of chains σ in D which intersects every curve joining A and the point at infinity in D. σ consists of a finite number of open arcs or a closed curve separating A from the point at infinity. Let $\Gamma^*(A)$ be the image of $\Gamma(A)$ under f. Then an extremal metric ρ^* for $\Gamma^*(A)$ exists and is equal to the gradient of a potential function G(w) which is harmonic in f(D) except for a logarithmic singularity at the point at infinity with the properties $G(w) \to 0$

as $w \to \zeta$, $\zeta \in f(A)$ except for a set of capacity zero and $dG^* = 0$ along the boundary part other than $f(A)$. To construct $G(w)$, let D_n be $D - A_n$. Then $\{D_n\}$ forms an exhaustion of D. Let $G_n(z)$ be a solution of the Dirichlet-Neumann problem satisfying

i) $G_n(z) = \log|z| + c_n + o(1)$ is harmonic in D_n except for the point at infinity,

ii) $G_n(z) \to 0$ as $z \to \zeta$, $\zeta \in \partial A_n$ and

iii) $\partial G_n/\partial n = 0$ on $\partial D_n - \partial A_n$(circular arcs).

Then $G_n(z)$ is monotone increasing as n increases and G_n tends to a potential function $G_0(z)$. $G(w)$ is given by $G_0(f^{-1}(w))$. Let $\hat{\rho}^*$ be an extension of ρ^* over Ω such that $\hat{\rho}^* = 0$ in $\Omega - f(D)$. Then we obtain from (5)

$$\|\hat{\rho}^* - \rho_0\|^2 \le a(\rho^*) - a(\rho_0) , \tag{6}$$

since $\hat{\rho}^*$ is admissible for $\Gamma(f(A))$. Let ρ_1 be $(2\pi)^{-1}|\text{grad } G_0(z)|$. Then by normalization (1), $a(\rho^*) = a(\rho_1)$. The function $G_0(z)$ can be expressed by the Green's function of A^c with pole at the point at infinity, denoted by $g_A(z)$, as $G_0(z) = g_A(z) + g_A(1/\bar{z})$ by symmetry. It is known that [1]

$$g_A(z) - g_A(1 / \bar{z}) = \log|z|.$$

We have near the point at infinity

$$G_0(z) = \log|z| + \log \frac{1}{(\text{Cap } A)^2} + o(1).$$

Hence $a(\rho_1) = -\pi^{-1} \log \text{Cap } A$ and we infer that

$$\|\hat{\rho}^* - \rho_0\|^2 \le \frac{1}{2\pi} \left(\log \frac{1}{\text{Cap } f(A)} - \log \frac{1}{(\text{Cap } A)^2} \right) \tag{7}$$

which implies the desired inequality (2).

We prove the equality statement in Theorem 1. In fact equality in (2) implies $\|\hat{\rho}^* - \rho_0\|^2 = 0$.

We infer that

$$g(w) = G(w)$$

in the image domain from which the equality statement follows.

References

[1] Ahlfors, L. V., Conformal invariants, McGraw-Hill Book Company (1973).

[2] Pommerenke, Ch., On the logarithmic capacity and conformal mapping. Duke Math. J. 35 (1968), pp. 321 - 326.

[3] Schiffer, M., Hadamard's formula and variation of domain functions. Amer. J. Math. Soc. 68 (1946), pp. 417 - 448.

[4] Suita, N., On radial slit disc mappings. Kōdai Math. Sem. Rep. 18 (1966) pp. 219 - 228.

Department of Mathematics
Tokyo Institute of Technology

PINCHING DEFORMATION OF ARBITRARY RIEMANN SURFACES AND VARIATIONAL FORMULAS FOR ABELIAN DIFFERENTIALS

Masahiko Taniguchi

Introduction.

Recently, Y. Kusunoki, F. Maitani and the author established a simple but powerful method of deriving variational formulas for square integrable harmonic differentials under quasiconformal deformation of arbitrary Riemann surfaces (cf. [2] and [4]). The prototype of this method can be found in Ahlfors' celebrated paper in 1960, and the method of orthogonal decomposition, which is classical but still fruitful, enabled them to revive the essence of Ahlfors' argument in such a general situation as stated above.

After slight modification, the same method can work even in more general situations. One way of formulating such a generalization is considering the pinching deformation family of arbitrary Riemann surfaces with possibly a finite number of nodes, which in turn can be regarded as an amalgamation of quasiconformal deformation and variation by opening up nodes.

In § 1, we state the definition of the real pinching deformation family and the complex one, and give a fundamental variational formula for abelian differentials under pinching deformation (Theorem 1). Some of typical applications of this formula will be discussed in § 2 and § 3 (cf. [9], [10], [11]).

In case of the real pinching deformation, taking the real part of the fundamental formula, we can derive several formulas for basic differentials and functions such as the period reproducers and Green's functions (cf. Theorem 2, [9] and [10]). Without variation by opening up nodes, these formulas reduced to those under quasiconformal deformation stated above, and without quasiconformal deformation, then can be regarded as a generalizations of Schiffer-Spencer's formulas for Green's functions on finite Riemann surfaces (Theorem 3).

In case of the complex pinching deformation, the fundamental formula gives ones for periods of the normalized meromorphic abelian differentials (cf. Theorem 4 and [11]). The case without variation by opening up nodes has already been discussed in [2] as a generalization of Ahlfors' results (which is the starting point of these series of investigations) and without quasiconformal deformation we have in particular the formulas for period matrices which generalize Fay-Yamada's ones for the case of compact surfaces. As a corollary, we can determine the singularity of period matrices at the center of the complex pinching deformation (Theorem 5).

§ 1. Pinching deformation and fundamental formulas.

Let R_0 be an arbitrary Riemann surface with a finite number of nodes (cf. [6, § 1]). Let $N(R_0) = \{p_j\colon j = 1, \cdots, n\}$ be the set of all nodes of R_0, and set $R_0' = R_0 - N(R_0)$. For every j, we fix a neighborhood U_j of p_j on R_0 such that each component, say $U_{j,k}$ $(k = 1, 2)$, of $U_j - \{p_j\}$ is mapped conformally onto $D_0 = \{0 < |z| < 1\}$ by a mapping $z=z_{j,k}(p)$. Also we suppose that $\{\overline{U}_j\}_{j=1}^{n}$ are mutually disjoint.

For every $t > 0$, let f_t be a quasiconformal mapping of R_0' onto another union R_t' of Riemann surfaces with the complex dilatation μ_t. Further we assume that

a) the support of μ_t is contained in $R_0 - U$, and

b) there is a bounded $(-1, 1)$–form μ on R_0' such that

$$\lim_{t\to 0} \| (\mu_t/t) - \mu \|_\infty = 0,$$

where $U = \cup_{j=1}^{n} U_j$, and $\| \cdot \|_\infty$ is the L^∞-norm on R_0'. For t=0, we denote by f_0 the identical mapping of R_0' onto itself.

Next for every given $t \geq 0$ and s_j with $0 \leq s_j < 1/2$ ($j = 1, \cdots, n$), let $R_{t,s}$ (with $s = (s_1, \cdots, s_n)$) be the Riemann surface obtained from R'_t by deleting two punctured disks $z_{j,t,s}^{-1}(\{0 < |z| < s_j\})$ ($k = 1, 2$) and identifying the borders by the mapping

$$z_{j,2,t}^{-1}(\eta_j \cdot s_j^2 / z_{j,1,t}(p))$$

for every j, where $z_{j,k,t} = z_{j,k} \circ f_t^{-1}$ and η_j is a constant with $|\eta_j| = 1$. We denote by $C_{j,t,s}$ the identified borders considering as a loop on $R_{t,s}$ (with a suitable orientation). Here in case that some of s_j are equal to 0, we regard that $C_{j,t,s}$ for such j collapses to a node corresponding to p_j.

For every (t,s) as above, let $J_{t,s}$ be the natural embedding of $R''_{t,s} = R_{t,s} - \cup_{j=1}^{n} C_{j,t,s}$ into R'_t, then $f_t^{-1} \circ J_{t,s}$ can be reformed to a well-behaved continuous surjection $f_{t,s}$ from $R_{t,s}$ to R_0. For instance, we set

$$\begin{aligned} f_{t,s}^{-1}(p) &= J_{t,s}^{-1} \circ f_t(p) \quad \text{on } R'_0 - V\text{, and} \\ &= (z_{j,k,t} \circ J_{t,s})^{-1}\left(\left[(1-2s_j) + \frac{s_j}{|z_{j,k}(p)|}\right] \cdot z_{j,k}(p)\right) \\ &\qquad \text{on } V_{j,k} \ (j = 1, \cdots, n;\ k = 1, 2), \end{aligned}$$

where $V_{j,k} = z_{j,k}^{-1}(\{0 < |z| < 1/2\})$, and $V = \cup_{j=1}^{n} (\cup_{k=1}^{2} V_{j,k})$.

We call the above family $\{R_{t,s}: t \epsilon [0, 1), s \epsilon [0, 1/2)^n\}$ of (marked) Riemann surfaces with nodes (with the markings induced by $\{f_{t,s}\}$) the real pinching deformation family with the center R_0 and the deformation data $(\{\mu_t\}, \{\eta_j\}, U)$.

Here we can generalize the real parameter t for quasiconformal deformation by a real vecto. Also we can modify the family for the case that both t and s are complex vewctor-valued, which we may formulate as follows.

Let $\{\mu_i\}_{i=1}^{m}$ be Beltrami differentials supported on $R_0 - \overline{U}$ with $\|\mu_i\|_\infty < 1/m$ for every i. For every $t = (t_1, \cdots, t_m)$ in $(B_t)^m$, where $B_t = \{t \epsilon \mathbb{C}: |t| < 1\}$, let f_t be the quasiconformal mapping of R'_0 onto another R'_t with the complex dilatation $\mu_t =$

$\Sigma_{i=1}^{m} t_i \cdot \mu_i$. Next for arbitrarily fixed $t\epsilon(B_t)^m$ and any given $s = (s_1, \cdots, s_n)$ in $(B_s)^n$, where $B_s = \{s\epsilon C: |s| < 1/2\}$, let $R_{t,s}$ be the Riemann surface with nodes obtained from R'_t by deleting two punctured disks $z_{j,k,t}^{-1}(\{0 < |z| < |s_j|^{1/2}\})$ $(k = 1, 2)$ and identifying resulting borders into $C_{j,t,s}$ by the mapping

$$z_{j,2,t}^{-1}(s_j/z_{j,1,t}(p))$$

for every j. Again regarding that $R_{t,s}$ has a node corresponding to p_j for every such j that $s_j = 0$, we have a family $\{R_{t,s}: t\epsilon(B_t)^m, s\epsilon(B_s)^n\}$, which we call the complex pinching deformation family with the center R_0 and the deformation data $(\{\mu_i\}, U)$. (Here we can reform $f_t^{-1} \circ J_{t,s}$ by the same way as above, using $|s_j|^{1/2}$ instead of s_j.

<u>Remark</u>. When R_0 has no nodes, then pinching deformation is nothing but quasiconformal deformation, and the following Theorem reduces to a technical step in a standard proof of variational formulas under quasiconformal deformation (cf. [2] and [4]).

The case that $\mu_t \equiv 0$ for every t, or equivalently, the case without quasiconformal deformation is considered as classical variation by opening up nodes (cf. [1], [3]).

Now fundamental variational formulas can be stated as follows.

<u>Theorem 1</u>. Let $\{R_{t,s}\}$ be the real pinching deformation family or the complex one with the center R_0 and the deformation data $(\{\mu_t\}, \{\eta_j\}, U)$ or $(\{\mu_j\}, U)$, respectively,

For every (t,s), let $\phi_{t,s}$ be a meromorphic differential on $R_{t,s}$ such that

1) $\phi_{t,s}$ converges to $\phi_{0,0}$ metrically on $K \cup (U \cap R'_0 - V)$, which means that

$$\lim_{|(t,s)| \to 0} \|\phi_{t,s} \circ f_{t,s}^{-1} - \phi_{0,0}\|_{K \cup (U \cap R'_0 - V)} = 0,$$

where and in the sequel,

$$|(t,s)| = t + \Sigma_{j=1}^{n} s_j \qquad \text{in the real case, and}$$

$$|(t,s)| = \Sigma_{j=1}^{m} |t_i| + \Sigma_{j=1}^{n} |s_j| \qquad \text{in the complex case,}$$

respectively, $\phi \circ f$ is the pull-back of ϕ by f, $\| \cdot \|_E$ is the Dirichlet norm on a Borel set E, and K is a closed subset of $R_0 - U$ such that $\mu_t \equiv 0$ on $R_0 - K$ for every t,

2) $$\int_{C_{j,t,s}} \phi_{t,s} = 0 \quad \text{for every j and (t,s), and}$$

there is a positive constant M such that

3) $$\| \phi_{t,s} \|_{U_{t,s} - N(R_{t,s})} < M \quad \text{for every (t,s),}$$

where $U_{t,s} = f^{-1}_{t,s}(U)$ and $N(R_{t,s})$ is the set of all nodes of $R_{t,s}$.

Next let ψ be a meromorphic abelian differential on R_0 such that

A) $\| \psi \|_{K \cup (U \cap R_0')}$ is finite, and

B) the (1, 1)-form $\overline{\phi_{0,0}} \wedge \psi$ is absolutely integrable on K and so is every $\omega_{t,s} = (\phi_{t,s} \circ f_{t,s}^{-1} \wedge {}^*\psi - \phi_{0,0})$ on $R_0 - U$.

Then in the real case, it holds that

$$\text{(R)} \quad \iint_{R_0'} \omega_{t,s} = t \cdot \iint_K \phi_{0,0} \cdot \mu \wedge {}^*\psi$$
$$+ 2\pi \cdot \Sigma_{j=1}^{n} \eta_j \cdot s_j^2 \cdot (a_{j,1}(0) \cdot b_{j,2}(0) + a_{j,2}(0) \cdot b_{j,1}(0))$$
$$+ o\,(t + \Sigma_{j=1}^{n} s_j^2),$$

and in the complex case, it holds that

$$\text{C)} \quad \iint_{R_0'} \omega_{t,s} = \Sigma_{i=1}^{m} t_i \cdot \iint_K \phi_{0,0} \cdot \mu_i \wedge {}^*\psi$$
$$+ 2\pi \cdot \Sigma_{j=1}^{n} s_j \cdot (a_{j,1}(0) \cdot b_{j,1}(0) + a_{j,2}(0) \cdot b_{j,1}(0))$$
$$+ o\,(| (t,s) |),$$

respectively, as | (t, s) | tends to 0.

Here we set $\phi_{0,0} = a_{j,k}(z_{j,k})dz_{j,k}$ and $\psi = b_{j,k}(z_{j,k})dz_{j,k}$ on $\overline{U}_{j,k} = \{ |z_{j,k}| < 1\}$ for every j and k.

Proof of Theorem 1 for the real case has been given in [9] and [10]. Also we find, by careful reading of the proof, that the condition B) in [9, Theorem 1] can be replaced by such one as stated in Theorem 1 (cf. [9, § 1, Remark]).

Proof for the complex case can be given exactly by the same argument as in that for the real case, hence is omitted.

<u>Remark</u>. From the proof, we can see that the reminder terms in the formula (R) and (C) can be bounded, for instance, by

$$\text{(R)}\quad M_0\{t(\|\phi_{t,s}\circ f_{t,s}^{-1} - \phi_{0,0}\|_K \cdot \|\psi\|_K + \|\mu_t/t-\mu\|_\infty \cdot \iint_K |\overline{\phi_{0,0}}\wedge\psi|)$$
$$+ \Sigma_{j=1}^{n} s_j^3 \cdot \|\phi_{t,s}\|_{U_{t,s}-N(R_{t,s})} \cdot \|\psi\|_U$$
$$+ \Sigma_{j=1}^{n} s_j^2 \cdot \| \phi_{t,s}\circ f_{t,s}^{-1} - \phi_{0,0} \|_{U\cap R_0' - V} \cdot \| \psi \|_U\}, \text{ and}$$
$$\text{(C)}\quad (\tfrac{1}{m} \cdot \Sigma_{i=1}^{m} | t_i |) \cdot \| \phi_{t,s}\circ f_{t,s}^{-1} - \phi_{0,0} \|_K \cdot \| \psi \|_K$$
$$+ M_0 \{\Sigma_{j=1}^{n} |s_j|^{3/2} \cdot \| \psi_{t,s} \|_{U_{t,s}-N(R_{t,s})} \cdot \| \psi \|_U$$
$$+ \Sigma_{j=1}^{n} |s_j| \cdot \|\phi_{t,s}\circ f_{t,s}^{-1} - \phi_{0,0}\|_{U\cap R_0'-V} \cdot \| \psi\|_U\}, \text{ respectively,}$$

with a constant M_0 independent of (t,s).

<u>§ 2. Application I; real cases.</u>

In this section, we will give some applications of (the proof of) the formula (R) in Theorem 1. For further applications of the formula (R), see [9, § 2] and [10, §§ 2-4]. In the sequel of this section, we denote by g(·, p; R) Green's function on a surface R with the pole p.

1) Assume that a component of R_0' admits Green's functions and every node of R_0 is essentially trivial, or equivalently, that every node divides R_0 and there is exactly one component, say S_0, of R_0' that admits Green's functions (cf. [9, § 2]). Take two mutually distinct points p and q in $S_0 - \bar{U}$, and let $\{R_{t,s}\}$ be the real pinching deformation family with the center R_0 and the deformation data $(\{\{\mu_t\}, \{\eta_j\}, U)$.

Set $g_{t,s} = g(\cdot, p_{t,s}; S_{t,s})$ for every (t,s), where $p_{t,s} = f_{t,s}^{-1}(p)$ and $S_{t,s}$ is the component of $R_{t,s} - N(R_{t,s})$ containing $f_{t,s}^{-1}(S_0)$. Then we have the following

Theorem 2. Under the above circumstances,

$$g_{t,s}(f_{t,s}{}^{-1}(q)) - g_{0,0}(q) = \frac{t}{2\pi}\cdot \mathrm{Re}\iint_{R_0'} i\cdot\phi_{0,0}\cdot\mu\wedge\psi$$

$$- \mathrm{Re} \mid \Sigma_{j=1}^{n} \eta_j \cdot s_j{}^2 \cdot (a_{j,1}(0)\cdot b_{j,2}(0) + a_{j,2}(0)\cdot b_{j,1}(0)) \mid$$

$$+ o(t + \Sigma_{j=1}^{n} s_j{}^2)$$

as $|(t,s)|$ tends to 0.

Here we set

$$\phi_{0,0} = {}^*dg_{0,0} - i\cdot dg_{0,0},$$

$$\psi = {}^*dg(\cdot, q; S_0) - i\cdot dg(\cdot, q; S_0),$$

and $a_{j,k}$ and $b_{j,k}$ are as in Theorem 1.

Remark. Note that we do not need the additional assumption that every $\mu_t \equiv 0$ on a neighborhood of {p, q}. The same argument as in the proof below enables us to get rid of the similar assumptions in [9, Theorems 3, 4] and [10, Theorem 3].

Proof. Set

$$\phi_{t,s} = {}^*dg_{t,s} - i\cdot dg_{t,s}$$

for every (t,s), and fix a mutually disjoint simply connected neighborhood W_1 and W_2 of p and q, respectively. For every (t,s), reform $f_{t,s}{}^{-1}$ to $S(f_{t,s}{}^{-1})$ as in the proof of [12, Theorem 2-I)] (with $U_j = W_j$) so that $S(f_{t,s}{}^{-1}) \equiv f_{t,s}{}^{-1}$ on $R_0 - W_1 \cup W_2$, $S(f_{t,s}{}^{-1})(p) = p_{t,s}$, $S(f_{t,s}{}^{-1})(q) = f_{t,s}{}^{-1}(q)$, and $S(f_{t,s}{}^{-1})$ is conformal in some neighborhood of {p, q}. Then we know that $\phi_{t,s}$ converges to $\phi_{0,0}$ strongly metrically with respect to $\{Sf_{t,s}\ (\equiv S(f_{t,s}{}^{-1})^{-1})\}$ ([7, Theorem 1]).

Also we can see by [9, Lemma 9] (cf. the proof of [9, Theorem 4]) that

$$I_{t,s} \equiv g_{t,s}(f_{t,s}{}^{-1}(q)) - g_{0,0}(q)$$

$$= \frac{-1}{2\pi}\cdot \mathrm{Re}\iint_{R_0'} (\phi_{t,s}\circ Sf_{t,s}{}^{-1} - \phi_{0,0})\wedge{}^*\psi$$

$$= \frac{1}{2\pi}\cdot \mathrm{Re}\iint_{R_0'} i\cdot\phi_{t,s}\circ Sf_{t,s}{}^{-1}\wedge\psi.$$

Now by the same argument as in the proof of Theorem 1, we can show that

$$I_{t,s} = \frac{1}{2\pi} \cdot \mathrm{Re} \iint_{R_0'} i \cdot \phi_{0,0} \cdot S\mu_t \wedge \psi$$
$$- \quad \mathrm{Re} \{ \Sigma_{j=1}^{n} \eta_j \cdot s_j^2 \cdot (a_{j,1}(0) \cdot b_{j,2}(0) + a_{j,2}(0) \cdot b_{j,1}(0)) \}$$
$$+ \quad o(t + \Sigma_{j=1}^{n} s_j^2)$$

as | (t,s) | tends to 0, where $S\mu_t$ is the complex dilatation of $Sf_{t,0}^{-1}$.

Since we can see by [12, Theorem 1] (cf. [12, § 2-(8)]) that

$$\iint_{R_0'} \phi_{0,0} \cdot S\mu_t \wedge \psi = \iint_{R_0'} \phi_{0,0} \cdot \mu_t \wedge \psi + 0(t^2),$$

we have the desired formula. q.e.d.

2). Next we consider the case that $t = 0$ and $n = 1$ (i.e. $s = s_1$), which can be regarded as the variation by opening up a node of Schiffer-Spencer's type ([3]) with a special choice of U.

We have treated in 1) the case that the single node p_0 of R_0 is essentially trivial, so here we consider the other cases. Suppose that R_0' admits Green's functions and p_0 is not essentially trivial, or equivalently, either (i) $R_0 - \{p_0\}$ is connected and admits Green's functions, or (ii) $R_0 - \{p_0\}$ consists of two components S_1 and S_2 both of which admit Green's functions.

Let p_k be the puncture of R_0' in $U_{1,k}$ for each k, and in the case (i), we set

$$G(p) = g(p, p_1 \bar{R}_0') - g(p, p_2; \bar{R}_0') \text{ on } R_0',$$

where $\bar{R}_0' = R_0' \cup \{p_1, p_2\}$. In the case (ii), assuming that $U_{1,k}$ is contained in S_k for each k, we set

$$G(p) = (-1)^{k-1} \cdot g(p, p_k; S_k \cup \{p_k\}) \text{ on } S_k \quad (k = 1, 2).$$

Next fix an $M_0 > 0$ such that $U_k = \{p \in R_0' : (-1)^{k-1} \cdot G(p) > M_0\} \cup \{p_k\}$ is a simply connected neighborhood of p_k in $\bar{R}_0'$ for each k. Take $((U_1 \cup U_2) \cap R_0') \cup \{p_0\}$ as U in the deformation data, and consider the real pinching deformation family $\{R_s\ (= R_{t,s})\}$ with the center R_0 and the deformation data $(\{0\}, \eta(= \eta_1), U)$. Then we have the following formula of Schiffer-Spencer's type.

Theorem 3 ([7, Theorem 2-3)] and [8, § 5]). Let p and q be distinct points on the same component S_0 of R'_0. Then

$$g(p, q; R_s) - g(p, q; S_0) = \frac{1}{2(\log s - M_0)} \cdot G(p) \cdot G(q)$$
$$- \quad s^2 \cdot \mathrm{Re}\ \{\ (\eta \cdot (a_{1,1}(0) \cdot b_{1,2}(0) + a_{1,2}(0) \cdot b_{1,1}\ (0)))\ \} + \epsilon(s^2)$$

as s tends to 0, where $a_{1,k}$ and $b_{1,k}$ are the same as in Theorem 1 with $\phi_{0,0} = {}^*dg$ $(\cdot, p; S_0) - i \cdot dg(\cdot, p; S_0)$ and $\psi = {}^*dg(\cdot, q; S_0) - i^*dg(\cdot, q; S_0)$.

Moreover, $\epsilon z(s^2) / s^2$ converges to 0 locally uniformly on $(R_0 - \bar{U}) \times (R_0 - \bar{U}) - \Delta$ with respect to (p, q) as s tends to 0, where $\Delta = \{(p, p):\ p \epsilon R_0 - \bar{U}\}$.

Remark. Note that

$$|\ z_{1,k}(p)\ | = \exp(\ (-1)^k \cdot G(p) + M_0)\quad (k = 1, 2).$$

Hence replacing $z_{1,k}$ and s by $z_k = \exp(-M_0) \cdot z_{1,k}$ (which maps $U_k - \{p_k\}$ onto $\{0 < |z| < \exp(-M_0)\}$) and by $t = \exp\ (-M_0) \cdot s$, respectively, we have the formula in [7, Theorem 2-3)].

Proof. For the sake of convenience, we include a proof using Theorem 1 (which has been suggested in [10, Remark 2]).

First note that $\phi = {}^*dG - i \cdot dG$ restricted on R''_s can be extended to a holomorphic differential, say ϕ_s, on R_s for every s, as is seen from the definition of G and U, and by [7, Lemma 4] we have

$$2(\log s - M_0) \cdot \int_{C_{1,s}} {}^*dg_s = 2\pi \cdot G(p), \tag{1}$$

where we set $g_s = g(\cdot, p; R_s)$ (cf. [7, § 4 -(13)]).

Now set

$$\phi_{0,s} = (^*dg_s - i\cdot dg_s) - \frac{1}{2\pi}\cdot\left(\int_{C_{1,s}} {^*dg_s}\right)\cdot\phi_s.$$

Then since $\int_{C_{1,s}} \phi_s = 2\pi$, $\int_{C_{1,s}} \phi_{0,s} = 0$, i.e. the condition 2) in Theorem 1 holds. Also we can see, again similarly as in the proof of [9, Lemma 9], that

$$\left(g_s(q) - \frac{1}{2\pi}\cdot\left(\int_{C_{1,s}} {^*dg_s}\right)\cdot G(q)\right) - g_0(q) \tag{2}$$
$$= \frac{-1}{2\pi}\cdot \mathrm{Re}\iint_{R_0'} (\phi_{0,s} \circ f_{0,s}{}^{-1} - \phi_{0,0}) \wedge {^*\psi}.$$

Now by [7, Theorem 1] and the definition of ϕ_s, it is easy to see that $\phi_{0,s}$ converges to $\phi_{0,0}$ strongly metrically with respect to $\{f_{0,s}\}$ as s tends to 0. Hence in particular, $\{\phi_s\}$ satisfies the condition 1) in Theorem 1. And [7, Lemma 2] implies that $\{\phi_{0,s}\}$ satisfies the condition 3) in Theorem 1. Since the conditions A) and B) in Theorem 1 clearly hold, where we can take the empty set as K, we have the first assertion by (1), (2) and Theorem 1-(R).

Moreover we know by Remark at the end of § 1 that the reminder term $\epsilon(s^2) / s^2$ is bounded by

$$(s \circ \|\phi_{0,s}\|_{U_{0,s}-N(R_{0,s})} + \|\phi_{0,s} - \phi_{0,0}\|_{U-V}) \cdot \|\psi\|_U.$$

Here $\|\psi\|_U$ is locally bounded on $R_0 - \bar{U}$ with respect to q ([8, Theorem 1]) and $\|\phi_{0,s} - \phi_{0,0}\|_{U-V}$ converges to 0 locally uniformly on $R_0' - \bar{U}$ with respect to p as s tends to 0 ([8, Lemma 7]). Also we can find a positive δ such that

$$M(p) = \sup_{s \epsilon [0,\, \delta)} \|\phi_{0,s}\|_{U_{o,s}-N(R_{o,s})}$$

is locally uniformly bounded on $R_0 - \bar{U}$.

In fact, fix p and a positive M so that $D_M = \{p' \epsilon S_0' \colon g(p', p; S_0') > M\}$ is simply connected and relatively compact in $R_0' - \bar{U}$. Then there is a neighborhood W of p such that for every $p' \epsilon$ W we can construct a 2-quasiconformal self-mapping $f_{p'}$ of R_0' such that $f_{p'}(p) = p'$ and $f_{p'}$ equals to the identity outside of D_M. Then applying [7, Lemma

1] with q=p, $R'=R_{0,s}$ and $f=f_{0,s}{}^{-1}\circ f_{p'}$ we can show the assertion by [6, Proposition 4] (cf. the proof of [7, Lemma 2]).

Thus we conclude the second assertion. q.e.d.

§ 3. Application II, a complex case.

In this section we treat the normalized abelian differentials with a given normal behavior. For the sake of simplicity, we consider only those of the first kind.

We recall some definitions. First let R^* be an ordinary Riemann surface obtained from the given R_0 by opening up all nodes of R_0. Then we can find a canonical homology base $\Xi = \Xi(R^*) = \{A_k, B_k\}_{k=1}^{g}$ of R^* modulo the ideal boundary (, where g may be infinite,) which satisfies the following condition;

(#) for every $p \in N(R_0)$, $f^{-1}(p)$ is (i) either freely homotopic to some A_k in Ξ (with a suitable orientation), or (ii) a loop on $R^* - \{A_k, B_k: A_k$ does not belong to $E_0\}$ dividing $R^* - \{A_k: A_k \in E_0\}$, where E_0 is the subset of all A_k in Ξ corresponding to nodes of R_0 as in (i) and f is a naturally induced allowable deformation (cf.[6,§2-1)])of R^* to R_0.

Note that we can consider $\Xi_0 \equiv \Xi(R_0) = \{A_k, B_k: A_k$ does not belong to $E_0\}$ as a homology base of R_0' modulo the ideal boundary.

Next fix a normal behavior space $\Gamma_0(R_0)$, i.e. a subspace of the Hilbert space $\Gamma_h(R_0)$ (consisting of all complex harmonic differentials square integrable on R_0') which satisfies the following conditions:

i) $\Gamma_0(R_0) \subset \Gamma_{hse}(R_0)$,

ii) $\int_{A_k} \omega = 0$ for every $\omega \in \Gamma_0(R_0)$ and $A_k \in \Xi_0$,

iii) $\Gamma_0(R_0) + {}^*\Gamma_0(R_0) = \Gamma_h(R_0)$, and

iv) $\Gamma_0(R_0) = \overline{\Gamma_0(R_0)}$.

And we say that a complex harmonic or meromorphic differential ϕ and R_0' has Γ_0

-behavior if there exists α in $\Gamma_0(R_0)$ and df in $\Gamma_{e0}(R_0)$ such that $\phi = \alpha + df$ outside of a compact set on R_0. (This condition imposes nothing on ϕ when R_0 is compact.)

For every A_k in Ξ, the k-th normalized abelian differential $\phi_k(R_0)$ of the first kind with Γ_0-behavior is, by definition, a holomorphic differential with Γ_0-behavior on R_0 uniquely determined by the conditions that

$$\int_{A_h} \phi_k(R_0) = \delta_{kh} \quad \text{for every } A_h \in \Xi,$$

Recall that, for every node p corresponding to a loop which has non-zero algebraic intersection number with B_k, $\phi_k(R_0)$ has simple poles at two punctures of R_0' corresponding to p, and is holomorphic at two punctures of R_0' that ϕ_k is the classical k-th normal differential of the first kind when the surface is compact and without nodes.

Now we consider the complex pinching deformation family $\{R_{t,s}\}$ with the center R_0 and the given deformation data $(\{\mu_i\}, U)$. Then we can obtain a normal behavior space $\Gamma_0(R_{t,s})$ on $R_{t,s}$ corresponding to $\Gamma_0(R_0)$ in a natural manner, and hence the k-th normalized abelian differential $\phi_k(R_{t,s})$ of the first kind with Γ_0-behavior on $R_{t,s}$, for every (t,s) and k (cf. the proof of [11, Theorem 1]). Also we can show the following

Proposition. For every k, $\phi_k(R_{t,s})$ converges to $\phi_k(R_0)$ strongly metrically (with respect to $\{f_{t,s}\}$) as (t,s) tends to (0,0).

Proof. In case that $A_k \in \Xi_0$, the assertion has been proved as [11, Corollary 2]. So assume that $A_k \in \Xi - \Xi_0$, then for every p_j, such that $C_{j,t,s} \times B_k$ is non-zero, we can show by [11, Corollary 4] that the characteristic ring domain of $\phi_k(R_{t,s})$ for $C_{j,t,s}$ is non-empty and contains $C_{j,t,s}$ for every (t,s) sufficiently near (0,0) (cf. [5, Proposition 1 and Lemma 5]). Hence we can show the assertion by the same argument as in the proof of [9, Theorem 6] (us ing [11, Corollary 3]). q.e.d.

Now applying Theorem 1, we can give another proof of some part of the following

Theorem 4 ([11, Theorem 5]). Fix $A_h \in \Xi$ and $B_k \in \Xi_0$ arbitrarily. The $F_{hk}(t,s) = \int_{B_k} \phi_h(R_{t,s})$ is holomorphic on $(B_t)^m \times (B_s)^n$, and

$$\frac{\partial}{\partial t_i} F_{hk}(t,s) \Big|_{(t,s)=(0,0)} = -\iint_{R_0'} \phi_h(R_0) \cdot \mu_i \wedge \phi_k(R_0) \ (i = 1, \cdots, n), \text{ and}$$
$$\frac{\partial}{\partial s_j} F_{hk}(t,s) \Big|_{(t,s)=(0,0)} = -2\pi i \cdot (a_{hj1} \cdot a_{kj2} + a_{hj2} \cdot a_{kj1}) \ (j = 1, \cdots, m),$$

where we set, for every v and j,

$$2\pi i \cdot a_{vjk} = \int_{\partial\{|z| \leq 1/2\}} \frac{1}{z} \cdot \phi_v(R_0) \circ z_{j,k} \quad (k = 1, 2).$$

In fact, further assume that A_h is in Ξ_0. Then the family $\{\phi_{t,s} \equiv \phi_h(R_{t,s})\}$ satisfies the conditions 1) and 2) in Theorem 1, by [11, corollary 2] and by the definition of Ξ_0, respectively, and it is clear that $\psi = -i \cdot \phi_k(R_0)$ satisfies the condition A) in Theorem 1. Also we know that $\{\|\phi_k(R_{t,s})\|_{R_{t,s}}\}$ is a bounded family for every $A_k \in \Xi_0$ (cf. the proof of [11, Corollary 2]), hence can see that $\{\phi_{t,s}\}$ and ψ satisfy the condition 3) and B) in Theorem 1.

Next fix a positive δ (< 1/2) and set $\alpha^\delta_{t,s} = F_\delta(\phi_{t,s}) - \phi_{0,0}$, where F_δ is defined as in the proof of [9, Lemma 5] with suitable $H_{t,s}$. Then, by [6, Lemmas 4, 5, and 7],

$$I_{t,s} \equiv F_{hk}(t,s) - F_{hk}(0,0) = (\alpha^\delta_{t,s}, \sigma(B_k, R_0))_{R_0'},$$

where $\sigma(B_k, R_0)$ is the period reproducer for B_k in $\Gamma_h(R_0)$.

Since $\alpha^\delta_{t,s}$ belongs to $\Gamma_0(R_0) + \Gamma_{e0}(R_0)$ by [11, Lemma 1], and since the projection of $\sigma(B_k, R_0)$ to $\Gamma_0(R_0)$ is equal to $\mathrm{Im}\ \phi_k(R_0)$, we can see by the condition iii) in the definition of a normal behavior space that

$$I_{t,s} = (\alpha^\delta_{t,s}, \mathrm{Im}\ \phi_k(R_0))_{R_0'}$$
$$= (\alpha^\delta_{t,s}, \overline{(-i \cdot \phi_k(R_0))})_{R_0'} = \iint_{R_0'} \alpha^\delta_{t,s} \wedge {}^*\psi.$$

Hence letting δ tend to 0, we conclude that

$$I_{t,s} = \iint_{R_0'} (\phi_{t,s} \circ f_{t,s}^{-1} - \phi_{0,0}) \wedge {}^*\psi.$$

Thus, in case that $A_k \epsilon \Xi_0$, the assertion follows by the formula (C) in Theorem 1. q.e.d.

Remark. When t = 0, n = 1 and the genus g is finite, we can obtain formulas for period matrices ([11, Corollaries 5 and 6]), which generalize Fay-Yamada's ones for the case of compact R_0 (cf. [1] and [13]).

Finally, we note that Theorem 4, [11, Corollaries 5 and 6] and the above Proposition gives the following

Theorem 5. For every h and k, set

$$f_{hk}(t,s) = \{\Pi_{j=1}^{n} (s_j)^{-n_{j,h} \cdot n_{j,k}}\} \cdot \exp(2\pi i \cdot F_{hk}(t,s)).$$

Then f_{hk} is holomorphic and zero points free on $(B_t)^n \times (B_s)^m$, where we set $n_{j,k} = C_{j,t,s} \times B_k$ for every j and k (, which is independent of (t,s)).

Also it holds that

1) $$\frac{\partial}{\partial t_i} (\log f_{hk}) \Big|_{(t,s)=(0,0)} = -2\pi i \cdot \iint_{R_0'} \phi_k(R_0) \cdot \mu_i \wedge \phi_h(R_0)$$ (i = 1, ⋯, n), and

2) $$\frac{\partial}{\partial s_j} (\log f_{hk}) \Big|_{(t,s)=(0,0)} = (2\pi)^2 \cdot (a_{kj1} \cdot a_{hj2} + a_{kj2} \cdot a_{hj1})$$ (j = 1, ⋯, m),

where a_{hjv} are as in Theorem 4.

Proof. Fix j, then by the assumption (#), $C_{j,t,s}$ is homologous to $\Sigma_{A_k \epsilon E_0} n_{j,k} \cdot A_k$ on any $R_{t,s}$ modulo the ideal boundary and every $n_{j,k}$ is either 0, or 1, or −1.

Fix h, and suppose that $n_{j,k}$ is non-zero. Change the canonical homology base Ξ to another one, say $\Xi' = \{A_k', B_k'\}$, such that $A_h' = C_{j,t,s}$ and $A_k' = A_k$ for every other k. Then B_h' should be homologous to $n_{j,h} \cdot B_h$, and every other B_k' should be homologous to

$B_k - n_{j,k} \cdot n_{j,h} \cdot B_h$, modulo the ideal boundary. Also the h-th normalized differential, say $\phi'_h(R_{t,s})$, of the first kind with Γ_0-behavior on $R_{t,s}$ with respect to Ξ' is coincident with $n_{j,h} \cdot \phi_h(R_{t,s})$.

Now Theorem 4 and [11, Corollary 6] implies that

$$(s_j)^{-1} \cdot \exp(2\pi i \cdot \int_{B'_h} \phi'_h(R_{t,s})) = (s_j)^{-1} \cdot \exp(2\pi i \cdot F_{hh}(t,s))$$

is holomorphic and zero points free on $\Omega_j = \{(t,s) \epsilon (B_t)^m \times (B_s)^n \colon \Pi_{v \neq j} s_v \neq 0\}$, and so is $(s_j)^{-n_{j,h} \cdot n_{j,k}} \cdot \exp 2\pi i \cdot F_{hk}$ also for every $k \neq h$, for $F_{hk}(t,s) = n_{j,h} \cdot \left\{ \int_{B'_k} \phi'_h(R_{t,s}) + n_{j,k} \cdot \int_{B'_h} \phi'_h(R_{t,s}) \right\}$ and $\int_{B'_k} \phi'_h(R_{t,s})$ is holomorphic on Ω_j for $k \neq h$.

Next suppose that $n_{j,h} = 0$. If $C_{j,t,s}$ corresponds to a dividing curve, then Theorem 4 and [11, Corollary 5] implies that F_{hk} (t,s) is holomorphic on Ω_j. If not, changing Ξ suitably as above, we can see by Theorem 4 and [11, Corollary 6] that $F_{hk}(t,s)$ is holomorphic on Ω_j.

Thus we conclude that $f_{hk}(t,s)$ is holomorphic and zero points free of $\Omega = \cup_{j=1}^{n} \Omega_j$. Since $E = (B_t)^m \times (B_s)^n - \Omega$ is a union of a finite number of complex submanifolds of $(B_t)^m \times (B_s)^n$ of codimension two, we have the first assertion.

Finally to show the second assertion, fix i. Then Theorem 4 implies that the equation 1) with the center $R_{0,s}$ instead of R_0 for any such s that every s_j is non-zero. (Recall that f_t restricted on R''_0 can be extended to a quasiconformal mapping of $R_{0,s}$ to $R_{t,s}$ with the complex dilatation μ_t.) Hence letting s tend to 0, we have the equation 1) by the first assertion proved above and Proposition. Next fix j. Then [11, Corollaries 5 and 6] imply and Theorem 4 that the equation 2) holds with the center $R_{0,s}$ instead of R_0 for any such s that $s_j = 0$ and every other component of s in non-zero. Hence again letting s tend to 0, we have the equation 2) by the first assertion proved above and [11, Corollary 4]. q.e.d.

References

[1] Fay, J., Theta functions on Riemann surfaces, Lecture Notes in Math. 352, Springer, (1973).

[2] Kusunoki, Y., and F. Maitani, Variations of abelian differentials under quasiconformal deformations, Math. Z., 181 (1982), pp. 435 - 450.

[3] Schiffer, M., and D. Spencer, Functionals of finite Riemann surfaces, Princeton, Univ. Press, Princeton, 1954.

[4] Taniguchi, M., On the variation of periods of holomorphic Γ_{h0}- reproducing differentials, Proc. Japan Acad. Ser. A, 56 (1980) pp. 315 - 317.

[5] Taniguchi, M., On convergence of holomorphic abelian differentials on the Teichmüller spaces of arbitrary Riemann surfaces, J. Math. Kyoto Univ., 24 (1984) pp. 305 - 321.

[6] Taniguchi, M., Square integrable harmonic differentials on arbitrary Riemann surfaces with a finite number of nodes, ibid., 25 (1985) pp. 597 - 617.

[7] Taniguchi, M., Continuity of certain differentials on finitely augmented Teichmüller spaces and variational formulas of Schiffer-Spencer's type, Tohoku Math. J., 38 (1986) pp. 281 - 295.

[8] Taniguchi, M., A note on continuity of Green's functions on Riemann surfaces, Kodai Math. J., 10 (1987) pp. 165 - 173.

[9] Taniguchi, M., Variational formulas on arbitrary Riemann surfaces under pinching deformation, J. Math. Kyoto Univ., 27 (1987) pp. 507 - 530.

[10] Taniguchi, M., Supplements to my previous papers; a refinement and applications, ibid., 28 (1988) pp. 81 - 86.

[11] Taniguchi, M., Abelian differentials with normal behavior and complex pinching deformation, ibid., 29(1989) pp. 45-56.

[12] Taniguchi, M., On first variation of Green's functions under quasiconformal deformation, ibid., to appear.

[13] Yamada, A., Precise variational formulas for abelian differentials, Kodai Math. J., (1980) pp. 114 - 143.

Department of Mathematics
Kyoto University

ON ALGEBROID SOLUTIONS OF SOME ALGEBRAIC DIFFERENTIAL EQUATIONS

Nobushige Toda and Masakimi Kato

1. Introduction

More than fifty years ago, K. Yosida ([14]) proved the following theorem.

Theorem A. If the differential equation with rational coefficients:

$$(w')^n = \left(\sum_{j=0}^{p} a_j w^j\right) / \left(\sum_{k=0}^{q} b_k w^k\right) \quad (a_p b_q \neq 0), \tag{1}$$

where n is a positive integer and $\Sigma a_j w^j$, $\Sigma b_k w^k$ are mutually prime in w, admits at least one transcendental finite valued algebroid solution w=w(z) in $|z| < \infty$, for which the ramification index ξ of the Riemann surface of w(z) is zero, then

$$q=0 \quad \text{and} \quad p \leq 2n.$$

Here we use the following notation. Let w=w(z) be a non-constant ν-valued algebroid function in $|z| < \infty$, R its Riemann surface and π the natural projection mapping from R to C. Let $\tilde{c}$ be a branch point of R which degree of ramification is λ-1 and put

$$n(r,R) = \sum_{|\pi(\tilde{c})| \leq r} (\lambda - 1)$$

$$N(r, R) = \frac{1}{\nu} \int_0^r \frac{n(t,R) - n(0,R)}{t} dt + \frac{1}{\nu} n(0,R) \log r$$

and

$$\xi = \liminf_{r \to \infty} \frac{N(r,R)}{T(r,w)} \quad \text{(the ramification index of R).}$$

It is known that $0 \leq \xi \leq 2(\nu - 1)$.

Some generalizations of this theorem are given in [2], [3], and [9].

On the other hand, the following theorem is well-known (see [1]).

Theorem B. Any meromorphic solutions in $|z| < \infty$ of the differential equation

$$(w')^n = \sum_{j=0}^{m} a_j w^j \quad (0 \le m \le n-1,\ a_j\text{: rational, } a_m \neq 0) \tag{2}$$

are rational.

As a generalization of this theorem it is natural that we consider the following problems.

Problem I. Are any algebroid solutions in $|z| < \infty$ for which $\xi = 0$ of the D.E. (2) algebraic?

Problem II. Are any algebroid solutions in $|z| < \infty$ of the D.E. (2) algebraic?

In this paper, we shall solve Problem I and give a partial answer to Problem II. The term "algebroid (resp. meromorphic)" will mean algebroid (resp. meromorphic) in the complex plane and we use the standard notation of the Nevanlinna theory of meromorphic functions ([5]) or algebroid functions ([6], [12], [13]).

2. Lemmas

Let $w=w(z)$ be a non-constant finite valued algebroid function and R be the Riemann surface of $w(z)$. We denote by M the set of meromorphic functions and by M(R) that of meromorphic functions on R. For a non-constant algebroid function v, $S(r,v)$ is any quantity satisfying

$$S(r,v) = o(T(r,v))$$

as $r \to \infty$, possibly outside a set of r of finite linear measure.

We shall give some lemmas for later use.

Lemma 1([6]). For any non-constant algebroid function v,

$$m(r,v'/v) = S(r,v)$$

The following two lemmas can be shown as in the case of meromorphic functions.

Lemma 2. Let f, g be in M(R) which are linearly independent over C and put

$$f + g = h.$$

Then,

$$T(r,f) \leq m(r,h) + N(r,f) + N(r,D) + S(r,f) + S(r,g),$$

where $D = g'/g - f'/f$ (cf. Lemma 1 in [8]}.

Lemma 3. Put

$$P(w) = \sum_{j=0}^{p} a_j w^j \quad (a_j \in M,\ a_p \neq 0).$$

Then,

(i) $$m(r,P(w)) \leq pm(r,w) + \sum_{j=0}^{p} m(r,a_j) + O(1),$$

(ii) $$T(r,P(w)) \leq pT(r,w) + \sum_{j=o}^{p} T(r,a_j) + O(1) \quad \text{(cf. [3])}.$$

3. Main Result

We use the same notation as in sections 1 and 2. Let w=w(z) be a non-constant ν-valued algebroid solution of the following D.E.

$$(w')^n = \sum_{j=0}^{m} a_j w^j \quad (1 \leq m \leq n-1,\ a_j \in M,\ a_m \neq 0) \tag{3}$$

and R be its Riemann surface over $|z| < \infty$. We rewrite (3) as follows.

$$(w')^n = a(w+b)^m + \sum_{j=0}^{m-2} b_j w^j, \tag{3'}$$

where $a = a_m$, $b = a_{m-1}/ma_m$ and $b_j = a_j - \binom{m}{j} a_m b^{m-j}$ (j=0, ⋯, m−2). Then, we have the following theorem.

Theorem. (I) When there is at least one j for which $b_j \neq 0$, for a constant K

$$\left(1 - \frac{m}{n}\right)T(r,w) \leq \min\left\{(\nu-1)\frac{m}{n}T(r,w),\ N(r,R)\right\} + K \sum_{j=0}^{m} T(r,a_j) + S(r,w) + \sum_{j=0}^{m} S(r,a_j).$$

(II) When all $b_j = 0$ and b is not constant, for a constant K

$$T(r,w) \leq \frac{n}{m}N(r,R) + K \sum_{j=0}^{m} T(r,a_j) + S(r,w) + \sum_{j=0}^{m} S(r,a_j)$$

and further if $\nu < n/m$

$$T(r,w) \leq K\left\{T(r,a_{m-1}) + T(r,a_m)\right\} + S(r,w) + S(r,a_{m-1}) + S(r,a_m).$$

(III) When all $b_j = 0$ and b is a constant,

$$T(r,w) \leq K(\alpha)\, T(\alpha r, a_m) + O(1)$$

where $\alpha > 1$ is any constant and $K(\alpha)$ is a constant depending only on α.

Proof. First we note that for a constant K

$$N(r,w) \leq K \sum_{j=0}^{m} N(r,a_j), \tag{4}$$

$$N(r,w') \leq K \sum_{j=0}^{m} N(r,a_j). \tag{5}$$

In fact, let s be the order of pole of w at $\tilde{c}$ and s_j that of a_j at c ($c=\pi(\tilde{c})$, $j=0, \cdots, m$).

(a) When $\tilde{c}$ is an ordinary point of R, the order of pole of $(w')^n = (s+1)n$ and the order of the pole of the right-hand side of (3) $\leq sm + \max s_j$.

(b) When $\tilde{c}$ is a branch point of R and w has the expansion near c:

$$w = (z-c)^{-s/\lambda} S((z-c)^{1/\lambda})$$

and then

$$w' = (z-c)^{-(s+\lambda)/\lambda} S_o((z-c)^{1/\lambda})$$

where $2 \leq \lambda \leq \nu$, $s \geq 1$ and $S(t)$, $S_o(t)$ are regular power series for which $S(0) \neq 0$, $S_o(0) \neq 0$, the order of pole of $(w')^n = (s+\lambda)n$ and the order of pole of the right-hand side of (3) $\leq sm + \lambda \max s_j$. From (a) and (b), we have

$$s \leq \frac{\nu}{n-m}(\max s_j - n) \leq \frac{\nu}{n-m} \sum_{j=0}^{m} s_j.$$

This inequality shows (4). Further, from (3)

$$nN(r,w') \leq mN(r,w) + \sum_{j=0}^{m} N(r,a_j)$$

and combining this inequality with (4) we have (5).

Now, we proceed to the proof.

(I) Put $f = -a(w+b)^m$, $g = (w')^n$ and $h=\sum_{j=0}^{k} b_j w^j$

where $k=\max\{j;b_j\neq 0\}$ $(0\leq k\leq m-2)$.

Then, from (3′)

$$f + g = h.$$

Case (a): h=0. As in the case of meromorphic solutions ([8]), we obtain

$$T(r,w)\leq \sum_{j=0}^{k} T(r,b_j) + O(1) \leq K \sum_{j=0}^{m} T(r,a_j) + O(1).$$

Case (b): $h\neq 0$ and f, g are linearly dependent over C. As in the case of meromorphic solutions ([8]), we obtain

$$T(r,w,) \leq K \sum_{j=0}^{m} T(r,a_j) + O(1).$$

Case (c): f and g are linearly independent over C. By lemma 2,

$$T(r,f) \leq m\ (r,h) + N(r,f) + N(r,D) + S(r,f) + S(r,g). \tag{6}$$

We estimate each term of (6).

$$mT(r,w)-T(r,a)-mT(r,b) + O(1) \leq T(r,f), \tag{7}$$

$$m(r,h)\leq km(r,w)+\sum_{j=0}^{k} m(r,b_j) + O(1) \quad \text{(by Lemma 3)}, \tag{8}$$

$$N(r,f)\leq N(r,a)+mN(r,w)+mN(r,b) \leq K\Big\{ \sum_{j=0}^{m} N(r,a_j)+N(r,0,a)\Big\} \quad \text{(by (4))}, \tag{9}$$

$$N(r,D)\leq N(r,0,w+b)+\bar{N}(r,0,w')+\min\Big\{(\nu-1)\bar{N}(r,0,w'),N(r,R)\Big\}$$
$$+ K\sum_{j=0}^{m} \Big\{N(r,0,a_j) + N(r,a_j)\Big\}. \tag{10}$$

As

$$N(r,D)\leq N(r,f'/f) + N(r,g'/g),$$

to prove (10) we estimate $N(r,f'/f)$ and $N(r,g'/g)$. First, as $f'/f = a'/a + m(w'+b')/(w+b)$,

$$N(r,f'/f) \leq N(r,a'/a) + N(r,(w'+b')/(w+b)),$$

$$N(r,a'/a) \quad \leq \overline{N}(r,0,a)+\overline{N}(r,a),$$

$$N(r,\ w'+b')/(w+b)) \leq N(r,0,w+b) + N(r,w+b)+N(r,w'+b') \tag{11}$$

by (4) and (5)

$$\leq N(r,0,w+b) + K \sum_{j=0}^{m} \{N(r,a_j) + N(r,0,a_j)\}.$$

Therefore we have

$$N(r,f'/f) \leq N(r,0,w+b) + K \sum_{j=0}^{m} \{N(r,a_j) + N(r,0,a_j)\}. \tag{12}$$

Next, as

$$g'/g = nw''/\ w',$$

$$N(r,g'/g) = N(r,w''/w'). \tag{13}$$

Let $\tilde{c}$ be a pole of w''/w' of order μ.

Case (i): $\tilde{c}$ is a pole of w'. Let p be its order. When $\tilde{c}$ is an ordinary point, μ=1.

When $\tilde{c}$ is a branch point and w' has the expansion near $c = \pi(\tilde{c})$:

$$w' = (z-c)^{-p/\lambda} S((z-c)^{1/\lambda})$$

where $2 \leq \lambda \leq \nu$ and S(t) is a regular power series for which $S(0) \neq 0$, $\mu = \lambda \leq \nu p$.

Case (ii): $\tilde{c}$ is a zero of w'. Let s be it order. When $\tilde{c}$ is an ordinary point, μ =1.

When $\tilde{c}$ is a branch point and w' has the expansion near $c = \pi(\tilde{c})$:

$$w' = (z-c)^{s/\lambda} S((z-c)^{1/\lambda})$$

where $2 \leq \lambda \leq \nu$ and S(t) is a regular power series for which $S(0) \neq 0$, $\mu = \lambda = 1+(\lambda-1)$.

Case (iii): $\tilde{c}$ is neither a pole nor a zero of w'. As

$$w'' = \Big(\sum_{j=0}^{m} a_j w^j\Big)'/n(w')^{n-1},$$

at least one of a_j has a pole at $c(=\pi(\tilde{c}))$ in this case and

$$\mu \leq \nu \max_j s_j$$

where s_j is the order of pole of a_j at c.

From these cases, we obtain

$$N(r,w''/w')\leq\overline{N}(r,0,w') + \min\{(\nu-1)\overline{N}(r,0,w'),N(r,R)] + N(r,w')+K\sum_{j=0}^{m} N(r,a_j). \quad (14)$$

Combining (5), (11), (12), (13) and (14), we have (10).

$$S(r,f)=S(r,w) + \sum_{j=0}^{m} S(r, a_j), \quad (15)$$

$$S(r,g)=S(r,w')=S(r,w)+ \sum_{j=0}^{m} S(r, a_j), \quad (16)$$

$$N(r,0,w+b)\leq T(r,w+b)+O(1)\leq T(r,w)+T(r.b)+O(1)$$

$$\leq T(r,w)+T(r,a_{m-1})+T(r,a_m)+O(1), \quad (17)$$

$$\overline{N}(r,0,w')\leq T(r,w')+ O(1)\leq\frac{m}{n}T(r,w)+ \sum_{j=0}^{m} T(r,a_j)+O(1) \quad \text{(by Lemma 3).} \quad (18)$$

From (6)-(10) and (15)-(18), we obtain the result.

(II) From

$$(w')^n = a(w+b)^m \quad (a\neq 0,\ b'\neq 0), \quad (19)$$

we have as in the case of meromorphic case ([7])

$$(w')^{n-m}(nw'' -\frac{a'}{a}w')^m = m^m a(w'+b')^m. \quad (20)$$

We want to prove

$$N(r,0,w')\leq N(r,R)+K\{N(r,0,a)+N(r,a)+N(r,0,b')+N(r,b')\} \quad (21)$$

and if $\nu<n/m$

$$N(r,0,w')\leq K\{N(r,0,a)+N(r,0,b')\}. \quad (22)$$

When $w'+b'=0$, this is trivial, so that we may suppose $w'+b'\neq 0$. Let $\tilde{c}$ be a zero of w', a, b' of order $\mu(>0)$, s, t respectively. (If c is a pole of a or b', s or t is negative.)

Case (i): $\tilde{c}$ is an ordinary point. As in the case of meromorphic case ([7]), using

$t^+= \max(0,t)$,

$$\mu \leq t^+ + \{s^+ - (s-1)^+\} + s^+/n.$$

Case (ii): $\tilde{c}$ is a branch point and w' has the expansion near c:

$$w' = (z-c)^{\mu/\lambda} S((z-c)^{1/\lambda}) \quad (c=\pi(\tilde{c}))$$

where $2 \leq \lambda \leq \nu$ and S(t) is a regular power series for which $S(0) \neq 0$. In this case,

$$w'' = (z\text{-}c)^{(\mu-\lambda)/\lambda} S_o((z-c)^{1/\lambda})$$

where $S_o(t)$ is a regular power series for which $S_o(0) \neq 0$.

(a) When $\mu \geq \lambda$, the order of zero of the left-hand side of (20) $\geq$ $\mu(n-m)+m(\mu-\lambda)=\mu n-\lambda m>0$. Then, $\mu \leq \lambda\, t^+$ or if $\mu > \lambda t^+$, $\mu n\text{-}\lambda m \leq s^+ + m\lambda t^+$. This means that $s^+ > 0$ or $t^+ > 0$ and $\mu \leq \lambda(t^+ + 1) + s^+/n$.

(b) When $1 \leq \mu \leq \lambda - 1$ and $\mu n - \lambda m > 0$, as (a) we have $s^+ > 0$ or $t^+ > 0$ and $\mu \leq \lambda(t^+ + 1) + s^+/n$.

(c) When $1 \leq \mu \leq \lambda - 1$ and $\mu n - \lambda m < 0$, a or b' must have a pole at c and $\mu \leq m\nu/n$.

(d) When $1 \leq \mu \leq \lambda - 1$ and $\mu n - \lambda m = 0$, if $\nu < n/m$ this case does not occur and if $\nu \geq n/m$, $\mu \leq \lambda - 1$.

From these cases, we have (21) and (22). Next, as in [7] we have

$$m(r,1/w') \leq K\{T(r,a)+T(r,b')\}+S(r,w')+S(r,a). \tag{23}$$

Combining (21), (22) and (23), we obtain (II).

(III) We can prove (III) of this theorem as in the case (IV) of Theorem 3 in [8].

<u>Corollary</u>. When the coefficients a_j of the D.E. (3) are all rational,

(i) if $\nu < n/m$, w is algebraic;

(ii) if $\xi = 0$, w is rational.

Added in proof. After we had sent the manuscript to the editors, we found that J. von Rieth ([11]) gave a result containing an answer to Problem I. The first author also proved a theorem containing an answer to Problem II in [9] applying a method used in [11].

References

[1] Bank, S.B., and R. P. Kaufman, On the growth of meromorphic solutions of the differential equation $(y')^m = R(x,y)$, Acta Math., 144, pp. 223-248 (1980).

[2] He, Y., and X. Xiao, Admissible solutions of ordinary differential equations, Contemporary Math., 25, pp. 51-61 (1983).

[3] Gackstatter, F., and I. Laine, Zur Theorie der gewöhnlichen Differentialgleichungen im Komplexen, Ann. Polon. Math., 38, pp. 259-287 (1980).

[4] Gol'dberg, A. A., and I. V. Ostrowskii, The distribution of values of meromorphic functions, Nauka, Moskow 1970 (in Russian).

[5] Nevanlinna, R., Le théorème de Picard-Borel et la théorie des fonctions méromorphes, Gauthier- Villars, Paris 1929.

[6] Selberg, H., Über die Wertverteilung der algebroiden Funktionen, Math. Z., 31, pp. 709-728 (1930).

[7] Toda, N., On the conjecture of Gackstatter and Laine concerning the differential equation $(w')^n = \sum_{j=0}^{m} a_j(z)w^j$, Kodai Math. J., 6, pp. 238-249 (1983).

[8] Toda, N., On the growth of non-admissible solutions of the differential equation $(w')^n = \Sigma_{j=0}^{m} a_j w^j$, Kodai Math. J., 7, pp. 293-303 (1984).

[9] Toda, N., On algebroid solutions of some binomial differential equations in the complex domain, Proc. Japan Acad., 64A, pp. 61-64 (1988).

[10] Toda, N., and M. Kato, On some algebraic differential equations with admissible algebroid solutions, Proc. Japan Acad., 61, Ser. A, No. 10, pp. 325-328 (1985).

[11] Rieth, J. von , Untersuchunggen gewisser Klassen gewöhnlichen differentialgleichungen erster und zweiter Ordnung im Komplexen, Doctoral Dissertation, Technische Hochschule, Aachen (1986).

[12] Ullich, E., Über den Einfluss der Verzweigtheit einer Algebroide auf ihre Wertverteilung, J. reine. und angew. Math., 167, pp. 198-220 (1931).

[13] Valiron, G., Sur la dérivée des fonctions algébróides, Bull. Soc. Math. France, 59, pp. 17-39 (1931).

[14] Yosida, K., On algebroid-solutions of ordinary differential equations, Japan. J. Math., 10, pp. 199-208(1934).

Nobushige Toda
Department of Mathematics
Nagoya Institute of Technology

Masakimi Kato
Department of Mathematics
Faculty of Liberal Arts
Shizuoka University

SOME ADVANCES IN THE THEORY OF FACTORIZATION OF ENTIRE OR MEROMORPHIC FUNCTIONS

Hironobu Urabe

Introduction and preliminaries.

The existence of entire solution g satisfying: $g(g(z)) = F(z)$ (for certain, given, entire function F(z)) has been extensively investigated by Baker [1] in 1958.

Subsequently, in 1968, F. Gross [4] and M. Ozawa [11] independently proved that a certain entire function F(z) does not have any factorization: $F(z) = f(g(z))$ with transcendental entire factors f and g. Since then, there have appeared a number of results (concerning primeness, uniqueness of the factorization etc.) obtained mainly by applying Nevanlinna theory. About these, one can consult Yang [28] (as well as Gross [6]).

In this note, mainly, several kinds of results (newly obtained by the author) will be summarized and some open questions are posed.

Firstly, we recall some definitions: a). primeness (etc.), b). unique factorizability (etc.), c). primeness in divisor sense (etc.), as follows.

a). Following Gross [4], we define a meromorphic function F(z) to have a factorization (under composition) with left-factor f and right-factor g, provided

$$F(z) = f(g(z)) = f{\circ}g(z), \quad (*)$$

where f and g are meromorphic. Note that when f is transcendental then g should be entire, since otherwise F(z) would have an essential singularity. F is called prime (pseudo-prime; left-prime; right-prime) if every factorization of the form (∗) implies that either f is bilinear or g is linear (f is rational or g is a polynomial; f is bilinear whenever g is transcendental; g is linear whenever f is transcendental, respectively).

Typically the function z + exp(z) is prime. Further, the function z + h (exp[exp(z)]) (h: entire) is shown to be prime, in view of Theorem 1 in [23].

b). Now suppose that a non-constant entire function F(z) has two factorizations:

$$F(z) = f_1 \circ \cdots \circ f_m(z) = g_1 \circ \cdots \circ g_n(z)$$

into non-linear entire factors. If m=n and if, with suitable linear polynomials $T_j(z)$ (j=1, ···, m−1), the identical relations:

$$f_1(z) = g_1 \circ T_1^{-1}(z), \quad f_2(z) = T_1 \circ g_2 \circ T_2^{-1}(z),$$

$$\cdots, \quad f_m(z) = T_{m-1} \circ g_m(z)$$

hold, then we say that the above two factorizations are equivalent (in entire sense). If any two factorizations of F into non-linear, prime, entire factors are equivalent, then F is called uniquely factorizable (cf. [21]). Of course, prime entire functions should be considered as uniquely factorizable. However, if it is impossible to (maximally) factorize F into non-linear, prime, entire factors (such as exp[z], cos z, cf. [17]), F should not be considered uniquely factorizable at all.

c). Further, in contrast to the primarity, we shall define the primeness in divisor sense for entire functions as follows; an entire function F(z) ($\not\equiv$ const.), with zeros, is called prime in divisor sense (pseudo-prime in divisor sense; left-prime in divisor sense; right-prime in divisor sense), if, for any identical relation of the form:

$$F(z) = f(g(z)) \cdot \exp[A(z)],$$

where f, g ($\not\equiv$ const.) and A are entire functions, one can deduce the following assertion: f has just one simple zero or g is a linear polynomial (f has only a finite number of zeros or g is a polynomial; f has just one simple zero whenever g is transcendental; g is a linear polynomial whenever f has an infinite number of zeros, respectively).

Note that the function z + exp[z] is also prime in divisor sense. However, we do not know whether the function $z + \exp(\exp[z])$ is prime in divisor sense or not.

1. Unique factorizability.

Early, the author proved the following result in [21].

Theorem A. Let $F(z) = (z + h(\exp[z]))\circ(z + Q(\exp[z]))$, where h is an entire function with the order $\rho(h(\exp[z])) < \infty$ and Q is a polynomial. Then F is uniquely factorizable.

Hence, in particular, the function $(z + \exp[z])\circ(z + \exp[z])$ is uniquely factorizable. But, we were unable to prove whether the function $(z + e_k(z)) \circ (z + e_m(z))$, with $e_k(z) = \exp[e_{k-1}(z)]$ $(k \geq 1)$ and $e_0(z) \equiv z$, is uniquely factorizable or not, provided that one of the natural numbers k and m is greater than one.

Relating these questions, we have been able to prove the following results ([26]) to be stated below. Here, Wittich's theorem ([27]) and a strong and beautiful theorem due to N. Steinmetz [19] are essential to our proofs.

Theorem 1.1. Let P, Q and S be non-constant polynomials. Then the composite function

$$(z + P(\exp[Q(\exp[z])]))\circ(z + S(\exp[z]))$$

is uniquely factorizable.

Theorem 1.2. Let P, Q and S be non-constant polynomials. Then the composite function

$$(z + S(\exp[z]))\circ(z + P(\exp[Q(\exp[z])]))$$

is uniquely factorizable, provided that either $\deg Q \neq \deg S$ or $P(\exp[Q(0)]) \neq P(0)$.

We don't know whether the latter condition in Theorem 1.2. is necessary or not for the validity of the theorem.

Theorem 1.3. Let P, Q and S be non-constant polynomials. Then $(z + P(\exp[z]))\circ(z + Q(\exp[z]))\circ(z + S(\exp[z]))$ is uniquely factorizable.

Theorem 1.4. Let P be a polynomial and h be an entire function with the order $\rho(h)<1$. Then the composite function $(z + P(\exp[z]))\circ(z + h(\exp[z]))$ is uniquely factorizable.

We don't know whether the function $(z + e_2(z)) \circ (z + e_2(z))$ is uniquely factorizable or not.

However, using a recent result obtained by Zheng Jianhua [30], we can prove the following fact: Let $H_j(z) = h_j(\exp[2\pi iz/\tau])$, where h_j are non-constant entire functions and τ is a non-zero constant, and let $\lambda\tau/\pi i$ is non-real or real but irrational. Set

$$F(z) = (H_1(z) + \exp[\lambda z + H_2(z)]) \circ (z + H_3(z)).$$

And assume that h_1 is a polynomial (hence H_1 is of exponential type) and that the order $\rho(h_3) < 1$. Then F is uniquely factorizable.

Here, in contrast to the above results, we would like to quote the following fact shown by G. Song and J. Huang [18].

Theorem B. Let $F(z) = \sin z \cdot \exp[\cos z]$. Then F is prime. But, $[F(z)]^2 = \{(1-z^2)\exp[z]\} \circ (\cos z)$ is not pseudo-prime.

Now, in this connection, we posed a question in [26]: Does there exist transcendental entire functions f and g such that f(g) is uniquely factorizable, but g(f) is not? We have answered this question affirmatively by proving the following result in [26].

Theorem 1.5. Let $F(z) = f(g(z))$, where $f(z) = z \cdot \exp[z^2/2]$ and $g(z) = z \cdot \sin z = (\sqrt{z} \cdot \sin \sqrt{z}) \circ (z^2)$. Then F is uniquely factorizable.

$$\text{Note } G(z) = g(f(z)) = (\sqrt{z} \cdot \sin \sqrt{z}) \circ (z^2) \circ (z \cdot \exp[z^2/2])$$
$$= (\sqrt{z} \cdot \sin \sqrt{z}) \circ (z \cdot \exp[z]) \circ (z^2).$$

Hence G(z) is not uniquely factorizable.

It would be desirable to answer the following fundamental question in [26]: If f and g are prime, transcendental, entire functions, does it follow that f(g) (and also g(f)) must be uniquely factorizable? But this seems to be a very difficult question to answer.

2. Iterative equation: $F = g(g)$.

2.1 . Uniqueness of the solution. In an earlier paper [23], we showed the following result.

Theorem C. Let f be an entire function of the form: $f(z) = z + p(\exp[z])$, where p is a non-constant entire function such that $\exp[k \cdot p(0)] \neq -1$ for any natural number k. Suppose that g is entire and satisfies the identity $g(g(z)) = f(f(z))$. Then we have necessarily that g and f are identical.

In this regard, we can prove the following facts [25].

Theorem 2.1. Let $f(z) = Q(\exp[z])$, where Q is a non-constant polynomial. Suppose that g is entire and satisfies the identity $g(g(z)) = f(f(z))$. Then $g(z) \equiv f(z)$.

Theorem 2.2. Let $f'(z) = P(z)\cdot\exp[Q(z)]$, where P ($\not\equiv 0$) and Q ($\not\equiv$ const.) are polynomials. Suppose that g is entire and satisfies the identity $g(g(z)) = f(f(z))$. Then we have that $g(z) = cf(z) + d$ for some constants c and d.

2.2 Non-existence of solutions. We have proved the following results ([25]).

Theorem 2.3. Let h be a non-constant entire function of finite order and H be also a non-constant, periodic, entire function with period b ($\neq 0$), that is $H(z + b) = H(z)$. Set $F(z) = h(z)\cdot\exp[H(z)]$. Then there exists no entire function g(z) which satisfies the identity $g(g(z)) = F(z)$.

Corollary. Let F be a non-constant, periodic, entire function with zeros such that $N(r, 0, F)$ has finite order, in other words, the exponent of convergence of the zeros of F is finite. Then there exists no entire function g(z) which satisfies the identical relation $g(g(z)) = F(z)$.

This is a generalization of a theorem due to I. N. Baker.

Theorem 2.4. Let P be a non-linear polynomial and h be a non-constant entire function of finite order. Set $F(z) = P(z) + h(\exp[z])$. Then there exists no entire function $g(z)$ which satisfies the identity $g(g(z)) = F(z)$.

2.3. Here, we want to study the condition that assures the existence of an entire solution g satisfying the equation $g(g(z)) = F(z)$ for certain specific entire functions $F(z)$. For example, we have the following result ([25]).

Theorem 2.5 Let h be a non-constant entire function and

$$F(z) = \int_0^z \exp[h(\exp[z])]\cdot dz.$$

Then there exists an entire function $g(z)$ which satisfies the identity $g(g(z)) = F(z)$ if and only if h can be expressed as

$$h(z) = L(z) + L(z^q\cdot\exp[k(z)]),$$

where

$$k(z) = \int_0^z (\exp[L(z)] - \exp[L(0)])/z \cdot dz + c$$

with a constant c and a non-constant entire function $L(z)$ satisfying $q=\exp[L(0)] \in \mathbb{Q}$ (the set of rational numbers).

In particular, if h is of hyper-order less than one, then no entire function $g(z)$ can satisfy the identity $g(g(z)) = F(z)$ for the above $F(z)$ as defined in Theorem 2.5.

Remark. If all the numbers $e_k = e_{k+1}(0)$ are irrational, then we can show that when $h(z) = e_k(z)$ there exists no entire function $g(z)$ which satisfies the identical relation $g(g(z)) = F(z)$ with $F(z) = \int_0^z e_{k+2}(z)\, dz$ ($k \in \mathbb{N}$, the set of natural numbers).

3. Primeness in divisor sense.

Using basic theorems in [3], [8] and [29], we have proved the following facts ([22]).

Theorem D. Let P be a polynomial. Then the function $z + P(\exp[z])$ is prime in divisor sense.

Theorem E. Let P and Q be non-constant polynomials. Then the function $F(z) = P(z) + Q(\exp[z])$ is right-prime in divisor sense. Further if, for any natural number k and constant c, the function $\exp[-kz]\cdot\{Q[\exp(z)] + c\}$ is non-constant, then F is prime in divisor sense.

Theorem F. Let P and Q be two non-constant polynomials. Assume that $Q(0) \neq 0$ and $Q(z)$ has only simple zeros. Then $F(z) = P(z)\cdot Q(\exp[z])$ is left-prime in divisor sense. Further if $\{P(z + c)\}^2$ is not even function for any constant c, then F is prime in divisor sense.

Theorem G. For a sequence of point-sets $\{A(n)\colon n \in \mathbb{N}\}$ with $A(n) = \{a_{nj}\}_{j=1}^{m_n}$ $(1 \leq m_n \leq \infty)$ and a sequence of mutually distinct prime numbers $\{p_n\}_{n=1}^{\infty}$ with $p_n \geq 3$, let $F(z)$ be an entire function whose zero-set be $\bigcup_{n=1}^{\infty} A_n$ (including multiplicities). We assume the following three conditions; (i) the order of the zero-point a_{nj} of F is all equal to p_n for $j = 1, \cdots, m_n$. (ii) there exists a half plane H such that $A_n \subset H$ $(n \geq 1)$. (iii) the order of $N(r, A_n)$ is less than one $(n \geq 1)$, while the order of $N\left(r, \bigcup_{n=1}^{\infty} A_n\right)$ is not less than one.

Then F is left-prime in divisor sense.

Remark. If we put a certain additional condition on A_n $(n \geq 1)$, the $F(z)$ in Theorem G becomes to be prime in divisor sense.

Relating these, we have proved the following results ([24]).

Theorem 3.1. Let Π_j $(j = 1, 2)$ be entire functions defined by infinite products such as

$$\Pi_1(w) = \prod_{n=1}^{\infty} (1 - w/a_n)\,, \qquad \Pi_2(w) = \prod_{n=1}^{\infty} (1 - w/b_n),$$

where $\{a_n\}_{n=1}^{\infty} \cap \{b_n\}_{n=1}^{\infty} = \emptyset$, $1 < |a_n| < |a_{n+1}|$, $1 < |b_n| < |b_{n+1}|$ $(n \geq 1)$ and $\{p_n\}_{n=1}^{\infty}$ and $\{q_n\}_{n=1}^{\infty}$ are two sequences of mutually distinct prime numbers with p_n

≥ 3 and $q_n \geq 3$ $(n \geq 1)$. Set

$$F(z) = \Pi_1(\exp[z])\cdot\Pi_2(\exp[-z])$$

and assume that, concerning the order $\rho(F)$ of F, $\rho(F)$ is finite and not integer (hence, $1 < \rho(F) < \infty$). Then F is prime in divisor sense.

To prove this result, a beautiful theorem due to Kobayashi [9] is essential (together with a basic theorem of Nevanlinna).

Concerning the primeness in divisor sense for functions with comparatively small growth, we have the following.

Theorem 3.2. Let F(z) be an entire function of order less than one which has at least two zero points with co-prime multiplicities such that all the zeros of F are contained in a half plane. Suppose that F is prime. Then F is prime in divisor sense.

Example. Let

$$F_1(z) = \prod_{n=1}^{\infty} (1 - z/n!),\ F_2(z) = \prod_{n=1}^{\infty} (1 - z/n^s)$$

with $s > 1$. Then F_j $(j = 1, 2)$ are prime (as shown by Ozawa [12]), and hence, in view of Theorem 3.2., they are prime in divisor sense.. Note that, when $s = 2$, $F_2(z) = \sin(\pi\sqrt{z}) / \pi\sqrt{z}$ is seen to be prime by virtue of Theorem 4.4. below.

Remark. It is an open question whether or not the functions $z + e_k(z)$ $(k \geq 2)$ are prime in divisor sense.

4. Prime meromorphic functions.

We have proved the following result ([23]).

Theorem H. Let

$$F(z) = [P(z) + Q(\exp[z])] / [S(z) + R(z)h(\exp[z])],$$

where P ($\not\equiv$ const.), Q($\not\equiv$ const.), R($\not\equiv 0$) and S are polynomials such that, for any $k \in \mathbb{N}$

and $c \in C$, the function $\exp[-kz] \cdot \{Q(\exp[z]) + c\}$ is non-constant, and h is an entire function such that the order of $h(\exp[z])$ is finite and not an integer and the lower order of it is larger than one. Assume that the numerator and denominator have no common zeros. Then F is prime.

In connection with the primeness in divisor sense and by applying the property concerning it, we have established the following result ([24]).

Theorem 4.1. Let G(z) be an entire function of finite non-integral order which is left-prime in divisor sense and let H be an entire function with at least one simple zero or two zeros with co-prime multiplicities such that $\rho(H) < \rho(G)$. Set $F(z) = H(z)/G(z)$. Then, assuming that G and H have no common zeros, F(z) is left-prime.

Corollary. Taking G(z) as F in Theorem 3.1. (i.e. $G(z) = \Pi_1(\exp[z]) \cdot \Pi_2(\exp[-z])$ for certain Π_j) and H(z) as $H(z) = Q(\exp[z])$, where Q is a polynomial with at least one simple zero or two zeros with co-prime multiplicities such that $Q(0) \neq 0$. Consider a periodic (non-entire) meromorphic function F defined by $F(z) = H(z)/G(z)$. Then, assuming that G and H have no-common zeros, F must be prime.

Remark. The above fact shows that there exists a prime function which is simply periodic and (no-entire) meromorphic. (By the way, the proof of the existence of such a function given by Ozawa in [14] seems to be incomplete.)

Concerning the factorization for meromorphic functions with only a finite number of poles, we have the following result.

Theorem 4.2. Let P, Q and R be polynomials such that P and Q are non-constant, deg R < deg P and that P and Q have only simple zeros with $Q(0) \neq 0$. Here we further assume that if R is constant then R must be equal to zero. Set $F(z) = [Q(\exp[z]) + R(z)]/P(z)$. Then, assuming that the numerator and denominator have no common zeros, F is prime.

Remark. Hence, if P_j (j = 1, 2) and Q are polynomials such that deg P_1 < deg P_2 and $P_2(z) - P_1(z)$ and Q(z) have only simple zeros, then the function $[P_1(z) + Q(\exp[z])]/[P_2(z) + Q(\exp[z])]$ is prime, provided that the numerator and denominator have no common zeros.

In addition, we have proved the following results ([24]).

Theorem 4.3. Assume that P is a non-constant polynomial, then $P(z)\cdot\cos(\sqrt{z})$ and $\cos(\sqrt{z})/P(z)$ are prime.

Theorem 4.4. Assume that P is a polynomial ($\not\equiv 0$), then $P(z)\cdot\sin(\sqrt{z})/\sqrt{z}$ and $\sqrt{z}\cdot\sin(\sqrt{z})/P(z)$ are prime.

Hence, in particular, the answers to some open questions of Gross ([5]) are affirmative (they are the cases where $P(z) = z^m$ for some natural number m).

Finally, we pose an open question: In general, can we show that, if f(z) is a prime transcendental entire function of order less than one, the meromorphic function f(z)/P(z) (f and P have no common zeros) is also prime for any non-constant polynomial P(z)?

Remark. In this regard, one can show that the above function is pseudo-prime (at least).

References

[1] Baker, I. N., Zusammensetzungen ganzer Funktionen, Math. Z. 69 (1958), pp. 121-163.

[2] Baker, I. N., and F. Gross, Further results on factorization of entire functions, Proc. Symposia Pure Math. (Amer. Math. Soc. Providence, R.I.) 11 (1968), pp. 30-35.

[3] Goldstein, R., On factorization of certain entire functions, J. London Math. Soc, (2) 2 (1970), pp. 221-224.

[4] Gross, F., On factorization of meromorphic functions, Trans, Amer. Math. Soc. 131 (1968), pp. 215-222.

[5] Gross, F., Prime entire functions, Ibid, 161 (1971), pp. 219-233.

[6] Gross, F., Factorization of meromorphic functions and some open problems, Complex Analysis (Kentucky), Lecture Note, Springer, No. 599 (1976), pp. 51-67.

[7] Hayman, W. K., Meromorphic functions, Clarendon Press, Oxford, 1964.

[8] Kobayashi, T., Distribution of values of entire functions of lower order less than one, Kōdai Math. Sem. Rep, 28 (1976), pp. 33-37.

[9] Kobayashi, T., On a characteristic property of the exponential function, Ibid, 29 (1977), pp. 130-156.

[10] Nevanlinna, R., Le théorème de Picard-Borel et la théorie des fonctions méromorphes, Gauthier Villars, Paris, 1929.

[11] Ozawa, M., On the solution of the functional equation $f \circ g(z) = F(z)$, I, II, III, IV and V, Kodai Math. Sem. Rep. 20 (1968), 159-162, 163-169, 257-263, 272-278 and 305-313, resp.

[12] Ozawa, M., On prime entire functions, I, II, Ibid, 22 (1970), 301-308, 309-312.

[13] Ozawa, M., On certain criterion for left-primeness of entire functions, Ibid, 26 (1975), pp. 304-317.

[14] Ozawa, M., On the existence of prime periodic entire function, Ibid, 29 (1978), pp. 308-321.

[15] Ozawa, M., On uniquely factorizable meromorphic functions, Kodai Math. J. 1 (1978), pp. 339-353.

[16] Pólya, G., On an integral function of an integral function, J. London Math. Soc, 1 (1926), pp. 12-15.

[17] Song, G., Factorization of exponential function and cosine function in composite sense, J. East China Normal Univ. Science Ed. No. 2 (1986), pp. 24-28.

[18] Song, G. D., and J. Huang, On pseudo-primality of th n-th power of prime entire functions, Kodai Math. J. vol. 10 (1987), pp. 42-48.

[19] Steinmetz, N., Über die faktorisierbaren Lösungen gewöhnlicher Differentialgleichungen, Math. Z. 170 (1980), pp. 168-180.

[20] Tichmarsh, E. C., The theory of functions, 2nd. ed. Oxford Univ. Press, 1939.

[21] Urabe, H., Uniqueness of the factorization under composition of certain entire functions, J. Math. Kyoto Univ. 18 (1978), pp. 95-120.

[22] Urabe, H., Primeness in divisor sense for certain entire functions, Ibid, 24 (1984), pp. 127-140.

[23] Urabe, H., On factorization of certain entire and meromorphic functions, Ibid, 26 (1986), pp. 177-190.

[24] Urabe, H., Primeness in divisor sense concerning entire functions and its application to the factorization of meromorphic functions, to appear.

[25] Urabe, H., On entire solutions of the iterative functional equation g(g(z)) = F(z), to appear.

[26] Urabe, H., and C. -C. Yang, On unique factorizability of certain composite entire functions, to appear.

[27] Wittich, H., Neuere Untersuchungen über eindeutige analytische Funktionen, Springer, 1955.

[28] Yang, C. -C., Progress in factorization theory of entire and meromorphic functions, Lecture Note, Pure and applied Math. Vol. 78 ed. by C. -C. Yang, Marcel Dekker Inc., 1982.

[29] Yang, C. -C., and H. Urabe, On permutability of certain entire functions, J. Lond Math. Soc. (2) 14 (1976), pp. 153-159.

[30] Zheng, J., On the factorization of several classes of entire functions, Kodai Math. J. 10 (1987), pp. 328-334.

Department of Mathematics
Kyoto University of Education

POLYNOMIAL DIFFERENCE EQUATIONS WHICH HAVE MEROMORPHIC SOLUTIONS OF FINITE ORDER

Niro Yanagihara

§ 1. Introduction

We consider here the difference equation

$$y(x+1)^m = P(y(x)), \tag{1.1}$$

in which P(y) is a polynomial of y. The equation (1.1) admits meromorphic solutions of finite order only if m = deg[P] ([5]). Further, (1.1) admits entire solutions of finite order if and only if

$$\text{either} \qquad P(y) = (ay + b)^m \tag{1.2}$$

$$\text{or} \qquad m = 2 \text{ and } P(y) = A^2 - y^2, \tag{1.3}$$

where a, b, and A are constants, and A ≠ 0. Obviously, we can take A = 1 ([6]).

What results would we obtain, if we permit solution to have poles? This is our concern in this paper.

The result is as follows:

Theorem 1. The equation (1.1) admits a meromorphic solution of finite order if and only if it is of the form either (1.2) or (1.3).

By the way, meromorphic solutions (of finite order) are determined. That is,

Theorem 2. Meromorphic solutions of the equation (1.1) with the form (1.3) (we take A = 1) are represented as

$$y(x) = \frac{1}{2}\left[\beta(x) + \frac{1}{\beta(x)}\right],$$

where $\beta(x)$ satisfies the difference equation

$$\beta(x+1) = i\,\beta(x) \ \text{or}\ \beta(x+1) = i/\beta(x).$$

§ 2. Proof of Theorem 1

The proof of theorem 1 is contained in the following Propositions 1-8, which will be proved in section below.

Proposition 1. Suppose the equation

$$y(x+1)^m = A(y(x)-c_1)^{i_1}(y(x)-c_2)^{i_2}\cdots(y(x)-c_h)^{i_h}$$

$$i_1, \cdots, i_h \geq 1, i_1 + \cdots + i_h = m,$$

admits a meromorphic solution. Then we have that $h \leq 3$.

Proposition 2. Suppose the equation

$$y(x+1)^m = A(y(x)-c_1)^{i_1}(y(x)-c_2)^{i_2}(y(x)-c_3)^{i_3}$$

$$i_1, i_2, i_3 \geq 1, i_1 + i_2 + i_3 = m, \tag{2.1}$$

admits a meromorphic solution. Then we have that

either (i) $m = 3, i_1 = i_2 = i_3 = 1$,

or (ii) $m = 4, i_1 = 2, i_2 = i_3 = 1$,

or (iii) $m = 6, i_1 = 3, i_2 = 2, i_3 = 1$.

Further, we have that $c_1c_2c_3 \neq 0$, and y(x) takes any of c_1, c_2, c_3.

Proposition 3. Suppose the equation of the type (i):

$$y(x+1)^3 = A\,(y(x)-c_1)\,(y(x)-c_2)\,(y(x)-c_3) \tag{2.2}$$

admits a meromorphic solution, then (2.2) reduces to the form

$$y(x+1)^3 + y(x)^3 = 1. \tag{2.2'}$$

Proposition 4. The equation (2.2′) does not admit any meromorphic solution of finite order.

Proposition 5. The equation of the type (ii):

$$y(x+1)^4 = A(y\,(x)-c_1)^2\,(y(x)-c_2)\,(y(x)-c_3) \tag{2.3}$$

does not admit any meromorphic solution.

Proposition 6. The equation of the type (iii):

$$y(x+1)^6 = A(y(x)-c_1)^3\,(y(x)-c_2)^2\,(y(x)-c_3) \tag{2.4}$$

does not admit any meromorphic solution.

Proposition 7. The equation

$$y(x+1)^m = A(y(x)-c_1)^{i_1}\,(y(x)-c_2)^{i_2},$$

$$m = i_1 + i_2 \geq 3 \tag{2.5}$$

does not admit any meromorphic solution.

Proposition 8. Suppose the equation

$$y(x+1)^2 = A(y(x)-c_1)\,(y(x)-c_2) \tag{2.6}$$

admit a meromorphic solution of finite order, the (2.6) reduces to the form

$$y(x+1)^2 + y(x)^2 = 1. \tag{2.6$'$}$$

For the proof, we make several use of the celebrated Theorem of Nevanlinna on deficiency and ramification [7].

§ 3. Proofs of Propositions 1 and 2

Proof of Proposition 1. At first we note the following fact:

Suppose $y(x)$ takes c_k and $y(x_0) = c_k$. Let ν_k be the order of zero point x_0 for $y(x) - c_k$. Let μ be the order of zero point x_0 for $y(x+1)$. Then

$$m\mu = i_k \nu_k.$$

Let t_k = G.C.D. (m, i_k), and $m_k = m/t_k$, $i_k{}' = i_k/t_k$. Then

$$m_k \mu = i_k{}' \nu_k,$$

hence $m_k | \nu_k$. Obviously $m_k \geq 2$. Therefore c_k is a totally ramified value of order at least m_k, for $y(x)$.

When $c_1 \ldots c_h = 0$. We may assume that $c_h = 0$. Suppose $y(x_0) = 0$. Then by (2.1) $y(x_0 + 1) = 0$. Let s_0 and s_1 be the orders of zero points x_0 and $x_1 = x_0 + 1$, respectively. Then we have

$$ms_1 = i_h s_0, \quad s_1 = (i_h/m)s_0.$$

By (2.1), $y(x_1 + 1) = 0$. Let s_2 be the order of $x_2 = x_1 + 1 = x_0 + 2$. Then

$$ms_2 = i_h s_1, \quad s_2 = (i_h/m)s_1 = (i_h/m)^2 s_0.$$

Repeating this procedure we obtain that

$$0 < s_k = (i_h/m)^k s_0 < 1$$

for sufficiently large k, which is a contradiction. Hence $y(x) \neq 0$. Then, as seen by (2.1), $y(x)$ cannot take any of $c_1, \dots c_{h-1}$, which implies that $h - 1 \leq 1$, $h \leq 2$.

It is to be noted that, as shown above, $c_1 \dots c_h \neq 0$ if $h \geq 3$.

When $c_1 \dots c_h \neq 0$. Suppose $y(x)$ does not take one of c_j, say c_h. If further, $y(x)$ does not take another c_j, say c_{h-1}, then $y(x)$ cannot admit any totally ramified value, which implies that $h \leq 2$. If $y(x)$ takes any value other than c_h, then all of $c_1, \cdots, c_{h-1}$ are totally ramified values, which means that $h - 1 \leq 2$, $h \leq 3$, since $y(x)$ can admit at nost two totally ramified values.

Suppose $y(x)$ takes any of $c_1, \dots, c_h$, which values are all totally ramified for $y(x)$. Hence $h \leq 4$, then c_j are all ramified to the order 2, which implies that the equation must be of the form

$$y(x + 1)^2 = A(y(x) - c_1)(y(x) - c_2)(y(x) - c_3)(y(x) - c_4),$$

which is impossible since degree of left side (= 2) $\neq$ degree of right side (= 4). Hence it cannot occur that $h = 4$. Q.E.D.

Proof of Proposition 2. As shown in the proof of Proposition 1, we have that $c_1c_2c_3 \neq 0$.

If $y(x)$ does not take c_3, then both of c_1 and c_2 are totally ramified at the order 2, by the theorem of Nevanlinna. Hence the equation must be of the form

$$y(x+1)^m = A(y(x) - c_1)^{m/2} (y(x) - c_2)^{m/2} (y(x) - c_3)^s$$

with a positive integer s, which is obviously impossible. Hence y(x) takes any of c_j and is totally ramified at c_j.

Let $(m, i_j) = t_j$ and $m_j = m/t_j$ for j = 1, 2, 3. Each c_j are m_j- fold ramified, hence we must have that

$$m_1^{-1} + m_2^{-1} + m_3^{-1} \geq 1.$$

The following possibilities may be considered:

1. $m_1 = m_2 = 2$, $m_3 =$ any.
2. $m_1 = 2$, $m_2 = 3$, $m_3 = 3$.
3. $m_1 = 2$, $m_2 = 3$, $m_3 = 4$.
4. $m_1 = 2$, $m_2 = 3$, $m_3 = 5$.
5. $m_1 = 2$, $m_2 = 3$, $m_3 = 6$.
6. $m_1 = 2$, $m_2 = 4$, $m_3 = 4$.
7. $m_1 = 3$, $m_2 = 3$, $m_3 = 3$.

Since $i_1 + i_2 + i_3 = m$, we see easily that cases 5, 6, 7 are only possible.

Q.E.D.

§ 4. Proof of Proposition 3

Lemma 4.1 Let f(x) and h(x) be meromorphic functions which satisfies

$$h(x)^2 = 4f(x)^3 - g_3, \tag{4.1}$$

in which g_3 is a non-zero constant.

Then there is an entire function $z = \varphi(x)$ such that

$$f(x) = \wp(\varphi(x)) \text{ and } h(x) = \wp'(\varphi(x)),$$

where $\wp(z)$ is the Weierstrass elliptic function which satisfies

$$\wp'(z)^2 = 4\,\wp(z)^3 - g_3. \tag{4.1'}$$

Proof. It is obvious that f(x) and h(x) admit no Picard exceptional values. Let z_0 be such that $\wp'(z_0) \neq 0$, and let x_0 be such that $f(x_0) = \wp(z_0)$. There is a neighborhood U of x_0 such that

$$\varphi(x) = \wp^{-1}[f(x)]$$

is defined and holomorphic in U. Obviously we have $h(x) = \wp'[\varphi(x)]$ in U. Suppose, at a point x_1, $f(x_1) = \wp(z_1)$ and $\wp'(z_1) = 0$. Then $(\wp')'(z_1) = 6\,\wp(z_1)^2 \neq 0$, hence $(\wp')^{-1}[h(x)]$ is defined in a neighborhood V of x_1. Thus $\varphi(x)$ can be continued analytically to x_1. Suppose $f(x_2) = \infty = \wp(z_2)$. Put $Q(z) = 1/\wp(z)$. By (4.1′),

$$Q'(z)^2 = 4Q(z) - g_3Q(z)^4, \text{ hence } Q''(z) = 2 - 2g_3Q(z)^3.$$

Thus $(Q')'(z_2) = Q''(z_2) = 2 \neq 0$, and

$$z = (Q')^{-1}[-h/f^2]\,(x)$$

is defined and holomorphic in a neighborhood W of x_2. Therefore $\varphi(x)$ can be continued analytically to x_2.

Thus $\varphi(x)$ can be continued everywhere in the complex plane, and is an entire function. Q.E.D.

Now we will prove Proposition 3. We consider the equation (2.2):

$$y(x+1)^3 = A(y(x) - c_1)\,(y(x) - c_2)\,(y(x) - c_3).$$

Put

$$y(x+1)\,/\,[y(x) - c_1] = A_1u(x) \quad \text{with } A_1 = [4A(c_2 - c_1)\,(c_3 - c_1)]^{1/3}$$

$$1\,/\,[y(x) - c_1] = v(x) + b,$$

where

$$b = \frac{1}{2} [(c_2 - c_1)^{-1} + (c_3 - c_1)^{-1}].$$

Then we have

$$v(x)^2 = 4u(x)^3 + b'^2, \text{ with } b' = \frac{1}{2} [(c_2 - c_1)^{-1} - (c_3 - c_1)]^{-1}. \tag{4.2}$$

Thus

$$y(x) = c_1 + [v(x) + b]^{-1}, \tag{4.3}$$

$$y(x + 1) = A_1 u(x) / [v(x) + b]. \tag{4.3'}$$

Therefore

$$A_1 \frac{u(x)}{v(x) + b} - c_1 = \frac{1}{v(x + 1) + b}. \tag{4.4}$$

Since zeros of $v(x) \pm b'$ are threefold ramified, we have that zeros of

$$A_1 \frac{u(x)}{v(x) + b} - c_1 - \frac{1}{b + b'} = \frac{b' - v(x + 1)}{(v(x + 1) + b) (b + b')}$$

hence zeros of

$$A_1 u(x) - c_2 v(x) - bc_2 \ \left(c_2 = c_1 + \frac{1}{b + b'}\right)$$

are threefold ramified.

By Lemma 4.1, we get

$$u(x) = \wp(\varphi(x)) \quad \text{and} \quad v(x) = \wp'(\varphi(x)). \tag{4.5}$$

with an entire function $\varphi(x)$. Therefore zeros of

$$G(x) = F(\varphi(x)) = A_1 \wp[\varphi(x)] - c_2 \wp'[\varphi(x)] - bc_2$$

are all of multiplicity 3. There are infinitely many zeros of

$$F(z) = A_1 \wp(z_0) - c_2 \wp'(z_0) bc_2, \tag{4.6}$$

and $\varphi(x)$ may admit at most two totally ramified values. Hence there is a zero point z_0 of $F(z)$ with multiplicity at least 3. Hence

$$F(z_0) = A_1 \wp'(z_0) - c_2 \wp'(z_0) - bc_2, \tag{4.6}$$

$$F'(z_0) = A_1 \wp' (z_0) - c_2 \wp'' (z_0) = A_1 \wp' (z_0) - 6c_2 \wp(z_0)^2 = 0, \tag{4.6'}$$

$$F''(z_0) = A_1 \wp''(z_0) - 12c_2 \wp(z_0) = 6A_1 \wp(z_0)^2 - 12c_2 \wp(z_0) \wp'(z_0) = 0. \tag{4.6''}$$

By (4.6) and (4.6″),

$$\wp'(z_0) = b, \qquad \wp(z_0) = 2bc_2/A_1.$$

By (4.6′),

$$A_1^{\,3} = 24\, c_2^{\,3}\, b. \tag{4.7}$$

If we consider zeros of

$$A_1 \frac{u(x)}{v(x) + b} - c_1 - \frac{1}{b - b'} = \frac{-(b' + v(x+1))}{(v(x+1) + b)(b - b')},$$

then similarly we obtain

$$A_1^{\,3} = 24\, c_3^{\,3}\, b. \tag{4.7'}$$

Therefore $c_2^{\,3} = c_3^{\,3}$.

Exchanging c_1 to c_2 or c_3, we get $c_1^{\,3} = c_2^{\,3} = c_3^{\,3}$. Hence we can write

$$c_2 = \omega\, c_1, \quad c_3 = \omega^2\, c_1,$$

with

$$\omega = (-1 + \sqrt{3}i) / 2 = e^{2\pi i/3}.$$

By (4.7) we get A = -1. Hence we obtain the following result:

The equation (2.2) must be of the form (taking $c_1 = 1$)

$$y(x+1)^3 = 1 - y(x)^3,$$

supposed it admits a meromorphic solution.

§ 5. Proof of Proposition 4

It is known that the functional equation

$$f(x)^3 + g(x)^3 = 1$$

admits meromorphic (not entire) solution, see e.g., [2], [3], [8]. However, we will show here that the equation (2.2′) cannot possess any meromorphic solution of finite order.

Putting $y(x+1) / [y(x)-1] = [-12]^{1/3} u(x)$ and $1/[y(x)-1] = v(x) - \frac{1}{2}$, we obtain from (2.2′)

$$v(x)^2 = 4u(x)^3 - \frac{1}{12}, \tag{5.1}$$

$$\frac{[-12]^{1/3}\ u(x)}{v(x)-(1/2)} = 1 + \frac{1}{v(x+1)-(1/2)} \tag{5.1'}$$

By Lemma 4.1, we have as in (4.5)

$$u(x) = \wp[\varphi(x)] \quad \text{and} \quad v(x) = \wp'[\varphi(x)], \tag{5.1''}$$

where $\wp[z]$ is the elliptic function (4.1′) with $g_3 = 1/12$. If $y(x)$ is of finite order, then $u(x)$ is also of finite order, therefore $\varphi(x)$ must be a polynomial, see [1] and [4, Theorem of Pólya].

Suppose x_0 is a pole of $v(x+1)$. Then by (5.1′)

$$[-12]^{1/3}\ u(x_0) - v(x_0) + \frac{1}{2} = 0.$$

By (5.1), we get

$$v(x_0) = -\frac{1}{2}, \quad u(x_0) = 1/\ [12]^{1/3}.$$

Further, let x_1 be such that $x_0 = x_1 + 1$, i.e.,

$$v(x_1+1) = -1/2, \quad u(x_1+1) = 1/\ [12]^{1/3}.$$

By (5.1′)

$$v(x+1) + \frac{1}{2} = \frac{[-12]^{1/3}\ u(x)}{[-12]^{1/3}\ u(x) - v(x) + 1/2} = 0 \quad \text{for } x = x_1$$

Thus either $u(x_1) = 0$ or $v(x_1) = \infty$.

Suppose $v(x_1) = \infty$. Put $z_0 = \varphi(x_1+1)$ and $z_1 = \varphi(x_1)$, where $\varphi(x)$ is a polynomial in (5.1″). Then

$$v(x) \;=\; \wp'(\varphi(x_1)) = -2/[\varphi(x) - \varphi(x_1)]^3 + \ldots$$

$$= (-2/\varphi'(x_1)^3)\ (x - x_1)^{-3} + \dots$$

$$u(x) = \wp(\varphi(x)) = 1/[\varphi(x) - \varphi(x_1)]^2 + \dots$$

$$= (1/\varphi'(x_1)^2)\ (x - x_1)^{-2} + \dots,$$

hence

$$\frac{[-12]^{1/3}\ u(x)}{[-12]^{1/3}\ u(x) - v(x) + (1/2)} =$$

$$= ([-12]^{1/3}/2)\ \varphi'(x_1)\ (x - x_1) + \dots \tag{5.2}$$

On the other hand,

$$\begin{aligned} v(x+1) + \tfrac{1}{2} &= \wp'(\varphi(x+1)) + \tfrac{1}{2} \\ &= \wp''(\varphi(x_1+1))\ \varphi'(x_1+1)\ (x - x_1) + \dots \\ &= 6\wp^2(\varphi(x_1+1))\ \varphi'(x_1+1)\ (x - x_1) + \dots \\ &= 6u\ (x_1+1)^2\ \varphi'(x_1+1)\ (x - x_1) + \dots \\ &= 6\ (3\sqrt{12}^{-2})\ \varphi'(x_1+1)\ (x - x_1) + \dots \end{aligned} \tag{5.2'}$$

Thus by (5.2) and (5.2′)

$$([-12]^{1/3}/2)\ \varphi'(x_1) = (6/([12]^{1/3})^2)\ \varphi'(x_1+1),$$

i.e.,

$$\varphi'(x_1 + 1)\ /\ \varphi'(x_1) = -1.$$

This is impossible, since $\varphi(x)$ is a polynomial and hence $\varphi'(x_1+1)\ /\ \varphi'(x_1)$ is nearly equal to 1 if x_1 is sufficiently large.

Suppose $u(x_1) = 0$. Then by (3.8), $v(x_1)^2 = -1/12$.

If $v(x_1) = i/\sqrt{12}$, then

$$[-12]^{1/3}\ u(x_1) - v(x_1) + (1/2) = -(i/\sqrt{12}) + (1/2).$$

Further

$$\begin{aligned} u(x) = \wp(\varphi(x)) &= \wp'(\varphi(x_1)\,(x - x_1) + \ldots \\ &= v(x_1)\, \varphi'(x_1)\,(x - x_1) + \ldots \\ &= (i/\sqrt{12})\, \varphi'(x_1)\,(x - x_1) + \ldots \end{aligned}$$

Thus

$$\frac{[-12]^{1/3}\; u(x)}{[-12]^{1/3} u(x) - v(x) + 1/2} =$$

$$= \frac{[-12]^{1/3}\; (i/\sqrt{12})}{-i/\sqrt{12} + 1/2}\, \varphi'(x_1)\,(x - x_1) + \cdots$$

Therefore by (5.2′)

$$6/\,[12]^{2/3}\; \varphi'(x_1 + 1) = \frac{[-12]^{1/3}\; (i/\sqrt{12})}{-i/\sqrt{12} + 1/2}\, \varphi'(x_1),$$

which is impossible since $\varphi'(x - x_1)\,/\,\varphi'(x_1)$ is nearly equal to 1 if x_1 is sufficiently large.

Similarly, the case $v(x_1) = -i/\sqrt{12}$ is also impossible.

Therefore we know that the equation (2.2) cannot possess any finite order meromorphic solution.

§ 6. Proof of Proposition 5

Lemma 6.1. Let f(x) and g(x) be meromorphic functions which satisfies

$$g(x)^2 = 1 - f(x)^4.$$

Then there is an entire function $z = \varphi(x)$ such that

$$f(x) = \mathrm{sn}[\varphi(x)] \;\; \underline{\text{and}} \;\; g(x) = \mathrm{sn}'[\varphi(x)],$$

where sn[z] is the Jacobi elliptic function with modulus −1, i.e.,

$$(\mathrm{sn}'[z])^2 = 1 - (\mathrm{sn}[z])^4 .$$

Proof is similar to the one of the Lemma 4.1. Q.E.D.

Now we will prove Proposition 5. We consider the equation

$$y(x+1)^4 = A(y(x) - c_1)^2 \, (y(x) - c_2) \, (y(x)) - c_3). \tag{6.1}$$

Put

$$\frac{y(x+1)}{y(x) - c_1} = A_1 u(x), \;\; A_1 = [-b'^2 A(c_2 - c_1)\,(c_3 - c_1)]^{1/4}$$

$$1/[y(x) - c_1] = b'v(x) + b, \tag{6.2}$$

with

$$b = \tfrac{1}{2}\,[(c_2 - c_1)^{-1} + (c_3 - c_1)^{-1}],$$

$$b' = \tfrac{1}{2}\,[(c_2 - c_1)^{-1} - (c_3 - c_1)^{-1}].$$

Then

$$v(x)^2 = 1 - u(x)^4 \tag{6.3}$$

and

$$\frac{A_1 u(x)}{b'v(x) + b} = c_1 + \frac{1}{b'v(x+1) + b} . \tag{6.3'}$$

Further

$$\frac{A_1 u(x) - c_2 b'v(x) - bc_2}{b' \, v(x) + b} = \frac{1 - v(x+1)}{(b'v(x+1) + b)\,(b' + b)} .$$

Since $v(x) - 1 = 0$ is fourfold ramified, we gat that zeros of

$$F(x) = A_1 u(x) - c_2 b'v(x) - bc_2$$

are also 4-fold ramified. By (6.3) and Lemma 6.1, we have

$$u(x) = \mathrm{sn}[\varphi(x)] \qquad \text{and} \qquad v(x) = \mathrm{sn}'[\varphi(x)]$$

with an entire function $\varphi(x)$. Put

$$G(z) = A_1 sn[z] - c_2 b' sn[z] - bc_2.$$

Suppose $v(x_0 + 1) - 1 = 0$, and put $z_0 = \varphi(x_0)$, $z_1 = \varphi(x_0 + 1)$. Then

$$G(z_0) = A_1 sn[z_0] - c_2 b' sn[z_0] - bc_2 = 0, \tag{6.4}$$

$$G'(z_0) = A_1 sn'[z_0] - c_2 b' sn''[z_0] = 0,$$

$$= A_1 sn'[z_0] + 2c_2 b' sn^3[z_0] = 0, \tag{6.5}$$

$$G''(z_0) = A_1 sn''[z_0] + 6c_2 b' sn^2[z_0] = 0$$

$$= -2A_1 \, sn^3[z_0] + 6c_2 b' sn^2[z_0] sn'[z_0] = 0,$$

i.e.,

$$-A_1 \, sn[z_0] + 3c_2 b' \, sn'[z_0] = 0. \tag{6.6}$$

$$\left\{G''(z)/sn^2[z]\right\}' |_{z = z_0} = -A_1 sn' [z_0] + 3c_2 b' sn''[z_0]$$

$$= -A_1 sn'[z_0] - 6c_2 b' sn^3 [z_0] = 0. \tag{6.7}$$

By (6.4) and (6.6)

$$sn[z_0] = 3c_2 b/(2A_1), \quad sn'[z_0] = b/(2b').$$

By (6.5)

$$-2A_1{}^4 + 27b^2 b'^2 c_2{}^4 = 0. \tag{6.8}$$

By (6.7)

$$-2A_1{}^4 + 81b^2 b'^2 c_2{}^4 = 0. \tag{6.8'}$$

From (6.8) and (6.8') we get $A_1 = 0$, i.e., $A = 0$, which is obviously impossible. Thus the equation (6.1) cannot possess any meromorphic solution.

§ 7. Proof of Proposition 6

We consider the equation

$$y(x + 1)^6 = A(y(x) - c_1)^3 (y(x) - c_2)^2 (y(x) - c_3). \tag{7.1}$$

Put

$$\frac{y(x + 1)}{y(x) - c_1} = A_1 \, v(x), \quad A_1 = [-A(c_2 - c_1)^2 (c_3 - c_1)]^{1/6}$$

$$1/[y(x) - c_1] = u(x) - b,$$

where

$$b = -\frac{1}{3}\left[\frac{2}{c_2 - c_1} + \frac{1}{c_3 - c_1}\right],$$

then

$$y(x+1) = \frac{A_1 u(x)}{u(x) - b}, \quad y(x) = c_1 + \frac{1}{u(x) - b} \tag{7.2}$$

and

$$\begin{aligned} v(x)^6 &= (u(x) - a)^2(u(x) + 2a) \\ &= u(x)^3 - 3a^2 u(x) + 2a^3, \end{aligned} \tag{7.3}$$

with

$$a = (c_2 - c_1)^{-1} + b = \frac{1}{3}\left[(c_2 - c_1)^{-1} - (c_3 - c_1)^{-1}\right] \neq 0.$$

By (7.2) we have

$$\frac{A_1 u(x)}{u(x) - b} - c_1 = \frac{1}{u(x+1) - b}. \tag{7.3'}$$

From (7.3), $u(x + 1)$ is 6-fold ramified over $-2a$, hence if $u(x_0 + 1) = -2a$, then we know that, noting $c_1 + \frac{1}{-2a - b} = c_3$,

$$F(x) = A_1 v(x) - c_3 u(x) + bc_3$$

has zeros of 6-th order at $x = x_0$. Therefore

$$F(x_0) = A_1 v(x_0) - c_3 u(x_0) + bc_3 = 0, \tag{7.4-0}$$

$$F'(x_0) = A_1 v'(x_0) - c_3 u'(x_0) = 0, \tag{7.4-1}$$

$$F''(x_0) = A_1 v''(x_0) - c_3 u''(x_0) = 0, \tag{7.4-2}$$

$$F^{(3)}(x_0) = A_1 v^{(3)}(x_0) - c_3 u^{(3)}(x_0) = 0, \tag{7.4-3}$$

$$F^{(4)}(x_0) = A_1 v^{(4)}(x_0) - c_3 u^{(4)}(x_0) = 0, \tag{7.4-4}$$

$$F^{(5)}(x_0) = A_1 v^{(5)}(x_0) - c_3 u^{(5)}(x_0) = 0, \tag{7.4-5}$$

On the other hand by (7.3)

$$6v^5v' = 3(u^2 - a^2)\, u', \tag{7.5-1}$$

$$2v^5v'' + 10v^4v'^2 = (u^2 - a^2)u'' + 2uu'^2, \tag{7.5-2}$$

$$2v^5v^{(3)} + 30v^4v'v'' + 40v^3v'^3 =$$

$$= (u^2 - a^2)u^{(3)} + 6uu'\, u'' + 2u'^3 \tag{7.5-3}$$

$$2v^5v^{(4)} + 40v^4v'v^{(3)} + 30v^4v''^2 + 240v^3v'^2v'' + 120v^2v'^4$$

$$= (u^2 - a^2)u^{(4)} + 8uu'u^{(3)} + 6uu''^2 + 12u'^2u''. \tag{7.5-4}$$

We will show later that there is an x_0 which satisfies (7.4)'s as well as (7.5)'s and $u'(x_0) \neq 0$. For such an x_0, we have by (7.5–1) and (7.4–1), writing $u_0 = u(x_0)$ and $v_0 = v(x_0)$,

$$2c_3v_0^5 = A_1(u_0^2 - a^2), \tag{7.6-1}$$

By (7.5–2), (7.4–2), (7.4–1) and (7.6–1), we get

$$5c_3^2\, v_0^4 = A_1^2\, u_0, \tag{7.6-2}$$

Similarly by (7.5–3), (7.4–3), (7.4–2), (7.4–1), (7.6–1), and (7.6–2)

$$20c_3^3\, v_0^3 = A_1^3\,. \tag{7.6-3}$$

Further by (7.5–4), (7.4–4), (7.4–3), (7.4–2), (7.4–1), (7.6–1), (7.6–2), and (7.6–3)

$$120v_0^2 = 0, \tag{7.6-4}$$

which obviously impossible. Hence the equation (7.1) cannot have any meromorphic solution.

Finally we will show that there is an x_0 with $u'(x_0) \neq 0$ which satisfies (7.4–0) and (7.3) for $x = x_0$.

Let u_0 and v_0 be such that

$$A_1v_0 - c_3u_0 + bc_3 = 0,$$

$$\text{and } v_0{}^6 - [u_0{}^3 - 3a^2u_0 + 2a^3] = 0.$$

There are at least three distinct values for such u_0. For

$$G(v) = v^6 - [(A_1/c_3)v + b)^3 + 3a^2[(A_1/c_3)v + b) - 2a^3$$
$$= v^6 - (A_1/c_3)^3v^3 - 3b(A_1/c_3)^2v^2 - 3(A_1/c_3)\,(a^2 - b^2)v$$
$$+ \ (3a^2b - 2a^3 - b^3) = 0$$

and

$$G'(v) = 6v^5 - 3(A_1/c_3)^3v^2 - 6b(A_1/c_3)^2v + (3A_1/c_3)\,(a^2 - b^2)$$

may have at most three common zeros, i.e., zeros of

$$6G(v) - vG'(v) = V^3 + 4bV^2 - 5(a^2 - b^2)\,V + 2(a - b)^2(2a + b)$$
$$= (V - (a - b))\,(V^2 + (a + 3b)\,V - 2(2a + b)\,(a - b))$$

with $V = (A_1/c_3)v$. Therefore, if $u'(x_0) = 0$ for any x_0 with $u(x_0) = u_0$, then $u(x)$ must be totally ramified over at least 6 distinct values, of which one is 6-fold ramified, which is impossible. Hence there must be an x_0 with $u'(x_0) \neq 0$.

§ 8. Proof of Proposition 7

We consider the equation

$$y(x + 1)^m = A(y(x) - c_1)^{i_1}\,(y(x) - c_2)^{i_2},$$
$$i_1 + i_2 = m \geq 3. \tag{8.1}$$

We may assume that $(m, i_1) = (m, i_2) = (i_1, i_2) = 1$.

Suppose $c_2 = 0$. As in the proof of Proposition 1, we see that $y(x) \neq 0$. Then $y(x) \neq c_1$ also. Put

$$y(x + 1)\,/\,y(x) = A_1u(x), \quad A_1 = (Ac_1{}^{i_1})^{1/m}$$
$$1/y(x) = v(x), \tag{8.2}$$

then $v(x) \neq \infty,\ 1/c_1$, and

$$u(x)^m = (c' - v(x))^{i_1}, \quad \text{where } c' = 1/c_1. \tag{8.2'}$$

Therefore $u(x) \neq \infty, 0$. By (8.2), we get

$$y(x + 1) = A_1 u(x)/v(x) = 1/v(x + 1).$$

Since $y(x + 1) \neq c_1$, we have that

$$A_1 u(x) - c_1 v(x) \neq 0. \tag{8.2''}$$

$w(x) = u(x)^{1/i_1}$ is meromorphic and $\neq 0, \infty$ because $u(x) \neq 0, \infty$. From (8.3), $v(x) = (1/c_1) - w(x)^m$. By (8.2″) we know that

$$A_1 w(x)^{i_1} + c_1\{w(x)^m - 1/c_1\} \neq 0,$$

hence $w(x)$ does not take values more than two, which is a contradiction. Thus c_2 cannot be 0, and $c_1 c_2 \neq 0$.

Suppose $y(x)$ does not take c_1. If $y(x) \neq c_2$ also, then $y(x) \neq 0$ by (8.1), which contradicts the theorem of Picard. Let $y(x_0) = c_2$. We denote by ν and μ orders of zero point x_0 for $y(x) - c_2$ and $y(x + 1)$, respectively. Then

$$m\mu = i_2 \nu,$$

hence $m \mid \nu$ and $i_2 \mid \mu$, thus 0 is a totally ramified value for $y(x)$ if $i_2 \geq 2$, which is impossible since $m \geq 3$. Thus $i_2 = 1$ and $i_1 = m - 1$ if $y(x) \neq c_1$.

Suppose $y(x)$ takes c_1 as well as c_2, at which values $y(x)$ is ramified to the order m. If $\mathrm{Min}(i_1, i_2) \geq 2$, then $y(x + 1)$ is ramified at 0 to the order at least 2. Hence

$$\left(1 - \tfrac{1}{m}\right) + \left(1 - \tfrac{1}{m}\right) + \left(1 - \tfrac{1}{2}\right) > 2$$

supposed $m \geq 5$. Thus we get $\mathrm{Min}(i_1, i_2) = 1$, if $m \geq 5$.

If $m = 4$ or 3, then obviously $\mathrm{Min}(i_1, i_2) = 1$.

Thus the equation (8.1) must be of the form:

$$y(x + 1)^m = A(y(x) - c_1)^{m-1}(y(x) - c_2). \tag{8.3}$$

Put

$$y(x+1) / [y(x) - c_1] = A_1 u(x), \quad A_1 = [\, A(c_2 - c_1)]^{1/m},$$

$$1/[y(x) - c_1] = v(x). \tag{8.4}$$

Then

$$y(x) = c_1 + 1/v(x),$$

$$y(x+1) = A_1 u(x) / v(x), \tag{8.5}$$

and

$$u(x)^m = b - v(x)\,, \quad b = 1/\,(c_2 - c_1), \tag{8.6}$$

$$\frac{A_1 u(x)}{v(x)} - c_1 = \frac{1}{v(x+1)}. \tag{8.6$'$}$$

We note that, as shown above, $y(x)$ takes c_2 hence $v(x)$ takes b.

Since $c_2 = c_1 + 1/b$, we get from (8.6′)

$$A_1 u(x) / v(x) - c_2 = [b - v(x+1)] / [v(x+1)\, b],$$

from which we see that, noting (8.6), zeros of

$$F(x) = A_1 u(x) - c_2 v(x) = A_1 u(x) + c_2 u(x)^m - c_2 b \tag{8.7}$$

are m-fold ramified. Hence, if $v(x_0 + 1) = b$,

$$F(x_0) = 0,$$

$$F'(x_0) = A_1 u'(x_0) - c_2 v'(x_0) = [A_1 + mc_2 u(x_0)^{m-1}]\, u'(x_0) = 0 \tag{8.7$'$}$$

$$F''(x_0) = [A_1 + mc_2 u(x_0)^{m-1}]\, u''(x_0) + m(m-1)\, u\, (x_0)^{m-2} u'(x_0)^2 = 0. \tag{8.7$''$}$$

Suppose $A_1 + mc_2 u(x_0)^{m-1} \neq 0$, then $u'(x_0) = 0$ hence $v'(x_0) = 0$. Suppose $A_1 + mc_2 u(x_0)^{m-1} = 0$, then $u(x_0) \neq 0$ and by (8.7″) we have also that $u'(x_0) = v'(x_0) = 0$.

Therefore, if $u_0 = u(x_0)$ and $v_0 = v(x_0)$ satisfy the system of equations

$$A_1 u_0 - c_2 v_0 = 0, \;\; u_0{}^m = b - v_0, \tag{8.8}$$

then $v(x)$ is totally ramified over v_0. On the other hand, there are at least $(m - 1)$ distinct values for such v_0, since if we put

$$G(v) = c_2^m v^m + A_1^m v - A_1^m b,$$

then $G(v) = 0$ and $G'(v) = mc_2^m v^{m-1} + A_1^m = 0$ admit at most one common zero point, which (if any) must be $v = [m / (m - 1)] b \neq b$ and hence cannot be a zero of $G''(v) = m(m - 1)c_2^m v^{m-2}$.

It is obvious from (8.6) that poles as well as b-points of $v(x)$ are m–fold ramified. Therefore $v(x)$ is totally ramified over at least $(m - 1 + 2) = (m + 1)$ distinct values, at least two of which are m–fold ramified. Since $m \geq 3$, this is impossible, which shows that the equation (8.1) cannot possess any meromorphic solutions.

§ 9. Proof of Proposition 8

We consider the equation

$$y(x + 1)^2 = A(y(x) - c_1)(y(x) - c_2). \tag{9.1}$$

By the same argument as in the head of § 8, we see that $c_1 c_2 \neq 0$.

Suppose $y(x)$ does not take c_1. Put

$$y(x + 1) / [y(x) - c_2] = A_1 u(x), \quad A_1^2 = A(c_1 - c_2), \quad 1/[y(x) - c_2] = v(x).$$

Then

$$u(x)^2 = c' - v(x), \qquad c' = 1/(c_1 - c_2),$$

$$y(x + 1) = A_1 u(x) / v(x) = c_2 + 1/v(x + 1) \neq c_1. \tag{9.2}$$

Since $y(x) \neq c_1$, $v(x) \neq c'$ hence $u(x) \neq 0$. By (9.2) we see

$$A_1 u(x) - c_1 v(x) \neq 0,$$

hence

$$A_1 u(x) + c_1[u(x)^2 - (c_1 - c_2)^{-1}] \neq 0,$$

which is impossible, since $u(x)$ cannot admit three exceptional values. Therefore, $y(x)$ takes both c_1 and c_2.

Now we will determine constants A, c_1 and c_2. Put

$$\frac{y(x+1)}{y(x)-c_1} = A_1\, u(x), \qquad A_1 = \sqrt{A(c_2 - c_1)}\,,$$

$$1/[y(x) - c_1] = v(x)\,, \qquad b^2 = 1/(c_2 - c_1), \tag{9.3}$$

then

$$y(x) = c_1 + \frac{1}{v(x)}\,,$$

$$y(x+1) = A_1\, u(x)\;/\;v(x), \tag{9.4}$$

and

$$u(x)^2 = b^2 - v(x), \tag{9.4$'$}$$

$$\frac{A_1 u(x) - c_2\, v(x)}{v(x)} = \frac{b^2 - v(x+1)}{v(x+1)}\, b^{-2}, \tag{9.4$''$}$$

since $c_2 = c_1 + 1/b^2$.

From (9.4$''$) we see that zeros of

$$F(x) = A_1 u(x) - c_2 v(x) = 0$$

are of multiplicity at least two. Hence if $F(x_0) = 0$, then

$$F(x_0) = A_1 u(x_0) - c_2 v(x_0) = 0, \tag{9.5}$$

$$F'(x_0) = A_1 u'(x_0) - c_2 v'(x_0) = [A_1 + 2c_2 u(x_0)]\, u'(x_0) = 0, \tag{9.5$'$}$$

by (9.4$'$). Let u_0 and v_0 be numbers such that

$$A_1 u_0 - c_2 v_0 = 0 \quad \text{and} \quad {u_0}^2 = b^2 - v_0. \tag{9.6}$$

Suppose that there is an x_0 such that $v(x_0) = v_0$ and $v'(x_0) \neq 0$. Then by (9.5$'$) $u'(x_0) \neq 0$. Then by (9.5$'$) $u'(x_0) \neq 0$. Hence by (9.5$'$) and (9.5) we have that

$$u(x_0) = -A_1/(2c_2) \text{ and } v(x_0) = -A_1^2/(2c_2^2),$$

which proves by (9.4′) that

$$A_1^2 = -4c_2^2\, b^2. \tag{9.6′}$$

On the other hand, suppose that for any x_0 such that $v(x_0) = v_0$, we have $v'(x_0) = u'(x_0) = 0$. Then $v(x)$ is totally ramified at such v_0. v_0 must satisfy the equation

$$c_2^2\, v_0^2 + A_1^2\, v_0 - A_1^2\, b^2 = 0\,, \tag{9.6″}$$

which has two distinct roots if $A_1^2 + 4_2^2\, b^2 \neq 0$. If so, $v(x)$ must be totally ramified at four distinct values ∞, b^2, and two v_0. On the other hand, the equation (9.1) can be written as

$$y(x+1)^2 = A[U(x)^2 - B^2] \tag{9.7}$$

where $U(x) = y(x) - c$, $c = \frac{1}{2}(c_1 + c_2)$, $B = \frac{1}{2}(c_2 - c_1) = 1/2b^2$. If we put

$$U(x) = \frac{B}{2}[\beta(x) + \frac{1}{\beta(x)}],$$

then $\beta(x)$ is easily seen to be meromorphic, by (9.7). We obtain

$$y(x+1)^2 = \frac{AB^2}{4}[\,\beta(x) - \frac{1}{\beta(x)}\,]^2\,,$$

$$\beta(x) - \frac{1}{\beta(x)} = \frac{2}{B\sqrt{a}}\, y(x+1) = \frac{2}{B\sqrt{A}}\left(c_1 + \frac{1}{v(x+1)}\right),$$

$$\beta^2 - \frac{2}{B\sqrt{A}}\left(c_1 + \frac{1}{v(x+1)}\right)\beta - 1 = 0. \tag{9.7′}$$

Hence the discriminant of the quadratic equation (9.7′) with respect to β,

$$[\left(4b^4c_1^2 + A\right) v(x+1)^2 + 8b^4c_1\, v(x+1) + 4b^4] \,/\, [A\, v(x+1)^2]$$

must be a square of some meromorphic function. Therefore roots of

$$\left(4b^4c_1^2 + A\right)v^2 + 8b^4c_1v + 4b^4 = 0 \tag{9.7″}$$

must be either b^2 or coincide with roots of (9.6″).

If b^2 is a root of (9.7''), then we have

$$\left(4b^4c_1^2 + A\right)b^4 + 8b^4c_1b^2 + 4b^4 = 0,$$

i.e.,

$$A_1^2 = A/b^2 = -4b^2\left(c_1^2 + 2c_1/b^2 + 1/b^4\right) = -4b^2c_2^2,$$

since $c_2 = c_1 + 1/b^2$. Hence we have either (9.6') or

the equation (9.7'') has the same roots as (9.6''), which means

$$\frac{4b^4c_1^2 + A}{c_2^2} = \frac{8b^4c_1}{A_1^2} = \frac{4b^4}{-A_1^2b^2}\,. \tag{9.7$^{\#}$}$$

Therefore

$$2c_1 = -1/b^2 = c_1 - c_2\,,\ c_1 = -c_2\,,$$

and

$$A^2 + A + 1 = 0.$$

Thus we have, taking $c_2 = 1$,

$$y(x + 1)^2 = \omega^2(y(x)^2 - 1), \tag{9.8}$$

where ω is the cubic root of 1.

We need the following Lemmas.

Lemma 9.1. The equation (9.8) does not admit any meromorphic solution of finite order.

Thus, if y(x) is a solution of finite order, then (9.7$^{\#}$) is not tenable, hence we must have (9.6'), i.e.,

$$A\,(c_2 - c_1)^2 = -4c_2^2.$$

Exchanging the role of c_1 to c_2, we get similarly

$$A\,(c_2 - c_1)^2 = -4c_1^2.$$

Thus we know that

$$c_2 = - c_1 \quad \text{and} \quad A = -1,$$

hence, taking $c_1 = 1$, the equation (9.1) must be of the form

$$y(x + 1)^2 = 1 - y(x)^2 \,, \tag{2.6'}$$

which proves Proposition 8.

Now we will prove the Lemma 9.1. By (9.8) we see that y(x) is totally ramified at $-i\omega$, 1, -1, $i\omega$. Hence

$$Y(x) = \frac{1-\eta}{1+\eta}\,\frac{y(x)+\eta}{y(x)-\eta}$$

is totally ramified at $\left(\frac{1-\eta}{1+\eta}\right)^2$, -1, 1, $-\left(\frac{1-\eta}{1+\eta}\right)^2$, where

$$\eta = \exp[\tfrac{7\pi}{12}\, i] \quad \text{and} \quad -\left(\frac{1-\eta}{1+\eta}\right)^2 = \tan^2\left(\frac{7\pi}{24}\right) > 1.$$

It is not difficult to see that Y(x) is written as

$$Y(x) = \mathrm{sn}[\varphi(x)]$$

with an entire function $\varphi(x)$, where $\mathrm{sn}[u] = \mathrm{sn}[u, \; -\left(\frac{1-\eta}{1+\eta}\right)^2]$. Thus

$$y(x) = \eta\,\frac{\mathrm{sn}[\varphi(x)] - [\eta - 1) \,/\, (\eta + 1)]}{\mathrm{sn}[\varphi(x)] + [(\eta - 1) \,/\, (\eta + 1)]}. \tag{9.9}$$

If y(x) is of finite order, then $\varphi(x)$ must be a polynomial [1], [4].

The equation (9.8) implies that $y(x + 3)^2 = y(x)^2$, hence $y(x + 3) = \pm\, y(x)$, therefore $y(x + 6) = y(x)$. Thus

$$\mathrm{sn}[\varphi(x + 6)] = \mathrm{sn}[\varphi(x)]$$

and

$$\varphi(x + 6) = \varphi(x) + K,$$

where K is a period of sn[u]. Differentiating, we have

$$\varphi'(x + 6) = \varphi'(x),$$

which implies that the polynomial $\varphi'(x)$ is a constant. Thus

$$\varphi(x) = hx + d, \text{ with some constants } h, d.$$

Suppose $sn[\varphi(x_0) + 1)] = 1$. By (9.9) $y(x_0) = 1$. Then by (9.8) $y(x_0 + 1) = 0$. Then $sn[\varphi(x_0 + 1)] = sn[\varphi(x_0) + h] = (\eta - 1) / (\eta + 1)$ by (9.9). On the other hand, if $sn[\varphi(x_1)] = [(\eta - 1) / (\eta + 1)]^2$, then $y(x_1) = -1$ by (9.9). Thus $y(x_1 + 1) = 0$ by (9.8), hence $sn[\varphi(x_1) + h] = (\eta - 1) / (\eta + 1)$. Therefore $sn[\varphi(x_1) + h] = sn[\varphi(x_0) + h]$, which implies $\varphi(x_1) - \varphi(x_0) = K$ with a period K. But $sn[\varphi(x_1)] \neq sn[\varphi(x_0)]$, a contradiction. Hence the equation (9.8) cannot possess any finite order solution.

Thus we have finished the proof of Lemma 9.1, hence of Proposition 8. Therefore the proof of Theorem 1 is complete.

§ 10. Proof of Theorem 2

Now we will determine meromorphic solutions of (2.6′):

$$y(x + 1)^2 = 1 - y(x)^2. \tag{2.6′}$$

Put

$$y(x) = \frac{1}{2} [\, \beta(x) + \beta(x)^{-1}]. \tag{10.1}$$

Then

$$\beta(x) = y(x) + \sqrt{y(x)^2 - 1} = y(x) + iy(x + 1),$$

which is meromorphic. By (2.6′) and (10.1) we obtain

$$\frac{1}{2} [\, \beta(x + 1) + \beta(x + 1)^{-1}] = \frac{1}{2i}[\beta(x) - \beta(x)^{-1}],$$

i.e.,